Quantum Networking

For my first teachers: my parents

Quantum Networking

Rodney Van Meter

Series Editor
Marcelo Dias de Amorim

iSTE

WILEY

First published 2014 in Great Britain and the United States by ISTE Ltd and John Wiley & Sons, Inc.

ISTE Ltd
27-37 St George's Road
London SW19 4EU
UK

www.iste.co.uk

John Wiley & Sons, Inc.
111 River Street
Hoboken, NJ 07030
USA

www.wiley.com

Library of Congress Control Number: 2014934407

British Library Cataloguing-in-Publication Data
A CIP record for this book is available from the British Library
ISBN 978-1-84821-537-5

Table of Contents

Notations xiii

Acknowledgments xv

Introduction xix

Chapter 1. Overview 1

1.1. Introduction 2
1.2. Quantum information 4
1.2.1. Principles 5
1.2.2. Imperfect quantum systems 7
1.2.3. Quantum computers 8
1.2.4. Applications of distributed quantum information 9
1.3. Quantum repeaters 10
1.3.1. Physical communication technologies 11
1.3.2. Multi-hop Bell pairs: quantum communication sessions 12
1.4. Network architectures 15
1.4.1. Semantics of distributed quantum information 16
1.4.2. Identifiers 17
1.4.3. Paths 17
1.4.4. Resource management discipline 18
1.4.5. A quantum internet 20
1.5. Conclusions 20

PART 1. FUNDAMENTALS 23

Chapter 2. Quantum Background 25

2.1. Introduction 26
2.2. Schrödinger's equation 28
2.3. Qubits 29
2.3.1. What is a qubit? 29
2.3.2. Quantum registers and weighted probabilities 30
2.3.3. Interference 32
2.3.4. Entanglement 33
2.3.5. Decoherence 34
2.3.6. Pure and mixed states and the density matrix 34
2.3.7. Fidelity 37
2.3.8. Measurement 38
2.3.9. The partial trace 39
2.4. Manipulating qubits 41
2.4.1. What is a quantum gate? 41
2.4.2. Single-qubit gates and the Bloch sphere 41
2.4.3. Global versus relative phase 44
2.4.4. Two-qubit gates 45
2.4.5. Quantum circuits 46
2.5. Bell pairs 47
2.5.1. The Bell basis 49
2.5.2. Measurement in the Bell basis 49
2.5.3. The Bell inequalities and non-locality 50
2.5.4. Experimental demonstration of violation of Bell's inequality . . . 52
2.6. The no-cloning theorem 53
2.7. Conclusion 54

Chapter 3. Networking Background 55

3.1. Concepts 56
3.1.1. Multihop communication: networks as graphs 56
3.1.2. Resources 59
3.1.3. Protocols 60
3.1.4. Naming and addressing 61
3.1.5. Security 62
3.2. Challenges in scaling up networks 63
3.2.1. Heterogeneity 63
3.2.2. Scale 64
3.2.3. Dealing with out-of-date information 64
3.2.4. Organizational needs 64
3.2.5. Misbehaving nodes 65

3.3. Design patterns . . . 65
3.3.1. Hierarchy . . . 65
3.3.2. Layering . . . 66
3.3.3. Narrow waist . . . 67
3.3.4. Multiplexing resources . . . 68
3.3.5. Smart versus dumb networks . . . 70
3.3.6. Distributed management and autonomy . . . 70
3.3.7. State machines . . . 71
3.3.8. Weak consistency and soft failure . . . 72
3.3.9. Distributed routing protocols . . . 73
3.3.10. Overlays, virtualization and recursion . . . 74
3.4. The Internet . . . 75
3.5. Conclusion . . . 77

Chapter 4. Teleportation . . . 79

4.1. The basic teleportation operation . . . 79
4.2. Experimental demonstration of teleportation . . . 82
4.3. State machines for teleportation . . . 84
4.4. Teleporting gates . . . 86
4.5. Conclusion . . . 88

PART 2. APPLICATIONS . . . 91

Chapter 5. Quantum Key Distribution . . . 93

5.1. QKD and the purpose of cryptography . . . 94
5.2. BB84: single-photon QKD . . . 97
5.3. E91: entanglement-based protocol . . . 100
5.4. Using QKD . . . 101
5.4.1. Campus-to-campus virtual private network . . . 101
5.4.2. Transport-layer security (TLS) . . . 103
5.4.3. Resilience of networks dependent on QKD . . . 104
5.5. Existing QKD networks . . . 105
5.6. Classical control protocols . . . 109
5.7. Conclusion . . . 111

Chapter 6. Distributed Digital Computation and Communication . . . 113

6.1. Useful distributed quantum states . . . 114
6.1.1. The stabilizer representation . . . 114
6.1.2. GHZ and W states . . . 115
6.1.3. Graph states . . . 116
6.2. Coin flipping . . . 118
6.2.1. The simplest multi-party distributed quantum protocol . . . 118

6.2.2. QKD-Based protocols 118
6.2.3. Practical, optimal quantum strong coin flipping 119
6.3. Leader election 119
6.3.1. The second simplest multi-party distributed quantum protocol . . 120
6.3.2. Tani *et al.*'s quantum protocol 120
6.4. Quantum secret sharing 121
6.4.1. Semi-classical, multi-party secret creation 121
6.4.2. The basic quantum secret sharing protocol 122
6.4.3. Verifiable quantum secret sharing and secure multi-party quantum computation 124
6.5. Byzantine agreement 126
6.5.1. The original problem 126
6.5.2. Ben-Or and Hassidim's quantum Byzantine agreement 127
6.6. Client-server and blind computation 128
6.7. Conclusion 130

Chapter 7. Entangled States as Reference Frames 131

7.1. Qubits in the environment 131
7.1.1. Precession 132
7.1.2. Quantum optical interference 133
7.2. Distributed clock synchronization 135
7.2.1. Chuang's algorithms 135
7.2.2. Jozsa *et al.*'s clock synchronization 138
7.2.3. Further work 140
7.3. Very long baseline optical interferometry 141
7.4. Conclusion 145

PART 3. LINES OF REPEATERS 147

Chapter 8. Physical Entanglement and Link-Layer Protocols 149

8.1. Creating entanglement using light 149
8.1.1. Quantum states of light 149
8.1.2. Emission 151
8.1.3. Transport 152
8.1.4. Detection 154
8.2. Memory and transceiver qubits 156
8.2.1. Gate noise 157
8.2.2. Single-qubit decoherence 158
8.2.3. Two-qubit decoherence 160
8.3. Link structure 161
8.4. State machines and protocol interactions 163

8.5. Managing density matrices in distributed software 164
8.5.1. Link-Level tracking of memory 167
8.5.2. Synchronizing higher layers 168
8.6. Examples 169
8.7. Conclusion 173

Chapter 9. Purification 175

9.1. Measurement revisited 175
9.2. Basic purification 177
9.2.1. Bit flip errors 178
9.2.2. Generalizing: incorporating phase flip errors and different Bell pairs 179
9.2.3. Multiple rounds and error redistribution 182
9.2.4. Resource consumption in multiple rounds 184
9.3. Scheduling purification 185
9.4. State machines and protocol interactions 187
9.5. More complex purification protocols 190
9.6. Experimental demonstrations 192
9.7. Conclusion 193

Chapter 10. Purification and Entanglement Swapping-Based Repeaters 195

10.1. Hardware architectures 195
10.2. Getting from here to there 197
10.2.1. Hop-by-hop teleportation 197
10.2.2. Basic entanglement swapping 200
10.2.3. Multi-hop swapping 202
10.3. Nested purification session architecture 203
10.3.1. Proof of polynomial resource growth 203
10.3.2. Problems to avoid 204
10.4. State machines and protocol interactions 206
10.5. Putting it all together 208
10.5.1. Simulating lines of repeaters 209
10.5.2. Greedy algorithm 211
10.5.3. Banded performance v. total distance 212
10.5.4. Finding the bands 212
10.5.5. Varying swapping thresholds 213
10.6. Considerations in the design of a simulator 215
10.7. Conclusion 217

Chapter 11. Quantum Error Correction-Based Repeaters 219

11.1. Quantum error correction 220
11.1.1. Steane code 221

11.1.2. Surface code 221
11.1.3. An early communication proposal 223
11.2. CSS repeaters 223
11.2.1. Protocols 225
11.2.2. Operational timing 227
11.2.3. Resources and performance 229
11.3. Surface code repeaters 230
11.3.1. Protocols 232
11.3.2. Operational timing 233
11.3.3. Resources and performance 234
11.4. Conclusion 235

Chapter 12. Finessing the Key Limitations 237

12.1. Quasi-asynchronous 238
12.1.1. Purification replacement operation 240
12.1.2. QEC-based operation 241
12.1.3. Timing variants 242
12.2. Memoryless 244
12.3. Summary: comparing quantum communication approaches 247
12.4. Conclusion 251

PART 4. NETWORKS OF REPEATERS 253

Chapter 13. Resource Management and Multiplexing 255

13.1. Simulated network and traffic 256
13.1.1. Network topology and simulator 256
13.1.2. Traffic load 258
13.1.3. Adjusting link target fidelity 258
13.2. Simulations 259
13.2.1. Circuit switching: upper and lower throughput bounds 259
13.2.2. Other multiplexing disciplines 260
13.3. Conclusion 263

Chapter 14. Routing 265

14.1. Introduction 265
14.2. Difficulties: differences between quantum and classical networks . . . 267
14.3. Problems and solutions 268
14.4. Simulation and results 270
14.4.1. The behavior questions 271
14.4.2. Simulated hardware and link costs 271
14.4.3. Simulated path candidates 274

14.4.4. Answering our behavior questions 275
14.4.5. Solving our problems 280
14.5. Conclusion 283

Chapter 15. Quantum Recursive Network Architecture 285

15.1. Review: network architecture 286
15.2. Recursive quantum requests 288
15.2.1. Processing in recursive networks 289
15.2.2. Naming a state 290
15.2.3. Defining quantum requests 291
15.3. Implementing recursion in quantum networks 294
15.3.1. Satisfying quantum requests 294
15.3.2. Paths and rendezvous points 294
15.4. Example 295
15.5. Conclusion 298

Chapter 16. Coda 301

16.1. Future development 301
16.1.1. Hardware 301
16.1.2. Making QRNA real 302
16.2. Open problems 303
16.3. Further readings for depth 304
16.3.1. Quantum repeaters and QKD 304
16.3.2. Optics and general quantum physics 304
16.3.3. Quantum computing 304
16.4. Further readings for breadth 305
16.4.1. Information theory 305
16.4.2. Dense coding 305
16.4.3. Quantum network coding 306
16.4.4. Entanglement percolation 306
16.5. Final thoughts 307

Bibliography 309

Index 331

Notations

$\|0\rangle$, $\|1\rangle$	Basis vectors for a qubit in the computational (Z) basis, written in Dirac's *ket* notation.
$\|+\rangle$, $\|-\rangle$	Basis vectors for a qubit in the X basis, written in Dirac's *ket* notation.
A, B	Names of nodes; abbreviations for Alice and Bob. Nodes are referred to with a numeric address in some places.
a, b, c, d	Diagonal elements of a two-qubit density matrix written in the Bell basis, corresponding to the probabilities of $\|\Phi^+\rangle$, $\|\Psi^+\rangle$, $\|\Phi^-\rangle$ and $\|\Psi^-\rangle$, respectively.
a_{AB}, etc.	Corresponding element of the d.m. of a two-qubit state (typically a Bell pair) shared between Alice and Bob.
$\mathbb{C}$	The set of complex numbers.
F	Generic for the fidelity of the state of one or more qubits, $F = \langle\psi\| \rho \|\psi\rangle$. $F = 1.0$ is a pure state. $F = 2^{-n}$ is the fidelity of a completely mixed state of n qubits.
l_0	Attenuation length in fiber.
$\|\psi\rangle$	Dirac's *ket* notation for a state vector. Generic for the state vector of a pure state of one or more qubits. It may be either a physical qubit or a logical one encoded using quantum error correction, depending on the context.
$\|\tilde{\psi}\rangle$	Dirac's *ket* notation for a qubit encoded using quantum error correction; a logical state, as opposed to a physical one.
$\|\bar{\psi}\rangle$	Dirac's *ket* notation for the NOT of a qubit.
H	Usually, the single-qubit Hadamard gate; occasionally, the Hamiltonian representing the physical evolution of a state.
P_b	Probability of success of the base-level physical entanglement operation.
P_{p1}	Probability of success of the first round of purification.
$\mathcal{P}_0^1$	Projective measurement operator for the value 0 on qubit 1.

T_1^A	T_1 energy relaxation time, or bit flip decay time, of the qubit in a Bell pair held at node A (Alice).
T_2^A	T_2 (phase relaxation time) of the qubit in a Bell pair held at node A (Alice).
$t_{\text{L1}}, t_{\text{LR}}$	Link-level one-way latency, round-trip time.
$t_{\text{E1}}, t_{\text{ER}}$	End-to-end one-way latency, round-trip time.
X, Y and Z	The single-qubit Pauli operators. Also written as σ_X, etc, in other texts and papers.
$\|\Psi^-\rangle^{(AB)}$	A Bell pair with one qubit held by node A and one qubit held by node B.
ρ	Generic for the density matrix for one or more qubits.
ρ_{AB}	Density matrix for a two-qubit state (typically a Bell state) shared between nodes A and B.
$O(\cdot)$	Asymptotic upper limit on growth in total number of computational operations, or execution time (circuit depth/algorithm steps) as problem size grows.
$\Theta(\cdot)$	Exact asymptotic growth in total number of computational operations, or execution time (circuit depth/algorithm steps) as problem size grows.
$\Omega(\cdot)$	Asymptotic lower limit on growth in total number of computational operations, or execution time (circuit depth/algorithm steps) as problem size grows.

Acknowledgments

> As no better man advances to take this matter in hand, I hereupon offer my own poor endeavours. I promise nothing complete; because any human thing supposed to be complete, must for that very reason infallibly be faulty.
>
> Herman Melville, *Moby Dick*

I owe more than I can say to my wife Mayumi and daughters Sophia and Esther. They have been patient throughout the writing of this book, accepting, "Sorry, I have to work on the book", as an excuse for everything from missing a soccer match to skipping my share of the housework. I love you three beyond all words.

My parents Doyle and Linda and sisters Sheila and Lera and their families have also been incredibly supportive. Lera's almost daily words of encouragement kept me going.

To borrow a phrase from Charlie Parker, Thaddeus Ladd is the other half of my heart. Without his patient teaching and guidance on the physics, in all probability I would not have been able to complete much of the research upon which my own share of the ideas in this book is founded, or even fully understand the impact of the giants of the field whose work I also attempt to explain here. I hope I have been able to return the favor at least in part by teaching him about systems and networks.

Besides Thaddeus, I owe a debt to Kohei Itoh, Mikio Eto, Eisuke Abe, Kae Nemoto, Bill Munro, Austin Fowler, Simon Devitt, Clare Horsman and Yoshihisa (Yoshi) Yamamoto for teaching me most of what I know about quantum information. The Core Research for Evolutionary Science and Technology (CREST) and Funding Program for World-Leading Innovative R&D on Science and Technology (FIRST) Quantum Summer Schools organized and taught by Yamamoto, Tarucha, Koashi, Nakamura, Tsai, Takeuchi, Imoto, Nemoto, Chuang, Wineland, Jozsa and others were immensely valuable; each time I attended, I learned a year's worth of new material.

For showing the way, being smart, or otherwise being inspirational: Ron Ayres, Charlie Bennett, Richard Feynman, Ed Stone and Wook.

For additional personal support on this book and related projects: Fred Baker, Thomas Clausen, Chip Elliott, Dave Farber, Bob Hinden, Kohei Itoh, Seth Lloyd, Paul Mockapetris, Jun Murai, Timo Jokiaho, Wook, Suzanne Woolf and Yoshi Yamamoto.

The first person I should thank with respect to the book itself is Marcelo Dias de Amorim, for suggesting this book in the first place, when we met at our semi-annual WIDE Camp in September 2012. The staff at ISTE have done an excellent job of keeping me at least somewhat on track; without them, the book would never have been finished.

For reviewing the book as a fairly complete entity, even as it was evolving: Kilnam Chon and Bill Manning. Shigeya Suzuki deserves a special mention for actually working on some of these topics in parallel with the development of the book; his patience as I said, "I think that's in the book... oh, wait, give me a day to write that..." in answer to many of his questions was extraordinary.

For reviewing parts of the book: Luciano Aparicio, Andi Frischknecht, Akira Furusawa, Jim Harrington, Thaddeus Ladd, Shota Nagayama, Sam Pottle, Yutaka Shikano, Shigeki Takeuchi, Seiichiro Tani, Todd Tilma, Joe Touch, Yidun Wan and Hideaki Yoshifuji.

For other advice on history and recent experimental work: Romain Alléaume, Thaddeus Ladd and Peter McMahon. For tidbits on radio interferometry: Min Yun.

For contributing to my modest share of the research covered in this book, and graciously allowing me to reuse large portions of several of our joint papers: Luciano Aparicio, Mourad Beji, Chia-Hung Chien, Byung-Soo Choi, Clare Horsman, Kaori Ishizaki, Hiroyuki Kusumoto, Thaddeus Ladd, Iori Mizutani, Bill Munro, Koji Murata, Shota Nagayama, Kae Nemoto, Takahiko Satoh, Shigeya Suzuki, Joe Touch, Jaw-Shien Tsai and Fumiki Yoshihara.

For photos and diagrams: Romain Alléaume, Chip Elliott, Akira Furusawa, Masahide Sasaki and Hajime Tazaki. Takaaki Matsuo and Shota Nagayama stepped in at the last minute and drew a stack of figures for the book.

Ultimately, I should thank the students in my Advancing Quantum Architecture (AQUA) "kenkyuukai" (research group) and my quantum information processing class, and the Murai Lab students and faculty in general, for bearing with me as I learned how to explain quantum computing and networking to classical systems folks.

My own share of the research presented here has been supported by three Kakenhi grants (21500020, 24102706, 25282197) from the Japan Society for the Promotion of Science (JSPS), including one through the Quantum Cybernetics program. This project has been made possible in part by a gift from the Cisco University Research Program Fund, a corporate advised fund of the Silicon Valley Community Foundation. This research is supported by the Cabinet Office, Government of Japan and the Japan Society for the Promotion of Science (JSPS) through the Funding Program for World-Leading Innovative R&D on Science and Technology (FIRST Program). My collaborators have been supported by the Ministry of Education, Culture, Sports, Science and Technology (MEXT) and the National Institute of Information and Communications Technology (NICT) in Japan, and the National Science Foundation (NSF) and other agencies in the United States. The generous and unrestricted support provided by the sponsors of the WIDE Project has enabled some of these collaborations. Thomas Clausen hosted me as a visiting professor in March 2011, as I got my start writing what would eventually become a string of survey and architecture papers that culminate in this book.

Although I have benefited immensely from the advice of a number of people who are more expert than me in many of the subfields covered in this book, I bear the ultimate responsibility for the contents; any misrepresentations of history, let alone actual technical mistakes, are my own. Comments are welcome; in this digital age, the print form of the first edition of a book is hardly the last word. I look forward to hearing from you.

This book consists in part of previously published material, used by permission of the copyright holders ACM, IEEE, National Institute of Informatics, SPIE and Springer. The material appeared in the following papers:

– Aparicio L., Van Meter R., "Multiplexing schemes for quantum repeater networks", *Proceedings of the SPIE*, vol. 8163, pp. 816308, August 2011.

– Aparicio L., Van Meter R., Esaki H., "Protocol design for quantum repeater networks", *Proceedings of Asian Internet Engineering Conference*, November 2011.

– Van Meter R., Ladd T.D., Munro W.J., *et al.*, "System design for a long-line quantum repeater", *IEEE/ACM Transactions on Networking*, vol. 17, no. 3, pp. 1002–1013, June 2009.

– Van Meter R., Touch J., Horsman C., "Recursive quantum repeater networks", *Progress in Informatics*, no. 8, pp. 65–79, March 2011.

– Van Meter R., Satoh T., Ladd T.D., *et al.*, "Path selection for quantum repeater networks", *Networking Science*, vol. 3, no. 1–4, pp. 82–95, December 2013.

– VAN METER R., HORSMAN C., "A blueprint for building a quantum computer", *Communications of the ACM*, vol. 53, no. 10, pp. 84–93, October 2013.

Some of the material appeared in my PhD thesis, "Architecture of a quantum multicomputer optimized for Shor's factoring algorithm," Graduate School of Science and Technology, Keio University, 2006.

Rodney VAN METER
Faculty of Environment and Information Studies
Keio University
March 2014

Introduction

We are going to need a quantum Internet, and to build it, we need quantum internetworking technology. This book is my contribution to both the technical and social work of getting there. It is based on my experiences during 15 years of work on classical computing systems and networks, followed by a decade of research on quantum computing systems and networks.

Quantum information, including both quantum computing and quantum communication, is poised to have a large and sustained impact on the fields of theoretical and experimental quantum physics, theoretical computer science (or informatics) and ultimately the information technology industry. One important subfield is quantum networking, especially using *quantum repeaters*, which are the focus of this tome. Quantum signals are weak and very fragile, and, in general, cannot be copied or amplified. Engineering quantum communication sessions that can reliably exchange data over long distances, in topologically complex networks built on heterogeneous technologies and managed by many independent organizations, requires an extraordinarily broad range of expertise, which few individuals anywhere have *in toto*. Over the next 300 or so pages, we will attempt to lay a common foundation on which each person can erect his or her contribution.

The primary audience of the book is two-fold:

– computer networking folks with no prior background in quantum information, who are curious and considering working in the field;

– quantum information experts who have yet to work in the area of repeaters and need an introduction, or those who have begun working in the area but need more background in networks.

Ideally, the book will produce a "meeting of minds" between the two communities. Networkers will find that quantum networking is less intimidating than it initially appears, and that there are breathtaking concepts underlying an emerging class of uses for distributed quantum information. Physicists will discover that networks are complex, artificial artifacts with emergent behaviors not immediately anticipated from the behavior of individual building blocks, and are built on some principles that are every bit as fundamental and beautiful as those they have been studying in physics. By the end of the book, readers from either community should be prepared to design a quantum repeater network, based on both classical network architecture and the existing literature on quantum repeaters. Readers should know enough to implement simulations of repeater networks that properly take into account (1) a reasonable abstraction of the physics, (2) the distributed, autonomous nature of decision-making and (3) the technical and operational heterogeneity of networks of networks such as the Internet.

The book is intended to be a readable introduction rather than a comprehensive, in-depth tome; each chapter is 10–20 pages, intended to be ingested in one sitting. Most chapters will use only basic linear algebra and probability theory. The approach emphasized throughout the book will be on the use of classical networking principles to build a sustainable, extensible, robust quantum repeater network architecture.

The overall flow of the book is an overview, three chapters on background (quantum information, networking concepts and teleportation), then three chapters on applications (QKD, distributed digital computation and entangled states as reference frames) to motivate the development of networking technology. In Part 3 of the book, the focus first shifts to the bottom of the stack, beginning with the physical entanglement experiments and link design. After working through purification, we come to the three major classes of communication session architecture for chains of quantum repeaters: the original entanglement swapping approach, the more recent error correction based approaches, and the recent work on asynchronous approaches. The book ends with a series of chapters on issues in multi-user, autonomous networks: multiplexing, routing and internetworking architecture, featuring the Quantum Recursive Network Architecture (QRNA).

The reader will find varying levels of mathematical and logical rigor in different chapters. In particular, a thorough discussion of physical implementations would fill a separate book, which we leave to the physicists. Likewise, at the highest level, the details of the security protocols and proofs for applications such as verifiable secret sharing are beyond the scope of this book; the applications are presented in just enough depth that casual readers will be able to understand why they are valuable, and what demands they make on the network itself.

Readers are assumed to be familiar with basic vector and matrix addition, multiplication and calculation of the determinant; exponentiation of matrices; complex numbers, including their exponentiation; and discrete probability. The mathematics presented here does not go beyond this level. Thus, although the concepts presented here are largely unfamiliar, abstract and sometimes counter-intuitive, the math itself is generally not particularly difficult. Chapter 2 includes explicit, worked examples of many of the mathematical principles. It is even possible for well-prepared first- and second-year undergraduates to work through the book.

For the advanced researcher, it is worth noting that this book lies halfway between the research monograph and the textbook on the spectrum. In the course of writing what I thought would be a relatively cut-and-dried presentation of some basics viewed from the point of view of a network engineer, I discovered a number of things that simply have not yet been done in the literature. Among them:

– distributed density matrix management (section 8.5);

– the "valley fold" timing for quasi-asynchronous repeaters (section 12.1);

– a moderately detailed analysis of network workloads imposed by applications of repeaters (Chapter 6);

– extended state machine-based designs for protocols.

Each of these likely will be a journal paper, perhaps more or less concurrent with the appearance of the book, but all but the last had their genesis in this writing project. (We began the state machine approach in a conference paper [APA 11b], but the book contains new material.) Each of these topics also deserves yet more attention than I have so far been able to give. I look forward to handing them off to my capable collaborators.

Chapter 1

Overview

"Teleportation" is a magic word, exotic and evocative, but it has been appearing in serious technical literature with increasing frequency. Both theoretically fascinating and experimentally demonstrated, teleportation is the key to quantum networks [GIS 07, KIM 08]. When used in discussions about quantum information, teleportation refers not to Captain Kirk stepping into a machine on the starship Enterprise, dissolving and reappearing on a planet's surface, but to an operation in which a quantum variable dissolves *here* and reappears *there*, on a different physical device. Only the quantum *state* moves; the electron or other physical device remains where it was, and the receiver can in fact be a very different form of physical device than the sender. The quantum state is destroyed at the sender in the process.

Classical networks communicate by physically copying data and transmitting the copy, but the rules of quantum mechanics forbid the creation of independent copies of an unknown, arbitrary quantum state. Instead of risking the loss of valuable, fragile quantum data by directly transmitting our only copy, networks will prepare generic states that are used to teleport data or to perform teleportation-derived operations on the data.

Quantum networks bring new capabilities to communication systems. Quantum physical effects can be used to detect eavesdropping, to improve the shared sensitivity of separated astronomical instruments or to create distributed states that will enable numerical quantum computation over a distance using teleportation. *Quantum communication* is the *exchange of quantum states* over a distance, generally requiring the support of substantial classical communication.

The quantum states that are exchanged may be "standalone" states, an individual element of quantum data. They may also be part of a larger quantum state, spanning devices or even network nodes in a way no shared classical state can. These latter states we refer to as *entangled* states, which we will study extensively in this book.

Applications running on classical computers will use these quantum states to accomplish one of the above tasks. The classical computer is connected to a quantum device, which may do no more than *measure* the quantum states to find a classical value (such as a bit of a secret key), or may store them for use in more complex quantum computers. A classical computer will treat a quantum computer as a type of coprocessor; likewise, the classical computer will see the quantum network through the eyes of a separate device.

Because quantum data is fragile and some quantum operations are probabilistic, errors and distributed calculations must be managed aggressively and perhaps cooperatively among nodes. Solutions to these problems will have both similarities to and differences from purely classical networks. Architectures for large-scale quantum networking and internetworking are in development, paralleling theoretical and experimental work on physical layers and low-level error management and connection technologies. Unentangled quantum networks have already been deployed, starting in the early 2000s; as of early 2014, entangled networks are not yet deployed, but may appear within the next few years and will form a vibrant research topic in the coming decade.

1.1. Introduction

The motivations for building networks are the same for both quantum and classical networks: the desire to connect people, devices such as computers or sensors, or databases that are in separate locations, for technical, economic, political, logistical, or sometimes purely historical reasons. What differs is the type of data and operation involved. Quantum computers, and quantum networks, use quantum variables rather than classical ones; the analogue of the classical bit is the quantum bit, or *qubit*.

Proper use of quantum information opens up new possibilities, making feasible solutions to some problems that are computationally intractable for classical computers (most famously, factoring large numbers) [SHO 97, LAD 10, NIE 00, VAN 13a] and adding new physical capabilities (most famously, detection of eavesdropping, leading to new, secure, distributed cryptographic key generation mechanisms) [BEN 84]. Other applications for distributed quantum systems include long-baseline optical interferometry for telescopes [GOT 12], high-precision clock synchronization [JOZ 00, CHU 00] and quantum forms of distributed tasks such as leader election [TAN 12] Byzantine agreement [BEN 05a] and coin flipping. Quantum and classical networks and computing systems will hybridize, allowing applications to select the most efficient mechanism for accomplishing a particular function.

Modern work on quantum communications can be said to have begun with Stephen Wiesner's quantum cryptography proposal, originating around 1970 [WIE 83], followed by Charlie Bennett and Giles Brassard's 1984 proposal for *quantum key distribution* (QKD) [BEN 84, DOD 09], which utilizes the new low-level quantum capability of eavesdropping detection to build a specific system function, namely the creation of shared, secret random numbers for keying of classical cryptographic systems. However, QKD in its basic form is limited in distance to a few hundred kilometers in optical fiber or perhaps more through free space, and is a single-application system.

Bennett *et al.*'s 1993 proposal for *quantum teleportation* made it possible to move data and execute simple calculations remotely, extending the feasible distance for QKD and vastly expanding the range of conceivable distributed quantum applications [BEN 93]. Teleportation involves local quantum operations at each end and classical messages from the sender to the receiver. It consumes a quantum state known as a *Bell pair* (introduced below), shared between the two end points, so, a key function of quantum networks is to replenish the supply of distributed Bell pairs as necessary. As with any physical operation, teleportation operates imperfectly, requiring an extensive system that labors to suppress errors. More than a goal in itself, teleportation serves as a building block for distributed quantum applications.

The need to deal with imperfect quantum states and to span multiple hops spurred the development of the concept of *quantum repeaters* [DÜR 07, SAN 11], which are a vibrant area of research in both experiment and theory. Classical repeaters amplify a signal at the physical level, or receive a weak, distorted or noisy signal then regenerate a clean, strong signal. Quantum repeaters, however, are prevented by the laws of physics from performing such operations directly. Instead, they support high-fidelity, long-distance quantum communication using teleportation over shorter distances and forms of error correction ranging from a simple parity check on a Bell pair to extraordinarily complex, full error correction schemes based on the mathematics of topology. Some repeater architectures manage data movement using computations distributed across all of the nodes in a path between source and destination, while others are more akin to the hop-by-hop packet forwarding used in the Internet; the best approach for a given set of physical capabilities remains an important open question. The basics of teleportation and simple forms of error correction have been experimentally demonstrated, and the race is on to build more complete repeaters.

Although QKD networks using trusted relays and optical switches are in use in medium-scale testbeds, the key architectural issues in large-scale repeater networks are only beginning to be addressed. Protocols to actually implement the repeater functionality must be developed. Path selection and resource management, both at the node level, where memory resources are precious, and the network level, including choosing who gets access to the network, will play a role in determining whether the networks actually work.

Beyond single networks lies the issue of *internetworking*. An individual network will be built and managed by a single organization. Initially, it will be built using a single quantum networking technology. What happens when we want to bring in a second technology? What happens when we want to connect our network to another organization's network? How do we get them to exchange quantum information? How do we manage the connection between the networks? Such a multi-network configuration is called an *internetwork*, or *internet*, for short. (Spelling it with an uppercase "I", and sometimes attaching the article "the", implies the primary, worldwide classical Internet we all use every day.)

Such an internet, of course, begins with the ability to recode quantum data from one form to another and physically connect heterogeneous technologies. Internetworking will require classical sharing of the correct abstraction for describing quantum states or computation requests and the ability to translate protocols for error management, as well as settling the issues of resource management and path selection.

Our goal, in this book, is to begin from scratch and build an understanding of quantum information, quantum repeaters and classical networking thorough enough to propose and evaluate a quantum internet architecture, including writing the classical software implementing the protocols.

1.2. Quantum information

To understand teleportation and distributed quantum information in principle, only a few concepts are required: superposition, measurement, interference, entanglement, no-signaling and no-cloning. To understand quantum networks in practice, it is equally imperative to study quantum systems in an imperfect world; all of the important behaviors of quantum networks arise from dealing with noise and loss using purification and quantum error correction. The primary mathematical tool for studying algorithms and basic concepts is the *state vector*, and for studying imperfect states, the primary tools are the *density matrix* and the *fidelity*, all of which we will see in the next chapter. Here, we restrict ourselves to a qualitative introduction to the key ideas.

1.2.1. *Principles*

Quantum computers have attracted interest because they are expected to asymptotically outperform classical computers on some important real-world problems [BAC 10, LAD 10, MOS 09, VAN 13a]. These gains in capability arise from the differences in storing and manipulating information using quantum states; here, we will restrict our discussion to qubits, though other forms of quantum information are possible. A qubit may be e.g. the direction of spin of a single electron, the direction of polarization of a single photon, or any of a large number of other proposed state variables. Like a classical bit, a qubit has two states, but unlike a classical bit, a qubit may be in a *weighted superposition* of the two states, allowing certain functions to be evaluated for *both* input values at the same time. A register of n qubits can, like a classical register, hold any of 2^n possible values. The quantum register can in fact hold a superposition of *all* of these values and can, in principle, be used to compute on all 2^n possible states at the same time.

The difficulty lies in extracting useful answers from a quantum computer. To read the results of a computation, dedicated hardware components *measure* the state of the system. The state of the quantum register *collapses* when the system is measured. It randomly picks one state out of the states that are part of the superposition, based on their relative weights. The other states go away, and it is as if they never existed.

A quantum algorithm manipulates the system to *reduce* the probability of undesirable states and *increase* the probability of desirable states, until the system has a high probability of measuring the quantum register and getting an answer to our problem, ideally in substantially fewer computational steps than a classical system would require. This is done by creating *interference* on the quantum states to reinforce good answers.

The concept of *entanglement*, in which the states of two or more quantum subsystems are correlated in a fashion that is not possible in classical systems, is the most difficult quantum concept to grasp. Two qubits can be entangled in a continuous spectrum of possible states; four types of entangled states known as *Bell states* or *Bell pairs* are commonly used. One such Bell state is a superposition of the state where both qubits are 0 and the state where both qubits are 1. In this state, when measured, each qubit has a 50% probability of being found in a 0 state and a 50% probability of being found in a 1 state. However, their probabilities are not independent; both values will be found to be the same.

Bell pairs form the basic communication and computation components for most distributed quantum computation, including teleportation, but are not the only form of entangled state. Bell states can be generalized into multi-party states called GHZ states or W states, and we also use entangled states known as *graph states*. Most distributed quantum computing algorithms will build around one or more of these key flavors of entangled state; so, the network must be able to create them efficiently. We will see Bell pairs in more mathematical detail in section 2.5, GHZ and W states in section 6.1.2 and graph states in section 6.1.3.

Basic teleportation is accomplished by first creating a Bell pair between the source and destination. The source entangles the qubit to be teleported with the source's half of the Bell pair; then, both qubits are measured, destroying the entanglement of the Bell pair and any superposition state of the data qubit. The measurement results in two *random* classical bits, uncorrelated with the state of the data qubit, which must be transmitted to the destination. Local quantum operations at the destination determined by those classical bits then recreate the original data qubit's state on the remaining Bell pair member. This sequence is illustrated in Figure 1.1. The latency of the classical information transmission prevents information from being transferred faster than the speed of light, and is known as the *no-signaling* constraint, and applies in many situations with quantum information.

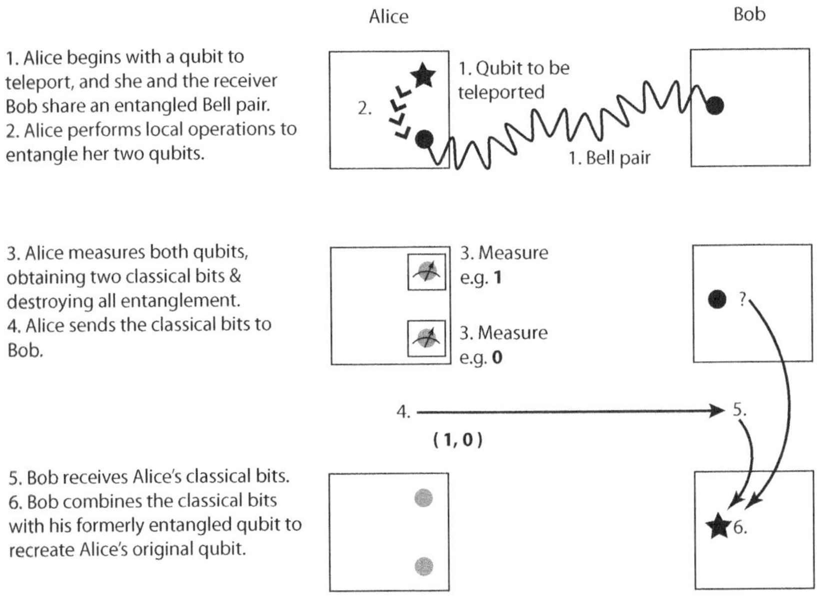

Figure 1.1. *Operations in teleporting a qubit from Alice to Bob*

The final concept required to understand both quantum computation and communication is the *no cloning* theorem, as we will see in more mathematical detail in section 2.6. Perfect *independent* copies of an unknown quantum state cannot be made. "Copies" of some states remain entangled with the original state. This entanglement is actually useful in many quantum algorithms, but an unentangled copy would be wildly more useful, allowing faster-than-light communication. It would be and is too good to be true.

A major consequence of the no-cloning theorem is that the system cannot copy and send precious quantum data when there is a risk of losing the data; loss of the in-transit copy would destroy even the copy kept due to the effects of entanglement and inadvertent measurement. This fact drives the common quantum networking approach of first building a high-quality, generic entangled state, then using that state to teleport or compute on our valuable data. We turn to handling these imperfections next.

1.2.2. *Imperfect quantum systems*

The central fact of all experimental quantum systems is this: the state of a quantum system is exceedingly fragile. Errors result in continuous degradation of our knowledge about the state of the quantum register. As the state drifts from its assigned value, the probabilities of the zero and one states change and the desired effects of interference may become muted or even incorrect. Beyond these errors that quickly accumulate, isolation of qubits from the environment is difficult, and qubits may be *accidentally* measured, destroying the valuable quantum state.

A measure known as the *fidelity* is one tool for tracking the quality of the state. Fidelity ranges from 0 to 1.0, with the latter being perfect. It is, essentially, the probability that our qubit or set of qubits is actually in the state we believe it ought to be in.

Various techniques for managing errors have been developed, some based on classical error correction and erasure correction techniques, others on uniquely quantum approaches [DEV 13, TER 13]. *Purification*, in which two or more multiqubit states are manipulated to form one higher-fidelity state, uses few quantum memory resources and simple quantum operations, but operates only on well-understood states such as Bell states rather than arbitrary application data. Purification is a type of error *detection*.

More complete protection of an arbitrary quantum state requires *quantum error correction*, in which we use a large number of physical qubits and add redundancy. It is possible to represent more than one qubit in an error correction block, as is done in classical error correction, but holding a single logical qubit is more common. The number of physical qubits can range from tens to possibly thousands, depending on the physical memory lifetime, quantum operation error rates and the performance required to successfully execute a given algorithm.

Besides errors involving the drift of the state, quantum communication systems are also subject to loss in the channel; for those systems expecting to use a single photon, this loss is fatal for that particular operation. Because losses in optical channels tend to be high, any communication system must be designed to manage this loss. Quantum optical states cannot be simply amplified without destroying the entanglement and superposition; so, other techniques must be used. Losses in the channel generally force a return message to be used acknowledging success or failure.

1.2.3. *Quantum computers*

Let us take a very short detour to look at quantum computers. After all, quantum networks will have some standalone applications, but a major goal is to use networks to connect computers!

The original concept goes back to the early 1980s, when Richard Feynman suggested that it was possible to simulate one quantum device using another, more efficiently than a classical computer could run such a simulation [FEY 02]. Paul Benioff suggested a quantum Turing machine [BEN 82]. David Deutsch explored some of the ideas behind such machines and proposed the first concrete quantum algorithm [DEU 85, DEU 92]. Seth Lloyd proposed the first plausible implementation of a real quantum computer in 1993 [LLO 93].

Theoretical approaches to organizing a computation using quantum effects include the gate model (similar to Boolean logic circuits), adiabatic quantum computation [AHA 04a, FAR 01], direct (analog) simulation, measurement-based quantum computation [RAU 03] and quantum random walks [AHA 93]. All have similar computational power, though the methods of creating algorithms for them are as different as classical digital and analog computers. To the extent that this difference affects quantum networks, in this book, we assume, and work with, the gate model. Measurement-based QC builds on top of a basic gate model and thus can benefit from the networks we describe here, but the adiabatic and direct models would need a very different form of network.

Peter Shor's 1994 announcement of his algorithm for factoring large numbers on a quantum computer generated huge excitement and an increase in research budgets [SHO 94]. The algorithm can factor composite numbers or take discrete logarithms in time polynomial in the number of bits, whereas the best known classical algorithm is superpolynomial [LEN 03]. Realization of such a speedup would dramatically affect the security of encryption algorithms such as RSA and the Diffie-Hellman key exchange used on the Internet, in e-commerce websites and site-to-site network encryption.

Numerous other algorithms have been developed. Lov Grover showed how to get a polynomial speedup on any combinatoric search problem, and it is known that it is impossible to get an exponential speedup on any arbitrary problem with no known structure [GRO 96, ZAL 99]. More recent algorithms cover various types of quantum chemistry calculations and simulations [BRO 10, BUL 09, KAS 11, LAN 11], certain classes of linear algebra problems [HAR 09], vector space problems [REG 02], graph problems [MAG 05], algebra [HAL 07], Boolean formula evaluation [AMB 07] and machine learning [LLO 13]. Bacon and van Dam [BAC 10] and Mosca [MOS 09] have published surveys which we recommend.

It is worth noting that the resource consumption of these algorithms is an area of ongoing research; how big, how fast and how accurate does a quantum computer have to be to solve interesting problems correctly [VAN 13a]? Current designs suggest that a computer will have to consist of many millions of qubits in order to apply error correction effectively [JON 12a, THA 06]. Execution times of algorithms on potentially buildable machines are also being investigated, although first-cut answers suggest that some algorithms with apparently attractive characteristics will in fact have dismayingly long run times [CLA 13, CLA 09, JON 12b].

A discussion of complexity classes and their application to quantum computation would fill a book, and we will not attempt to delve into it here. Scott Aaronson's PhD thesis is a good survey [AAR 04]. Key ideas here are again due to Charlie Bennett and to Ethan Bernstein and Aaronson's adviser Umesh Vazirani [BEN 97, BER 97].

All of this would have remained purely an exercise in theory, if not for the development of methods for suppressing errors, as discussed in the last section. John Preskill, Peter Shor, Andrew Steane, Charlie Bennett, Manny Knill and others contributed key insights to fault tolerant operation of a quantum computer [BEN 96c, KNI 96, PRE 989b, SHO 95, STE 96]. Excellent surveys on this topic have proliferated in the last few years [DEV 13, GRA 09, RAU 12, TER 13].

For a non-mathematical treatment of the ideas, the book by Williams and Clearwater is excellent [WIL 99]. Nielsen and Chuang [NIE 00] is the canonical text and covers algorithms as well as the underlying technology. The collection edited by Bouwmeester, Ekert and Zeilinger in 2000 [BOU 00] remains an excellent introduction to the technology.

1.2.4. *Applications of distributed quantum information*

The concepts of quantum communication are inherently fascinating, worthy of basic research by anyone's definition. However, as engineers striving to build networks, we must understand how the networks will be used, in order to evaluate our design decisions. Moreover, systems will only be deployed in large numbers when a compelling economic case appears. Thus, the study of quantum networks involves an equal measure of studying applications for distributed quantum states.

Earlier, we introduced QKD as an application. Implementations of QKD are well beyond the experimental phase [ELL 03, DOD 09]. A few commercial products are available, and metropolitan-area testbed networks exist in Boston, Vienna, Geneva, Barcelona, Durban, Tokyo, several sites in China and elsewhere throughout the world. In fact, the BB84 technique deployed in most links in these networks does not use entangled quantum states, although another approach, developed by Artur Ekert, does [EKE 91]. QKD is certainly the most practical, commercially attractive use of quantum networks in the near term. QKD has been integrated into custom encryption suites and the Internet standard IPsec suite and has been proposed for use with the TLS protocol common on the World Wide Web [ELL 02, MIN 09, NAG 09].

Other security-related functions have been proposed, including leader election and Byzantine agreement under assumptions of very powerful adversaries [BEN 05a, TAN 12]. Executing these algorithms would require nodes with more functionality than the ability to measure qubits for QKD, but likely would not require a fully functional, large-scale quantum computer.

We can reason that, like classical systems, one quantum computer is useful, but two are even more so, and connecting them together brings immediate benefits. Especially given that quantum algorithms (such as Shor's algorithm for factoring large numbers) are security-related, it seems reasonable to suppose that clients would like to be able to use remote quantum servers securely. A form of computation known as *blind computation* would allow a client to use the services of a remote machine, without revealing the algorithm, input or output data [BRO 09]. This will require *very* high rates of teleportation, low residual error rates and a powerful server; various schemes proposed alter the demands made of the client [MOR 13].

We can view QKD as a type of sensor network in which the interaction between the physical world and our quantum information devices figures prominently. Even more directly, distributed quantum states can be used as a form of *reference frame*, so that physical measurements can be conducted over a distance, more accurately or efficiently than using purely classical means. For example, synchronization of clocks is a common, critical use of communication signals and quantum algorithms have been proposed that will converge with asymptotically fewer operations than a classical method requires [JOZ 00, CHU 00]. A mechanism for improving the resolution of optical interferometry for astronomy has been proposed [GOT 12]. All of these will be very demanding applications with respect to both Bell pair production rates and the precision of those states.

1.3. Quantum repeaters

Quantum networks, like classical networks, will involve nodes and links and a layered communication architecture with individual protocol modules communicating vertically up and down a protocol stack and horizontally with peers. This section focuses first on the physical components that make up a link, before the discussion moves to arranging multiple links into a chain, then a network.

1.3.1. *Physical communication technologies*

Quantum communication channels are implemented by sending states of light down a physical channel. These states may be single photons, or other quantum optical states with either large or small numbers of photons. A channel may be a waveguide such as an optical fiber, or free space. It may involve a single transmitter and receiver, or multiple receivers that can individually be enabled or disabled in a shared bus configuration. A link uses a quantum channel and associated classical channel to connect two or more nodes.

A node may have quantum memory that can be used to store a qubit that is entangled with the pulse as it is sent out. When receiving a pulse, a node may either directly measure the pulse using, for example, an avalanche photodiode (APD), or may transfer its quantum state to a memory for later use or analysis. The pulses may come from weak lasers, flourescing atoms, or emission of single photons from a quantum dot, a structure created to exhibit some of the behavior of an atom.

One of the most promising hardware approaches for entangled networks uses microscopic pieces of diamond. When a carbon atom in the diamond lattice is replaced with a nitrogen atom, a positive electrical potential in the lattice capable of trapping a single electron is created. This approach, called *nitrogen vacancy (NV) centers in diamond*, may work at room temperatures, in contrast to most other solid-state quantum systems, which require cryogenic temperatures. Other promising experimental approaches include various forms of quantum dots. Ion traps that hold individual atoms in a vacuum are perhaps the most experimentally advanced approach. Entanglement of up to fourteen qubits in a single trap has been accomplished.

All of these experimental approaches have drawbacks; most do not operate at telecom wavelengths, which will dramatically shorten feasible link distances, though wavelength conversion schemes are also under development. They suffer from short memory lifetimes to differing extents, and the probability of correctly transferring the optical state to the static qubit remains inadequate for reasons ranging from low optical coupling efficiency to basic physics. None of these approaches is ready for mass production, and currently, all require hand-tuning and complex experimental setups.

The necessary classical messages include heralding at the physical layer to coordinate timing of the quantum pulses and many messages for coordinating the higher-level error management, data movement, distributed state creation and application functionality. Researchers often assume that the classical messages follow the same path through a network as the quantum messages, though except for the physical herald, this is not strictly necessary. When the classical messaging uses a different network topology, analysis of the communication efficiency must be done with care.

1.3.2. *Multi-hop Bell pairs: quantum communication sessions*

The purpose of the technologies just described is to create link-level entanglement. Interesting communication requires extending that entanglement across multiple hops while maintaining adequate fidelity. The *quantum repeater*, building on basic entanglement functionality with purification and teleportation, lays the foundation for quantum networks. Here, we discuss direct transfer of quantum information, and the generation of long-distance Bell pairs over a fixed chain of links and nodes. Below, we will take up the more general question of how such a chain can be part of an actual network.

The most obvious method of moving quantum data from place to place is direct hop-by-hop transmission by transferring the qubit state onto a photon and firing that photon down the link toward a node at the far end. If that link does not reach all the way to the destination, then it is received and forwarded on to the next node, relay fashion. However, as noted above, this places the valuable quantum data at unacceptable risk of loss. Instead, we could build Bell pairs over each link and teleport our qubit one hop at a time. Disappointingly, hop-by-hop teleportation is only marginally better than direct transmission because each hop degrades the fidelity of the data qubit.

Alternatively, what about creating a Bell pair at the source and performing hop-by-hop teleportation not on our valuable data qubit, but on one of the two Bell qubits instead? This will extend the length of the Bell pair, as shown in Figure 1.2. If the qubit being sent is lost before reaching the destination, the Bell pair can be discarded and restarted. Once the qubit reaches the destination, the Bell pair can be used to teleport the important data qubit, without fear of loss. However, the same problem arises: the fidelity of the Bell pair degrades with each teleportation operation as well as over time if the system keeps the qubit in memory.

One solution is executing purification in a distributed fashion, as in Figure 1.3. When purifying Bell pairs, node A holds one half of each of two pairs, and node B holds the other half of each. Using local quantum operations, including measurement of one of the Bell pairs, A and B can probabilistically improve the fidelity of the other Bell pair. Note that, because purification does not require direct quantum communication, it can operate over any distance, provided the requisite Bell pairs and a classical communication channel are available. The biggest drawback to distributed purification is that it requires that each end convey the results of its local measurement to the far end. Assuming that both nodes can independently identify the set of operations to perform, the minimum time for completion of purification is the one-way classical message latency between the two nodes.

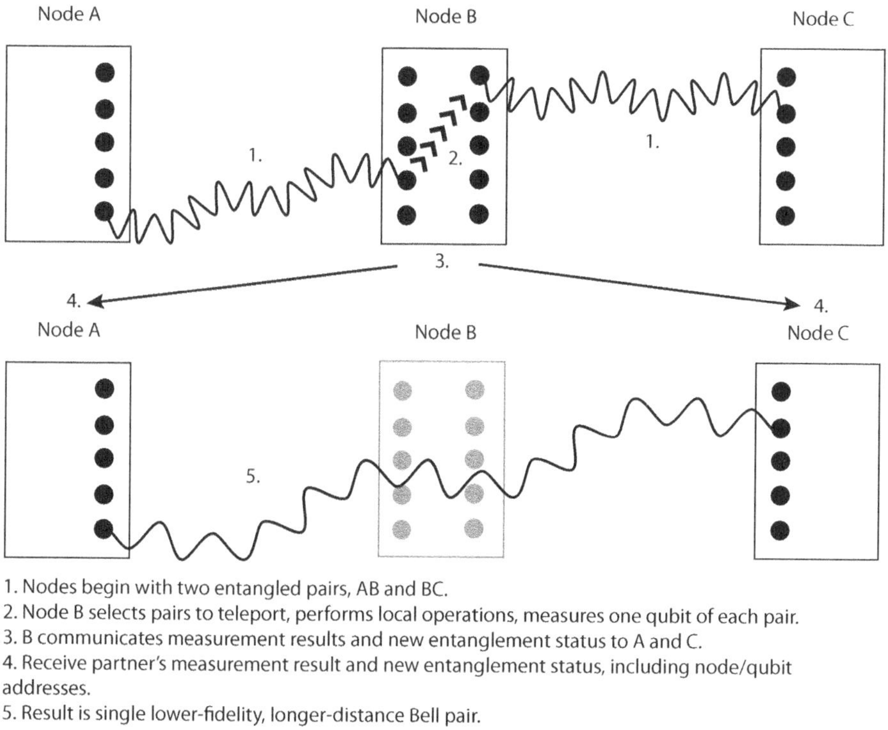

Figure 1.2. *Teleportation can lengthen one Bell pair using another*

The one-hop-at-a-time extend-purify-extend approach will work, but fails to take full advantage of the fact that *the distributed states being created are generic*, which allows the network to effectively build the Bell pair in parallel. The network can choose to build from both ends of the needed connection, or from the middle; note that the teleportation operation shown in Figure 1.2 operates independently of the length of the two Bell pairs. In the late 1990s, Wolfgang Dür, Hans Briegel and their collaborators proposed "nesting" Bell pair purification and teleportation so that the length of entanglement *doubles* in each round, allowing a logarithmic-depth number of rounds to create end-to-end entanglement over a large number of short hops: eight one-hop Bell pairs become four two-hop Bell pairs, then two four-hop Bell pairs and finally one eight-hop Bell pair [DÜR 99]. This has become the benchmark approach to repeaters, with much research assuming a power of two number of hops.

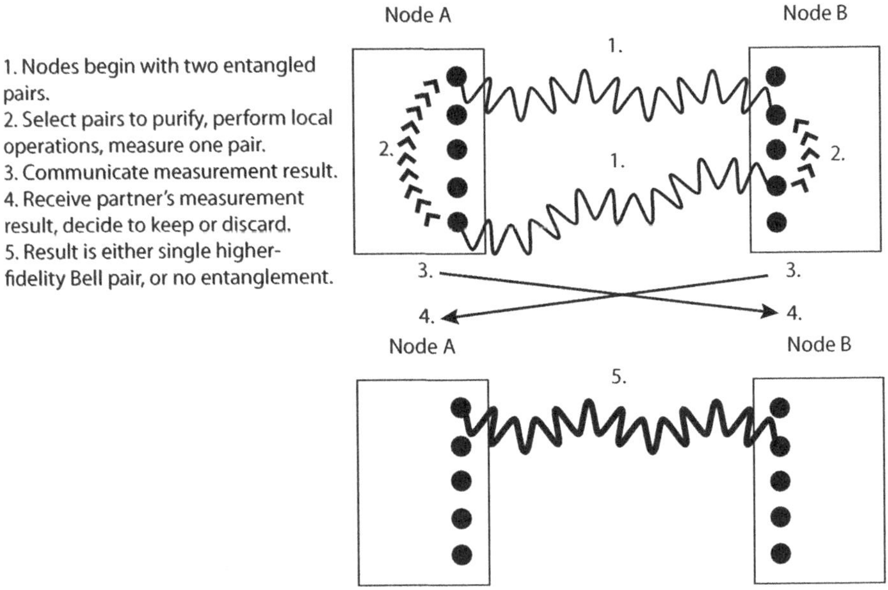

Figure 1.3. *Steps involved in purification of Bell pairs*

This purify-and-teleport architecture is not the only known approach; quantum error correction (QEC) can replace purification. Managed properly, QEC protects the data more completely, reducing the need for multi-hop purification and its associated need for round-trip delays. These advantages come at the expense of substantially more memory and computation resources at each node. Overall, in this book, we will study five approaches to connecting a source to a destination via a chain of links and nodes.

Layered communication describes how protocol functions are vertically composed within a communications node to provide increasingly complex capabilities. Layered quantum communication relies on five key functions that are unique to quantum networks, but only the first of these is actual quantum communication; the rest are classical functions for managing quantum states.

– *Physical layer*: we rely on a quantum physical layer using light to encode quantum state. Many technologies for this layer are under development.

– *Link-level entanglement*: because most physical entanglement mechanisms are probabilistic, the link layer will include an acknowledgment to the sender indicating which attempts succeeded.

– *Remote state composition*: in the Internet, links are composed into paths by copying packets from one link to the next. In a quantum network, links are less readily composed due to the no-cloning theorem. Quantum paths thus either establish end-to-end entanglement from entangled links or use that entanglement to teleport the quantum state from one end to the other. This layer is very sensitive to the link-layer capabilities as well as the error management mechanism.

– *Error management*: in the classical Internet, errors are managed using redundancy (e.g. forward error correction) or error detection and retransmission. As noted earlier, the no-cloning theorem prevents straightforward use of either of these mechanisms. The fidelity of quantum states is critical in reducing the need for error management.

– *Application*: the application may be QKD, or a physical reference frame for an instrument, or a numeric computation or decision algorithm based on shared state. The application will determine if end-to-end entanglement is required, or if the QKD-like model of direct measurement is adequate. Some applications may desire quantum states other than Bell pairs, including any of several common forms of three-party or larger states, such as the W and GHZ states we mentioned above.

One of the most difficult tasks in quantum networking is for the software running in separate nodes to maintain an adequately consistent idea of the density matrix of a quantum state that spans two or more nodes. The fidelity of the state will change, depending on the quality of the hardware at each node and the actions taken by the node. Unless the nodes exchange information about such matters, it is impossible to accurately assess the current state and the need for further action such as error management.

1.4. Network architectures

The physical entanglement, error management and communication session technologies we have discussed will get us to laboratory-scale demonstrations, but do not form a complete, deployable network architecture. Real-world networks must accommodate heterogeneous links arranged in complex topologies, managed by many autonomous organizations; traffic sources competing for use of the network; and ongoing network events such as a node or link suddenly becoming unavailable. An architecture for a large-scale network must support independent decision making by the nodes in a manner that will result in robust, efficient operation of the network as a whole. Although classical design principles can be applied to quantum networking, the resulting architectures can be quite different due to the radical restrictions and unique capabilities of the quantum domain.

The chosen communication session architecture will be executed over a *path* composed of links and nodes. Each node, link and software service, and even the quantum states themselves, must have an *identifier*, a name that software can use when composing those paths and tracking states as they are fabricated. Because the global network does not exist solely for the exclusive use of one communication session, a *resource management discipline*, either explicit or implicit, will govern. All of these interlocking aspects of the architecture depend on an understanding of the *semantics* of the requests on the network; so, let us begin there.

1.4.1. *Semantics of distributed quantum information*

Before constructing actual networks, we must decide how we want nodes to communicate across a chain of links. Above, we assumed that our goal was to move quantum data from place to place, either directly or by building Bell pairs. Now that we have some idea of what is in our quantum toolbox, we can ask a sophisticated question: what should the semantics of a request across the network be?

The request model depends first on what the network is designed to *do*. In classical networks, traditionally the network layer only sends data, via unicast, multicast or broadcast, with other functions delegated to higher-layer protocols. To support the applications we have discussed, a quantum network can operate in one of three modes: (a) it can teleport data from place to place, (b) it can execute certain computational operations over a distance (a technique known as *teleporting gates*) or (c) it can create distributed quantum states. Each of these options results in a different form of contract between the requesting end node and the network.

Perhaps the most fundamental operation, at the network layer, would be creation of high-fidelity distributed Bell pairs, which alone are adequate for building more complex distributed states, teleporting data or executing remote operations. Conservative engineering practice would suggest that, as with operating system APIs and IP packet semantics, simple is best, favoring this as the lone network-supplied operation. However, data movement as a primitive may improve performance by operating more asynchronously. Providing remote computation requests, or a richer set of state-creation services in the network, may reduce application complexity or improve overall system performance by reducing the total number of operations that must be performed. With appropriate network protocols, all three modes can be mixed in the same network.

Above, we noted that most of the functionality in a quantum network is actually classical. The same is true of the applications that consume the services provided by the network, whether QKD, or distributed digital computation, or use as a reference frame in instrumentation. Perhaps the most commonly-used approach in software for sending or receiving data over a network is the *sockets* interface, developed at UC Berkeley in the early 1980s [WRI 95a, WRI 95b]. Software engineers studying quantum networks have begun to ask, "What does a quantum socket look like?" The answer to this question will, of course, be different depending on which of the above options is chosen.

1.4.2. *Identifiers*

Because we must identify where we want to send data, networks naturally require names for the nodes. These names may be symbolic and easy for humans to remember or numeric *addresses* that are used more directly by internal systems and help to guide the choice of *route* to the destination.

On the Internet, end points of the communication must be able to name each other, but in general know nothing about the other nodes involved in completing their communications. For quantum communication, end nodes using a purify-and-swap approach must communicate directly with nodes along the path and so must be able to name and discover those nodes.

On the Internet, a node can generally determine where to next send a packet using only the destination address (a type of name) carried in the packet itself. When the packet arrives at the destination, further information carried in the packet (other types of names) is used to distinguish which software program (process) is waiting for the packet. In other types of networks, this discrimination can be done either explicitly or implicitly.

In quantum networks, the entangled states built within the network must have some sort of identifier, to facilitate the software work of management and delivery to applications. This is complicated by the fact that a Bell pair in the middle of the network might not yet be assigned to serve a particular end-to-end session.

1.4.3. *Paths*

Multihop networks require a means of selecting a path through the network [PER 00]. The need to understand the network topology manifests itself both in the choice of path through the static, physical topology, and in the dynamic operation of the network, as in Figure 1.5. This is a complex operation in networks of thousands of nodes and links with as much as eight orders of magnitude difference in bandwidth, made vastly more difficult when tens of thousands of such networks are interconnected.

On the Internet, each network is run by an independent organization, and the internal structure of each of those networks is kept hidden from the outside. Path selection is therefore a two-level process, enhancing scalability and autonomy. Moreover, calculation of the best route from a given node to anywhere else is an ongoing task, but a complete path for a given Internet connection is not determined before data is sent. Information held locally allows globally consistent decisions to be made for each data packet independently inside each node.

One common approach to path calculation in medium-scale networks is a distributed form of Dijkstra's shortest path first algorithm [DIJ 59, GOV 02, MOY 97]. From network to network, a separate protocol is used to minimize the number of networks that a packet passes through.

Quantum repeater networks and internetworks will undoubtedly be organized in a similar fashion. Dijkstra's algorithm can be adapted to medium-scale repeater networks [VAN 13b], but no concrete proposal yet exists for routing at the global level.

The path selection mechanism will be affected by the choice of communication session architecture and in turn makes demands of the identifier and naming architecture. In the purify-and-swap approach, nodes in a path must communicate with more than just their immediate neighbors, as in Figure 1.4. In its most natural form, this requires that any node be able to discover and name any other node in the entire internetwork, but we have just seen that a network typically keeps its own internal structure hidden from the outside. A scalable, robust solution to this problem is imperative.

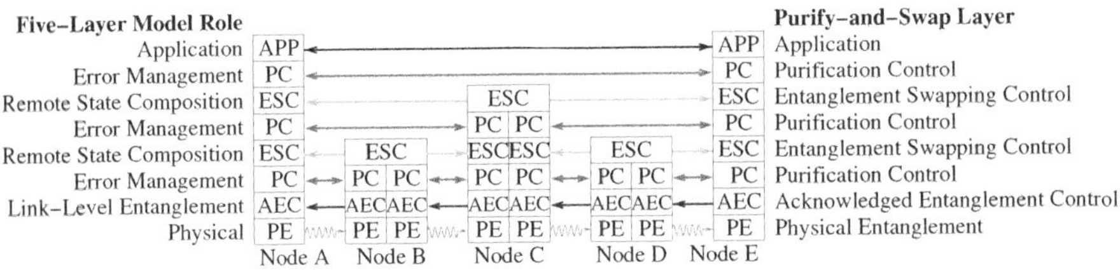

Figure 1.4. *Protocol layers and their interaction in purify-and-swap repeaters, in a five-node, four-hop chain. The labels on the left indicate the model layer represented, and the labels in the boxes and on the right indicate the protocol name for purify-and-swap repeaters. Double-headed arrows indicate bidirectional classical communication is required. The only quantum portion of the stack is the physical layer, shown with all links propagating left to right*

1.4.4. *Resource management discipline*

The network architecture must specify resource management for data requests: are the qubit memories in a node and the Bell pairs created across a link committed to the exclusive use of a single quantum communication session, or are they shared? This choice is affected by the communication session architecture, and in turn affects the construction of network paths and the naming of quantum states that currently span more than one node.

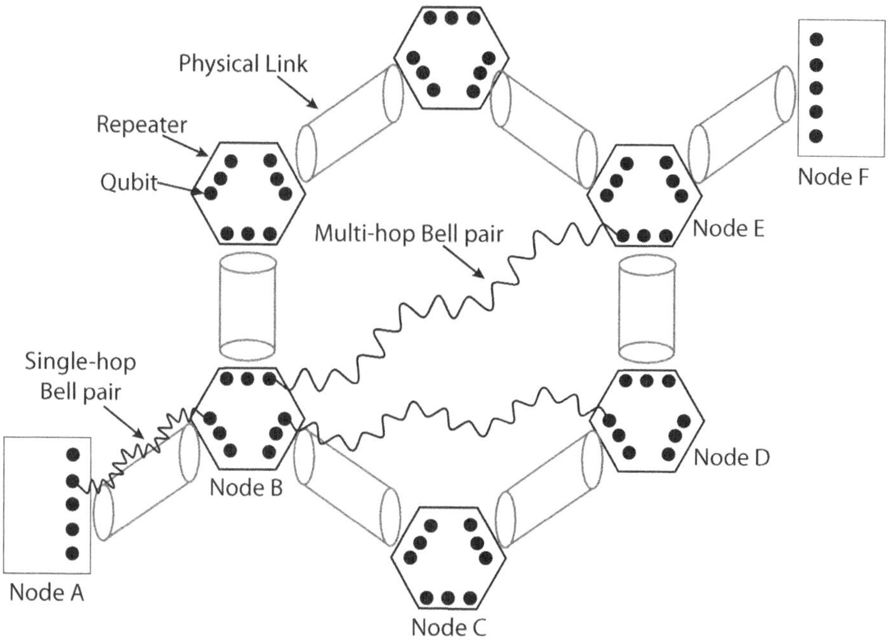

Figure 1.5. *Even when Node B knows that A is trying to build a Bell pair with F, B may be uncertain whether its Bell pair connected to Node D or Node E is "closer" to the destination*

Completely dedicating all of the resources along a path to a single session is the most obvious approach, especially as fragile quantum memories impose fairly stringent real-time constraints. However, this approach blocks other connections from using the same links and may reduce the overall rate at which work is completed across the network. Our recent research suggests that sharing the resources using one of several forms of *multiplexing* may raise the aggregate throughput of the network if implemented carefully.

The generic states generated throughout the network could serve the needs of many pending requests, much as production from a factory can be sent to serve any of a large number of customers. An interesting dynamic resource assignment problem arises: does an entangled Bell pair between two nodes "belong" to a specific end-to-end request, or are all Bell pairs "up for grabs"? Bell pairs that span long distances along routes carrying a lot of traffic may be especially desirable commodities.

1.4.5. *A quantum internet*

We have discussed some of the difficult problems in networks as they scale up in the number of nodes and links: heterogeneity, autonomy of management, naming, resource management and path selection and control. Distributed management of density matrices requires complex protocols and real-time decisions based on inevitably out-of-date information. The choice of quantum communication session architecture, or the design decision to allow nodes to choose one of several session architectures, affects all of these architectural choices.

All of these problems are in turn made more difficult in an internetwork. My research group and our collaborators have begun developing a quantum internetwork architecture called quantum recursive network architecture (QRNA) [VAN 11], which we hope will provide a structure for addressing these problems at global scale. QRNA may or may not ultimately become the architecture for the Quantum Internet, but we will present it in some detail in Chapter 15 to provide a platform for discussing the issues in a concrete manner.

QRNA is based on the classical recursive network architecture (RNA) developed by Joe Touch and collaborators [TOU 08]. A recursive network architecture can be viewed as a natural fit for quantum repeater internetworks. The naming structure, network topology, path composition and even the creation of multi-hop entangled states can all be simplified by judicious use of recursion.

QRNA provides a general-purpose request mechanism abstracted from underlying layers to accommodate any of the communication session architectures presented above. It supports requests for creation of distributed states (including both two-party Bell pairs and multi-party states) and operations on those states. Requests may be recursively decomposed and distributed throughout the network in order to build the end-to-end state requested by an application, meshing smoothly with the protocol layering model in Figure 1.4.

Like the Internet, nodes need not understand the topology of the entire network or even the names of all of the nodes involved in a communication session. This is accomplished by allowing a link as seen by a node to be either a physical link or a recursively organized network.

1.5. Conclusions

Quantum networks come in both entangled and unentangled forms. QKD networks are already up and running in various metropolitan areas throughout the world, although the coupling at intermediate nodes is strictly classical at the moment. In contrast, entangled networks remain rudimentary, existing only at small scales in laboratories and have yet to demonstrate all necessary functionality in a single experiment [SAN 11].

For entangled networks, by far, the most important ongoing research is on the physical layer – if quantum memories and local operations do not reach sufficiently high fidelities and entanglement success probabilities do not rise, quantum networks will remain a laboratory exercise. Applications for distributed quantum states, whether numerical computation or sensor networks, will drive the need for quantum networks; without them, no one will buy and deploy quantum networking equipment. Both of these areas are being addressed in depth, the first by experimental physicists, the second by theorists in both computer science and physics. It bears pointing out that the performance required for some of these applications remains several orders of magnitude beyond even optimistic hardware predictions for the next several years.

To bridge the considerable gap between theoretical large-scale, wide-area applications and small-scale experiments, an overarching network architecture and matching protocols must be developed. These protocols must emphasize optimized use of quantum memory, both spatially, by reducing the number of qubits that must be stored, and temporally, by reducing the length of time a qubit must be held in an intermediate state, e.g. by eliminating round-trip messaging where possible. The real-time factors in physical memory decay, or the high resource requirements of error correction-protected memories, must be managed properly. Moreover, the form of requests within the network is critical to efficiency. Improvements in the request model can alter the demands on the size, quality and capabilities of the physical system. Ultimately, the problems exemplified in Figures 1.5 and 10.6 are matters of giving each node enough information to make high-quality, autonomous decisions. The design of robust, efficient classical protocols usable in multi-user, multi-technology quantum internetworks will demand the technical skills of the data networking community.

The Quantum Internet, once realized, will allow us to exploit entanglement over long distances for new computational capabilities and for new physical capabilities such as eavesdropping detection. More speculatively, we can imagine uses such as sensor networks for quantum-enhanced telescopes and tests of the fundamental correctness of quantum mechanics. Within a few years, quantum networking and teleportation will move out of the physics laboratory and into the network engineering domain, offering some of the most exciting and intellectually challenging research and development topics of the coming decade.

PART 1

Fundamentals

Chapter 2

Quantum Background

The focus of this book is quantum networks, but in order to understand their operation and importance, we must learn about the general principles upon which quantum computers are founded as well. A *quantum computer* is a device that takes advantage of quantum mechanical effects to perform certain computations asymptotically faster than a purely classical machine can. It relies on *quantum parallelism*, using physical phenomena that can be held, like Schrödinger's cat, in more than one state at once. This allows us to compute on all of those states at the same time using a single operation. Quantum parallelism is best understood in the context of the concepts of *superposition* (section 2.3.1), *entanglement* (section 2.3.4), *measurement* (section 2.3.8), *interference* (section 2.3.3) and *no-cloning* (section 2.6).

Of course, to discuss quantum information in precise terms, we must learn how quantum data is represented and manipulated (section 2.4). In this book, we will use the *state vector* representation (section 2.2) and the *density matrix* representation (section 2.3.6) extensively. We will occasionally use the *stabilizer* representation, which we defer introducing until section 6.1.1.

A quantum computer performs, in principle, exponentially many computations simultaneously; however, exponentially many *results* of those computations cannot be read out, leaving us with the fascinating problem of how to use such a machine to accelerate computations that interest us. In fact, a common construction for a quantum algorithm is "Evaluate $f(x)$ for all $0 \leq x < 2^n$, and return a value x_0 such that $f(x_0) = a$." This is achieved by using interference so that measuring the value of x is likely to give x_0 and not a value x_1 for which $f(x_1) \neq a$. The most famous result in quantum computing to date, Shor's algorithm for factoring large numbers [SHO 94, SHO 97], offers an asymptotic superpolynomial speedup over the current champion classical algorithms. A few instances of exponential speedup over classical algorithms have been discovered, but it is known that quantum computers do not automatically give an exponential improvement; on completely unstructured search problems, only a square root improvement is possible [GRO 96]. For more on quantum computation, see the additional reading list in section 16.3.3.

This chapter reviews the basic concepts needed to understand quantum computing and communication, with emphasis on manipulating qubits and grasping the key idea of entanglement through the example of Bell pairs (section 2.5). We begin with quantum mechanics, presenting Dirac's *ket* notation, with a few notes on linear algebra, then state Schrödinger's equation and Hamiltonian dynamics as a fact; deep derivation or analysis of quantum mechanics are deferred to q.m. textbooks. We then informally define a qubit, discuss its state-vector and Bloch sphere representations and corresponding manipulations. Two-qubit gates are explained, along with the three-qubit Toffoli and Fredkin gates. The latter part of the chapter covers Bell pairs, their experimental demonstration and the no-cloning theorem. Readers are also referred to both popular [NIE 03, WIL 99] and technical [KIT 02, NIE 00] texts on quantum computing for more breadth and depth.

2.1. Introduction

First, let us introduce the linear algebraic notation commonly used in quantum computing. We will not give rigorous definitions, instead limiting ourselves to a few of the practical matters that a working engineer needs to understand. Readers are assumed to be familiar with basic vector and matrix addition, multiplication and calculation of the determinant; exponentiation of matrices; complex numbers, including their exponentiation and discrete probability.

Paul Dirac was one of the founders of quantum mechanics. He introduced the *bra-ket* notation. $|\psi\rangle$ is Dirac's *ket* notation for a column vector, corresponding to but more expressive than the notation $\overrightarrow{v}$ or **v** seen in many math and computer science (especially computer graphics) books. The ket is defined as

$$|\psi_A\rangle = \begin{bmatrix} a_0 \\ a_1 \\ \vdots \\ a_{N-1} \end{bmatrix}. \qquad [2.1]$$

We will find it convenient in this book to subscript both vectors and matrices starting from 0, rather than 1. $\langle\psi|$ is the *bra* corresponding to the ket. The bra is a complex-conjugate row vector,

$$\langle\psi_A| = \begin{bmatrix} a_0^* & a_1^* & \cdots & a_{N-1}^* \end{bmatrix}, \qquad [2.2]$$

recalling that a^* is the complex conjugate of a. That is, if $a = x+iy$, then $a^* = x-iy$.

$\langle\psi_A|\psi_B\rangle$ is the dot product of the two vectors ψ_A and ψ_B,

$$\langle\psi_A|\psi_B\rangle = \sum_{i=0}^{N-1} a_i^* b_i, \qquad [2.3]$$

and $|\psi_A\rangle\langle\psi_B|$ is their outer product, defined as

$$|\psi_A\rangle\langle\psi_B| = \begin{bmatrix} a_0 b_0^* & \cdots & a_0 b_{N-1}^* \\ \vdots & \ddots & \vdots \\ a_{N-1} b_0^* & \cdots & a_{N-1} b_{N-1}^* \end{bmatrix}. \qquad [2.4]$$

Most of the quantum operations we will deal with in this book can be written as *unitary* transforms. (The two major exceptions are measurement, which we will see in sections 2.3.8 and 9.1, and decoherence, which we will see in section 8.2.) A unitary matrix obeys the equation $U^\dagger U = UU^\dagger = I$, where $U^\dagger$ is the adjoint, or conjugate transpose, of U. In keeping with normal matrix multiplication rules, a series of gates or transforms applied to a register can be written as

$$U_k \cdots U_3 U_2 U_1 |\psi\rangle, \qquad [2.5]$$

where U_1 is the first gate applied, U_2 is the second, etc. This can be confusing as we draw circuit diagrams with time flowing left to right. Circuit diagrams will be introduced in section 2.4.5 below.

We will need the know the *eigenvectors* and *eigenvalues* of a matrix. The direction of an eigenvector of a operator matrix is left unchanged by the application of the operator, though its magnitude and sign may change by a scalar factor, the corresponding eigenvalue. We can write this as

$$U\,|\psi\rangle = \lambda\,|\psi\rangle\,, \qquad [2.6]$$

and read it as λ is the eigenvalue corresponding to the eigenvector $|\psi\rangle$ of the operator U. Most often, we will see eigenvalues of ± 1 and eigenvectors corresponding to the basis states below.

To compose the state of two or more qubits into one vector, or operations on multiple qubits into a single operator, we will use the *tensor product*. For two $N \times N$ matrices A and B, $A \times B$ is the $N^2 \times N^2$ matrix

$$A \otimes B = \begin{bmatrix} a_{0,0}B & a_{0,1}B & \cdots & a_{0,N-1}B \\ a_{1,0}B & a_{1,1}B & \cdots & a_{1,N-1}B \\ \vdots & \vdots & \ddots & \vdots \\ a_{N-1,0}B & a_{N-1,1}B & \cdots & a_{N-1,N-1}B \end{bmatrix}. \qquad [2.7]$$

If A and B are both 2×2 matrices,

$$A \otimes B = \begin{bmatrix} a_{0,0} \begin{bmatrix} b_{0,0} & b_{0,1} \\ b_{1,0} & b_{1,1} \end{bmatrix} & a_{0,1} \begin{bmatrix} b_{0,0} & b_{0,1} \\ b_{1,0} & b_{1,1} \end{bmatrix} \\ a_{1,0} \begin{bmatrix} b_{0,0} & b_{0,1} \\ b_{1,0} & b_{1,1} \end{bmatrix} & a_{1,1} \begin{bmatrix} b_{0,0} & b_{0,1} \\ b_{1,0} & b_{1,1} \end{bmatrix} \end{bmatrix}$$
$$= \begin{bmatrix} a_{0,0}b_{0,0} & a_{0,0}b_{0,1} & a_{0,1}b_{0,0} & a_{0,1}b_{0,1} \\ a_{0,0}b_{1,0} & a_{0,0}b_{1,1} & a_{0,1}b_{1,0} & a_{0,1}b_{1,1} \\ a_{1,0}b_{0,0} & a_{1,0}b_{0,1} & a_{1,1}b_{0,0} & a_{1,1}b_{0,1} \\ a_{1,0}b_{1,0} & a_{1,0}b_{1,1} & a_{1,1}b_{1,0} & a_{1,1}b_{1,1} \end{bmatrix}. \quad [2.8]$$

Finally, we will need the *trace* of a matrix, which is simply the sum of the diagonal elements.

2.2. Schrödinger's equation

Erwin Schrödinger, another of the founders of quantum mechanics, is famous in the popular press for his *gedankenexperiment* of the cat in a box, possibly alive and possibly dead, depending on the state of a decaying atom. This thought experiment brilliantly captures the essence of superposition and decoherence, two of the key concepts in quantum information. It remained a philosophical paradox for much of the history of quantum mechanics, while Schrödinger helped build the mathematical foundations of q.m.

Schrödinger's most important contribution to the early development of q.m. is his famous equation,

$$i\hbar \frac{\partial|\psi\rangle}{\partial t} = H|\psi\rangle, \quad [2.9]$$

describing the dynamics of a quantum system. In this equation, $|\psi\rangle$ is the *state vector*, H is an operator (represented as a square matrix) known as the *Hamiltonian* of the system. This equation is the heart of all quantum mechanics. Solutions matching the time evolution of the system are of the form

$$|\psi\rangle \to e^{-iHt/\hbar}|\psi\rangle = U|\psi\rangle, \quad [2.10]$$

where U is the unitary transform corresponding to applying H for a length of time t. Experimentalists usually describe the behavior of the system in terms of its Hamiltonian to emphasize the temporal nature of the evolution, but we are interested in specific types of behavior achieved by using fixed time intervals; so, it will be easiest for us to use the unitary operators. Unitary operators can, in turn, be expressed as gates, which we will use throughout this book.

Solutions to the Schrödinger equation can be weighted, linear combinations of any of the possible solutions such that the weights all add up to 1, which leads us to discussion of qubits and superposition.

2.3. Qubits

We have already been using the term "qubit" informally. In this section, we will look at the characteristics of qubits and how they are represented.

2.3.1. *What is a qubit?*

A classical bit is a data element with two values, 0 and 1. It can be represented using an almost endless array of physical phenomena; classical computers typically use charge in active CMOS circuits or the direction of a tiny magnetic field on a disk drive. A *qubit* is the quantum equivalent of a bit. It is represented using either a true two-level system, such as the direction of polarization of a photon or the direction of spin of an electron, or a pseudo-two-level system, such as two energy levels of an atom that can be treated as a two-level system. Of course, an electron spins in either the "up" or "down" direction, not zero and one; therefore, we chose to label the two states as our zero and one states, much as we choose e.g. +5 volts to be a logical one and ground to be a logical zero in classical circuits. The difference between a classical bit and a qubit is that a qubit can be in a *superposition* of the two states; it can be partially zero and partially one. The state of a qubit can be written as

$$|\psi\rangle = \alpha|0\rangle + \beta|1\rangle, \qquad [2.11]$$

where α and β are complex numbers, $|\alpha|^2$ is the probability of finding the qubit in the state 0 and $|\alpha|^2 + |\beta|^2 = 1$: the qubit must be found to be in one state or the other.

The above expression can also be written as

$$|\psi\rangle = \begin{bmatrix} \alpha \\ \beta \end{bmatrix} \qquad [2.12]$$

showing the same probabilities for finding the states 0 and 1, implicit in the position within the vector. The top element of the vector corresponds to the zero state and the bottom element, to the one state. Technically, the 0 and 1 inside the ket are labels for the states; we could choose to represent any two basis vectors by $|0\rangle$ and $|1\rangle$, but in this book, we will always use the convention that

$$|0\rangle \equiv \begin{bmatrix} 1 \\ 0 \end{bmatrix}, |1\rangle \equiv \begin{bmatrix} 0 \\ 1 \end{bmatrix}. \qquad [2.13]$$

The state of a single qubit is often thought of in terms of the *Bloch sphere* representation, in which the state of a qubit is a unit vector, as shown in Figure 2.1 (this sphere is often called the Poincaré sphere by researchers working in optics, when coupled with a specific choice of mapping polarization states to points on the sphere). If the vector points at the north pole, our qubit is in the $|0\rangle$ state, and if it points at the south pole, the qubit is in the $|1\rangle$ state. The north-south axis is the Z axis, the positive X axis is toward the reader (out of the page or screen, for a 2-D representation) and the Y axis is right-left. When the unit vector points toward you, that is the $(|0\rangle + |1\rangle)/\sqrt{2}$ state; when it points away from you, that is the $(|0\rangle - |1\rangle)/\sqrt{2}$ state. These two states are called the $|+\rangle$ and $|-\rangle$ (read "ket plus" and "ket minus") states. The positive Y axis is $(|0\rangle + i\,|1\rangle)/\sqrt{2}$, and the negative Y axis is $(|0\rangle - i\,|1\rangle)/\sqrt{2}$. The *phase* is the position of our vector about the Z axis (the angle ϕ in the figure).

Unfortunately, visualizing the state of more than one qubit is more complicated than a set of spheres, one per qubit. If it were that easy, there would be no exponential growth in the complexity of our states, and quantum computation would be uninteresting. It is possible to visualize the state of more than one qubit as a *set* of points on the Bloch sphere, in what is called the *Majorana representation.* Its utility is limited to pure states; there are not enough degrees of freedom to represent mixed states [MAR 04].

2.3.2. *Quantum registers and weighted probabilities*

For a single qubit, our state vector is two-dimensional, and we will use $|0\rangle \equiv \begin{bmatrix} 1 \\ 0 \end{bmatrix}$ as the zero state and $|1\rangle \equiv \begin{bmatrix} 0 \\ 1 \end{bmatrix}$ as the one state. For a n-qubit register, the state vector is $N = 2^n$-dimensional. We will often write the binary expansion of the state inside the ket, in the form $|0111\rangle$ (a four-qubit state with the value seven). This state can also be written $|0\rangle|1\rangle|1\rangle|1\rangle$ or $|0\rangle \otimes |1\rangle \otimes |1\rangle \otimes |1\rangle$, emphasizing that it is the tensor product of four separate two-level systems. This notation will be explained in more detail below.

Sometimes, we will write $|7\rangle$ as the state of the set of qubits. Although the number may be written base ten for convenience, it is represented in binary in the quantum register (many physical phenomena, such as the energy levels of an atom, may have more than two levels and therefore may use e.g. $|2\rangle$ to represent the third level, but we will confine ourselves to two-level qubits in this book). The size of the register will usually be understood from context, and if the integer is small, the high-order bits are of course understood to be zero. Occasionally, it may be necessary to write $|0\rangle^{\otimes k}$ to indicate a set of k qubits all in the zero state.

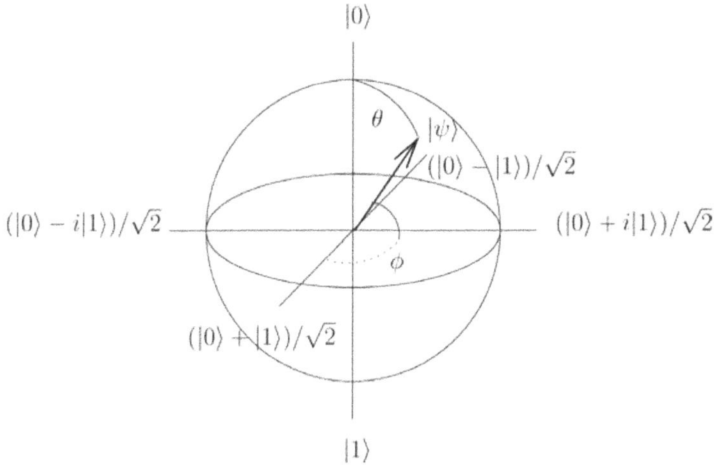

Figure 2.1. *The Bloch sphere. In this view, the positive Z axis points up, the positive X axis projects forward from the page, and the positive Y axis points to the right*

We will refer to a related set of two or more qubits as a *quantum register*. Two classical bits can be in any of the four states 00, 01, 10 and 11. Two qubits can be in a weighted combination of all four states at the same time. For two qubits, we can write

$$|\psi\rangle = \alpha\,|00\rangle + \beta\,|01\rangle + \gamma\,|10\rangle + \delta\,|11\rangle = \begin{bmatrix} \alpha \\ \beta \\ \gamma \\ \delta \end{bmatrix} \qquad [2.14]$$

where $|\alpha|^2 + |\beta|^2 + |\gamma|^2 + |\delta|^2 = 1$ and the amplitudes are complex numbers, $\alpha, \beta, \gamma, \delta \in \mathbb{C}$. For example, if $\alpha = \delta = 1/\sqrt{2}$ and $\beta = \gamma = 0$, we have a 50% probability of finding $|00\rangle$ and a 50% probability of finding $|11\rangle$, but no chance of finding the other states. This gives us the equation

$$|\psi\rangle = \frac{|00\rangle + |11\rangle}{\sqrt{2}} \qquad [2.15]$$

Similarly, three qubits can be in eight states, and n qubits can be in all 2^n possible states at once,

$$|\psi\rangle = \sum_{i=0}^{2^n-1} \alpha_i |i\rangle, \qquad [2.16]$$

subject to the constraint that their total weights, α_i must sum to 1,

$$\sum |\alpha_i|^2 = 1. \qquad [2.17]$$

For a four-qubit state, the complete state vector is $2^4 = 16$ entries long. Fortunately, many (indeed, up to $2^n - 1$) of the α_i may be, and often are, zero, allowing us to leave them out when we write out a state such as

$$|\psi\rangle = \begin{pmatrix} \frac{\sqrt{3}}{2} \\ 0 \\ 0 \\ 0 \\ 0 \\ 0 \\ 0 \\ 0 \\ 0 \\ 0 \\ 0 \\ 0 \\ 0 \\ 0 \\ 0 \\ \frac{1}{2} \end{pmatrix} \begin{matrix} \leftarrow \text{amplitude of } 0000 \\ \leftarrow \text{amplitude of } 0001 \\ \leftarrow \text{amplitude of } 0010 \\ \leftarrow \text{amplitude of } 0011 \\ \leftarrow \text{amplitude of } 0100 \\ \leftarrow \text{amplitude of } 0101 \\ \leftarrow \text{amplitude of } 0110 \\ \leftarrow \text{amplitude of } 0111 \\ \leftarrow \text{amplitude of } 1000 \\ \leftarrow \text{amplitude of } 1001 \\ \leftarrow \text{amplitude of } 1010 \\ \leftarrow \text{amplitude of } 1011 \\ \leftarrow \text{amplitude of } 1100 \\ \leftarrow \text{amplitude of } 1101 \\ \leftarrow \text{amplitude of } 1110 \\ \leftarrow \text{amplitude of } 1111 \end{matrix} = \frac{\sqrt{3}}{2}|0000\rangle + \frac{1}{2}|1111\rangle. \qquad [2.18]$$

Clearly, the more compact representation on the right is preferable to the more tedious and error-prone form in the middle!

2.3.3. *Interference*

The state of a quantum system is a wave function that matches Schrödinger's equation. As with classical wave mechanics, two waves can *interfere*, depending on the relative phases of the waves. That interference can be positive, enhancing the amplitude (hence, probability) of a particular state, or negative, decreasing the probability. Because the phase of a state is actually complex, the addition of phases is also complex.

As a simple example, let us introduce our first quantum gate, the *Hadamard gate*. The Hadamard gate is

$$H = \frac{1}{\sqrt{2}} \begin{bmatrix} 1 & 1 \\ 1 & -1 \end{bmatrix}. \quad [2.19]$$

Consider the state created by application of a Hadamard to the $|0\rangle$ state

$$|\psi\rangle = H|0\rangle = \frac{1}{\sqrt{2}} \begin{bmatrix} 1 \\ 1 \end{bmatrix} = \frac{|0\rangle + |1\rangle}{\sqrt{2}} = |+\rangle. \quad [2.20]$$

The state now consists of two terms, a superposition of two states, and is often referred to as the $|+\rangle$ state. Applying a second Hadamard gate, will return the system to its original state by interfering the two terms,

$$H|\psi\rangle = \frac{1}{2} \begin{bmatrix} 1 & 1 \\ 1 & -1 \end{bmatrix} \begin{bmatrix} 1 \\ 1 \end{bmatrix} = \frac{1}{2} \begin{bmatrix} 1+1 \\ 1-1 \end{bmatrix} = \begin{bmatrix} 1 \\ 0 \end{bmatrix} = |0\rangle. \quad [2.21]$$

The top element in the array exhibits constructive (positive) interference $(1 + 1)$, and the bottom element shows destructive (negative) interference $(1 - 1)$.

2.3.4. *Entanglement*

Two quanta can be in a shared state in which operations on one affect the other. The quanta are said to be *entangled*. One consequence is that the probabilities of two entangled qubits are not independent (but see section 2.3.6 below for an important caveat). If the state of the system is e.g. $(|00\rangle + |11\rangle)/\sqrt{2}$ ($\alpha = \delta = 1/\sqrt{2}$, in the above notation), when we measure the system, we will find either that both qubits are zero or that both qubits are one. Although each qubit has a 50% probability of being zero and a 50% probability of being one, their state is not independent. Starting from this state, we will never find one qubit to be zero and the other qubit to be one.

Entanglement is a continuous phenomenon, not discrete. There are numerous measures of the amount of entanglement present in a system, but they all use a scale running from zero to one, where zero is completely unentangled and one is fully entangled (see Munro *et al.* and references therein [MUN 01]). For the purposes of this book, our primary interest will be in fully-entangled and fully-unentangled pairs of qubits. We will see entanglement in more detail when we discuss Bell pairs below.

2.3.5. *Decoherence*

Quantum states are very fragile: excited atoms decay and spins of electrons and atomic nuclei spontaneously flip. Any quantum system can be affected by interacting with its environment, leaking information about its state out into the environment where we cannot recover or use the information. We call this gradual decay of the state of a system *decoherence*. When decoherence sets in, measurement of the system probably will not produce the desired results, causing the failure of our quantum algorithm. The two key measures of decoherence are the T_1 and T_2 times. T_1 is the energy relaxation time, and T_2 is the phase relaxation time. Both processes are memoryless, with probabilistic behavior. The amount of time we can count on the state of a qubit remaining in a usable state is a function of the minimum of T_1 and T_2. Researchers determine these values experimentally, and an important area of device research is extending these times by careful engineering of the environment and control system. Decoherence is discussed in depth in section 8.2.

2.3.6. *Pure and mixed states and the density matrix*

Quantum states can be either *pure* or *mixed*. So far, we have discussed only pure states. "Pure" does not mean that the superposition, when written out in state-vector form, contains only one term; pure means that it is *possible* to write the state in state-vector form. For example, $|\psi\rangle = |0\rangle$ and $|\psi\rangle = (|0\rangle + |1\rangle)/\sqrt{2}$ are both pure states. However, not all quantum states can be written out completely in the state-vector form. Experimentalists often prefer to write the state using the $2^n \times 2^n$ *density matrix* form, which can represent a more complex state of the system. In particular, the density matrix representation allows us to write down a representation of the state of the system when the complete state cannot be known, such as when part of the information in the quantum state has leaked out into the environment. Using the example of our basic entangled state, $|\psi\rangle = (|00\rangle + |11\rangle)/\sqrt{2}$, our density matrix is

$$\rho = |\psi\rangle\langle\psi| = \frac{1}{\sqrt{2}}(|00\rangle + |11\rangle)\frac{1}{\sqrt{2}}(\langle 00| + \langle 11|) \qquad [2.22]$$

$$= \frac{1}{2}|00\rangle\langle 00| + \frac{1}{2}|00\rangle\langle 11| + \frac{1}{2}|11\rangle\langle 00| + \frac{1}{2}|11\rangle\langle 11| = \begin{bmatrix} \frac{1}{2} & 0 & 0 & \frac{1}{2} \\ 0 & 0 & 0 & 0 \\ 0 & 0 & 0 & 0 \\ \frac{1}{2} & 0 & 0 & \frac{1}{2} \end{bmatrix}. \qquad [2.23]$$

The entries along the diagonal of the density matrix, $\rho_{j,j}\,|j\rangle\langle j|\,, 0 \leq j < 2^n$, correspond to the probability of finding the system in a particular state and hence must be real. The off-diagonal elements, $\rho_{j,k}\,|j\rangle\langle k|\,, 0 \leq j, k < 2^n, j \neq k$, are the *quantum coherences* and may be imaginary.

To be a valid density matrix, the *trace* (the sum of the diagonal) must be one, written $\mathrm{Tr}(\rho) = 1$. The trace must be one because, when measured, the system will be found to be in *some* state. For pure states, the square of the density matrix also has trace one, $\mathrm{Tr}(\rho^2) = 1$. If the density matrix is diagonalized (achieved via an appropriate change of basis), a pure state will have only a single non-zero element. The eigenvector corresponding to this eigenstate is the state of the system.

In Figure 2.1, we introduced the Bloch sphere for visualizing the state of a single pure qubit. It can also be used to visualize mixed states of a single qubit. A pure state is a vector of length 1.0, touching a point on the Bloch sphere, while mixed states are shorter, lying as points inside the sphere.

In section 2.3.4 above, we referred to a caveat on our definition of entanglement; with this understanding of the difference between pure and mixed states, we are now ready to discuss it. The state of two qubits can, in fact, be dependent, without being entangled, if the state is mixed. In contrast to the state in equation [2.23], we can also have the state

$$\frac{1}{2}(|00\rangle\langle 00| + |11\rangle\langle 11|) = \frac{1}{2}\begin{bmatrix} 1 & 0 & 0 & 0 \\ 0 & 0 & 0 & 0 \\ 0 & 0 & 0 & 0 \\ 0 & 0 & 0 & 1 \end{bmatrix}. \quad [2.24]$$

In this mixed state, the state of the two qubits is not independent, but they are not entangled; actions on one qubit cannot affect the state of the other. In this particular case, the density matrix now represents classical dependent probabilities.

More formally, the density matrix is

$$\rho \equiv \sum_i p_i |\psi_i\rangle\langle\psi_i|, \quad [2.25]$$

where p_i is the classical probability of finding the system in the pure state $|\psi_i\rangle$. We can see the difference in power between the state vector and density matrix representations and understand the difference between classical probabilities and quantum superposition a little better by looking at a simple example. The state $|+\rangle = (|0\rangle + |1\rangle)/\sqrt{2}$ is *not* the same as a qubit whose state is either $|0\rangle$ or $|1\rangle$, but is simply unknown. The former is a true superposition of waves, the latter is equivalent to the hidden state of a flipped coin before it is revealed. Writing those in density matrix form, we have

$$\rho_+ = |+\rangle\langle+| = \left(\frac{|0\rangle + |1\rangle}{\sqrt{2}}\right)\left(\frac{\langle 0| + \langle 1|}{\sqrt{2}}\right) \quad [2.26]$$

$$= \frac{1}{2}(|0\rangle\langle 0| + |0\rangle\langle 1| + |1\rangle\langle 0| + |1\rangle\langle 1|) \quad [2.27]$$

$$= \frac{1}{2}\begin{bmatrix} 1 & 1 \\ 1 & 1 \end{bmatrix} \quad [2.28]$$

for the superposition state and

$$\rho_{50/50} = \frac{1}{2}(|0\rangle\langle 0| + |1\rangle\langle 1|) \quad [2.29]$$

$$= \frac{1}{2}\begin{bmatrix} 1 & 0 \\ 0 & 1 \end{bmatrix} \quad [2.30]$$

for the classical probability distribution.

In equation [2.21], we saw how the Hadamard gate applied to the $|+\rangle$ state via the equation $H|+\rangle$ creates interference that cancels the $|1\rangle$ component, leaving us with $|0\rangle$. To apply a unitary U to a density matrix, we use

$$\rho' = U\rho U^\dagger, \quad [2.31]$$

which in these two cases gives us

$$\rho'_+ = H\rho_+ H^\dagger = H\frac{1}{2}\begin{bmatrix} 1 & 1 \\ 1 & 1 \end{bmatrix}H^\dagger = \begin{bmatrix} 1 & 0 \\ 0 & 0 \end{bmatrix} = |0\rangle\langle 0| \quad [2.32]$$

and

$$\rho'_{50/50} = H\rho_{50/50}H^\dagger = H\frac{1}{2}\begin{bmatrix} 1 & 0 \\ 0 & 1 \end{bmatrix}H^\dagger = \frac{1}{2}\begin{bmatrix} 1 & 0 \\ 0 & 1 \end{bmatrix} = \frac{1}{2}|0\rangle\langle 0| + \frac{1}{2}|1\rangle\langle 1|. [2.33]$$

Applying a Hadamard on the $|+\rangle$ state, gives us $|0\rangle$, as we expect, but doing it on the 50/50 probability state, gives us the exact same mixed state back! The superposition is two actual wave components together, whereas the mixed state is simply an unknown state. If you have a coin that has been flipped and is covered and you reach under the cover and turn it over before revealing it, you do not change the probability; it is still 50/50 heads or tails. This demonstrates the distinction between superposition components in the wave function and purely classical probabilities.

It is worth noting that the completely mixed state is the same in any basis. If it is 50/50 in the $\{|0\rangle, |1\rangle\}$ basis, it is 50/50 in $\{|+\rangle, |-\rangle\}$ or any other basis:

$$\begin{aligned}
\rho_{50/50} &= \frac{1}{2}(|+\rangle\langle+| + |-\rangle\langle-|) && [2.34]\\
&= \frac{1}{2}\begin{bmatrix} 1 & 1 \\ 1 & 1 \end{bmatrix} + \frac{1}{2}\begin{bmatrix} 1 & -1 \\ -1 & 1 \end{bmatrix} && [2.35]\\
&= \frac{1}{2}\begin{bmatrix} 1 & 0 \\ 0 & 1 \end{bmatrix} && [2.36]\\
&= \frac{1}{2}(|0\rangle\langle 0| + |1\rangle\langle 1|). && [2.37]
\end{aligned}$$

Both here and in our earlier discussion of the state vector, our basis vectors have been $\{|00...0\rangle, |00...1\rangle, ..., |11...1\rangle\}$. Rewriting a state into the density matrix using this basis, any entangled state will have at least one nonzero off-diagonal element. In section 2.5.1, we will see that it is also possible to write a state using a different basis set, in which this constraint may not hold.

2.3.7. *Fidelity*

Mixed states described using the density matrix are states about which we have imperfect information. We can quantify this as the *fidelity* of the state with respect to some desired state. We will define the fidelity as

$$F = \langle\psi|\, \rho\, |\psi\rangle, \qquad [2.38]$$

where $0 \leq F \leq 1$ is the fidelity, $|\psi\rangle$ is the state we think we have created and ρ is the density matrix of the actual state. The fidelity can also be described as the *overlap* of our actual state with the desired state. The fidelity is 1.0 for a pure state and declines as noise in the system degrades the quality of the state. For an n-qubit state, the *completely mixed* state in which all qubits are random, we have $F = 2^{-n}$.

The fidelity is often defined as $F = \sqrt{\langle\psi|\, \rho\, |\psi\rangle}$*, but we will dispense with the square root in this book, in keeping with Jozsa's definition [JOZ 94].* This definition simplifies some mathematics. If $|\psi\rangle$ contains only a single term, then the fidelity becomes the corresponding diagonal entry of the density matrix. For example, if $|\psi\rangle = |00\rangle$, then the fidelity is the upper left element of the density matrix, $\rho_{0,0}$.

Let us look at one example of manipulating kets and the density matrix. Imagine we are initializing a two-qubit register to the $|00\rangle$ state, but that the initialization process is imperfect. To learn *how* imperfect, we repeat the process a number of times and measure the state, to build up a statistical picture of our ability to create the desired state. (We will ignore imperfect measurement here.) We might find that each qubit has an independent 1% chance of being mis-set as $|1\rangle$ instead of $|0\rangle$. The d.m. would be

$$\rho = \begin{bmatrix} 0.9801 & 0 & 0 & 0 \\ 0 & 0.0099 & 0 & 0 \\ 0 & 0 & 0.0099 & 0 \\ 0 & 0 & 0 & 0.0001 \end{bmatrix} \quad [2.39]$$

$$= 0.9801\,|00\rangle\langle 00| + 0.0099\,|01\rangle\langle 01| + 0.0099\,|10\rangle\langle 10| + 0.0001\,|11\rangle\langle 11|\,. \quad [2.40]$$

(Because we have asserted that the initialization process on each qubit is independent, there is no entanglement and no off-diagonal elements here.) To calculate the fidelity with our desired state $|\psi\rangle = |00\rangle$, we have

$$\begin{aligned} F &= \langle\psi|\,\rho\,|\psi\rangle & [2.41] \\ &= \langle 00|\,\rho\,|00\rangle & [2.42] \\ &= \langle 00|\,(0.9801\,|00\rangle\langle 00| + 0.0099\,|01\rangle\langle 01| \\ &\quad +0.0099\,|10\rangle\langle 10| + 0.0001\,|11\rangle\langle 11|)\,|00\rangle \\ &= 0.9801\langle 00|00\rangle\langle 00|00\rangle + 0.0099\langle 00|01\rangle\langle 01|00\rangle \\ &\quad +0.0099\langle 00|10\rangle\langle 10|00\rangle + 0.0001\langle 11|00\rangle\langle 11|00\rangle \\ &= 0.9801. & [2.43] \end{aligned}$$

In general, for quantum networking, our error detection schemes will only work if $F > 0.5 + \epsilon$, where ϵ bounds the fidelity substantially away from one half.

2.3.8. *Measurement*

Measurement of a qubit causes the collapse of the wave function, forcing the state of the system into just one term of the superposition. In the famous thought experiment of Schrödinger, measurement is opening the box containing his cat and finding out if the cat is dead or alive. Until measurement takes place, the state of the system can be in the superposition state, with various histories and outcomes only determined probabilistically. When we measure the system, the state and history pick one consistent "storyline" that the system must have followed, in effect choosing among possible pasts based on their relative probabilities. If we measure such that more than one history remains possible, the system remains in a state that is consistent with all of them as in the double-slit quantum interference experiment (see, for example, volume I, Chapter 37 of the Feynman Lectures [FEY 63]). We will discuss interference further in sections 2.3.3 and 7.1.2.

In our basic example of $|\psi\rangle = |0\rangle$, we know the system is 100% in the zero state. Measurement of the qubit's state will definitely produce a zero [1]. For $|\psi\rangle = (|0\rangle + |1\rangle)/\sqrt{2}$, zero and one each have a 50% probability of being found. Once our measurement determines the state (e.g. 0), the entire system will be forced to a state consistent with the idea that our qubit has been zero all along.

For two or more qubits, we can measure either the entire system or only part. Measuring a single qubit can alter the state of the system. For example, consider our two-qubit state $|\psi\rangle = (|00\rangle + |11\rangle)/\sqrt{2}$. If we measure the low-order bit (the right-hand one of our pair), we have a 50% probability of each outcome, and our result will force the system to a matching state. We can write the measurement outcome and the resulting state as

$$0: \quad |\psi\rangle \rightarrow |0\rangle \qquad [2.44]$$

$$1: \quad |\psi\rangle \rightarrow |1\rangle. \qquad [2.45]$$

In this case, measuring one qubit has determined the state of the other. For the state $|\psi\rangle = (|00\rangle + |10\rangle)/\sqrt{2}$, we can factor the state as $|\psi\rangle = (|0\rangle + |1\rangle)|0\rangle/\sqrt{2}$. Measuring the low-order qubit will clearly always yield the result 0. The state of the system then moves to $(|0\rangle + |1\rangle)/\sqrt{2}$; the high-order qubit (now our only qubit) has not changed. We can say that two qubits were *separable*; there was no entanglement between them.

Measurement is a complex and sometimes counter-intuitive topic. It is important and deep enough that books and conferences are devoted to it [ALT 01]. One good place to start studying this topic is Preskill's lecture notes [PRE 98a]. We will also return to a slightly more mathematical treatment in section 9.1.

2.3.9. *The partial trace*

We are now ready to discuss the *partial trace* of a system. We use the partial trace for various purposes, including expressing the loss of a photon in optical quantum computing or the "leaking" of information about the state out into the environment.

---

1 Assuming the measurement is performed along the Z (0/1) axis. We will deal with measurements in other bases below.

We can discuss the state of a system in terms of the *system* and the *reservoir*, where system in this case refers to the qubits we are interested in and have control over, and reservoir refers to the rest of the world. Initially, the system and the reservoir are not entangled; that is, they are separable, and the state can be written as

$$\rho = \rho_S \otimes \rho_R, \tag{2.46}$$

where ρ is our overall state, ρ_S is the state of the quantum system and ρ_R is the state of the reservoir (which we can never know fully). Over time, information leaks out of the quantum system into the larger world or the reservoir. If $\rho(t)$ is the state at time t,

$$\rho = \rho_S \otimes \rho_R \cdot \rho_S(t) = \mathrm{Tr}_R(\rho(t)), \tag{2.47}$$

where Tr_R is the partial trace with respect to the reservoir.

For a two-qubit system, numbering our qubits 0 and 1, in keeping with normal computer architecture convention, we will let ρ^0 be the density matrix for the system traced out over qubit 1, and ρ^1 be traced out over qubit 0. Defining the partial trace as

$$\rho^0 = \mathrm{Tr}_1(\rho) = \langle {}_1 0|\rho|0_1\rangle + \langle {}_1 1|\rho|1_1\rangle, \tag{2.48}$$

where $|0_1\rangle$ is the basis vector for the zero state for qubit one. Noting that $\langle 0|0\rangle = \langle 1|1\rangle = 1$ and $\langle 0|1\rangle = \langle 1|0\rangle = 0$ and that the trace is linear, the partial trace for the example in equation [2.23] is

$$\begin{aligned}
\rho^0 &= \mathrm{Tr}_1(\rho) = \frac{1}{2}\mathrm{Tr}_1(|00\rangle\langle 00|) + \frac{1}{2}\mathrm{Tr}_1(|11\rangle\langle 00|) + \frac{1}{2}\mathrm{Tr}_1(|00\rangle\langle 11|) \\
&\quad + \frac{1}{2}\mathrm{Tr}_1(|11\rangle\langle 11|) \\
&= \frac{1}{2}\langle {}_1 0|00\rangle\langle 00|0_1\rangle + \frac{1}{2}\langle {}_1 0|11\rangle\langle 00|0_1\rangle + \frac{1}{2}\langle {}_1 0|00\rangle\langle 11|0_1\rangle + \frac{1}{2}\langle {}_1 0|11\rangle\langle 11|0_1\rangle \\
&\quad + \frac{1}{2}\langle {}_1 1|00\rangle\langle 00|1_1\rangle + \frac{1}{2}\langle {}_1 1|11\rangle\langle 00|1_1\rangle + \frac{1}{2}\langle {}_1 1|00\rangle\langle 11|1_1\rangle + \frac{1}{2}\langle {}_1 1|11\rangle\langle 11|1_1\rangle \\
&= \frac{1}{2}|0\rangle\langle 0| + \frac{1}{2}|1\rangle\langle 1| \\
&= \begin{bmatrix} \frac{1}{2} & 0 \\ 0 & \frac{1}{2} \end{bmatrix}.
\end{aligned} \tag{2.49}$$

$\mathrm{Tr}((\rho^0)^2) = 1/2$, indicating that our state is now a mixed state. Our pure state has become mixed with the environment, and we can no longer write down a definitive description of the quantum register alone.

2.4. Manipulating qubits

Quantum computation proceeds by taking a set of qubits, modifying their state such that a "computation" of some interest is performed and reading out the result so that we learn what happened. Feynman originally conceived of quantum computers as systems designed to simulate the physical behavior of many-body systems, which are hard to examine experimentally or in classical simulation, solving quantum mechanical problems directly in an analog fashion rather than via numerical calculation of properties of the wave function [FEY 02, LLO 96, ABR 97, BYR 06]. This approach is similar to e.g. simulating a set of mechanical resonators using a set of electrical resonators, as is done in analog computing [KOR 56, GIL 64, MEA 89]. However, this is not the only way to use quantum phenomena to solve problems. A quantum computation can be defined as a circuit, in which the system is built and programmed and behaves roughly analogously to a classical digital computer. Recent advances include adiabatic quantum computing [FAR 01, STE 03, AHA 04b] and cluster-state computing [RAU 03, NIE 05, WAL 05]. All of these are equivalent in computational power, but are believed to be very different in how useful algorithms are found. In this book, we will deal almost exclusively in terms of the circuit model, which is the basis for Shor's factoring algorithm and most of the other important quantum algorithms discovered to date.

2.4.1. *What is a quantum gate?*

In the circuit model, quantum computations are decomposed into separate gates and can be organized more or less along the lines of classical circuits. In order for our computational capabilities to be "universal", we must be able to reach any point on the Bloch sphere for a single qubit, and we must be able to entangle two qubits. First, we discuss the individual gates that compose a quantum computation, and in the next subsection, we discuss larger circuits in more detail.

Physicists, especially theorists, occasionally refer to a large unitary transform as a *quantum gate*, but in this book, we will restrict the use of the term to smaller units, which for most proposed implementations will be more physically realistic. Our gates will be one-, two-, and three-qubit transforms only.

2.4.2. *Single-qubit gates and the Bloch sphere*

Only one interesting single-bit operation, the NOT gate, exists in the classical world (ignoring setting and resetting the bit). In the quantum world, a single-qubit operation can be any rotation on the Bloch sphere. Rotations about the axes of the Bloch sphere can be described in terms of the *Pauli matrices*. The transforms for 180° rotations are

$$X = \sigma_x = \begin{bmatrix} 0 & 1 \\ 1 & 0 \end{bmatrix} \qquad [2.50]$$

$$Y = \sigma_y = \begin{bmatrix} 0 & -i \\ i & 0 \end{bmatrix} \qquad [2.51]$$

$$Z = \sigma_z = \begin{bmatrix} 1 & 0 \\ 0 & -1 \end{bmatrix}. \qquad [2.52]$$

For rotation of an angle θ about each axis, the transforms are (from Nielsen and Chuang [NIE 00]):

$$R_x(\theta) = e^{-i\theta X/2} = \begin{bmatrix} \cos\frac{\theta}{2} & -i\sin\frac{\theta}{2} \\ -i\sin\frac{\theta}{2} & \cos\frac{\theta}{2} \end{bmatrix} \qquad [2.53]$$

$$R_y(\theta) = e^{-i\theta Y/2} = \begin{bmatrix} \cos\frac{\theta}{2} & -\sin\frac{\theta}{2} \\ \sin\frac{\theta}{2} & \cos\frac{\theta}{2} \end{bmatrix} \qquad [2.54]$$

$$R_z(\theta) = e^{-i\theta Z/2} = \begin{bmatrix} e^{-i\theta/2} & 0 \\ 0 & e^{i\theta/2} \end{bmatrix}. \qquad [2.55]$$

These rotations correspond to the unitary U in equation [2.10] The angle $\theta = \omega t$, for some ω determined by the physical system, including some factor that drives the rotation of the qubit, such as a magnetic field.

Universal quantum computation requires that we be able to reach any location on the Bloch sphere starting from any other. Naturally, we do not need arbitrary rotations about all three axes in order to achieve this; two will do. Moreover, arbitrary rotations can be approximated using a small set of fixed rotations [DAW 06]. Figure 2.2 shows us such a set of gates, with their graphic representations and unitary transform matrices. The particular set shown is technically redundant; the control-Z and SWAP gates can be constructed from the others.

As a simple example, consider the state created by application of a Hadamard gate to the $|0\rangle$ state, as in equation [2.20]. The state now consists of two terms, a superposition of two states. Likewise, applying the Hadamard to the $|1\rangle$ state, we have

$$|\psi\rangle = H|1\rangle = \frac{1}{\sqrt{2}}\begin{bmatrix} 1 & 1 \\ 1 & -1 \end{bmatrix}\begin{bmatrix} 0 \\ 1 \end{bmatrix} = \frac{1}{\sqrt{2}}\begin{bmatrix} 1 \\ -1 \end{bmatrix} = \frac{|0\rangle - |1\rangle}{\sqrt{2}} = |-\rangle, \qquad [2.56]$$

which is often referred to as the $|-\rangle$ state. Geometrically, we visualize the Hadamard gate as a 180° (π) rotation about the Z axis, followed by a 90° ($\pi/2$) rotation about the Y axis. The rotation about the Z axis does not directly affect the probability of finding either a 0 or a 1 if the state is measured right away, but this two-step manipulation clearly shows the importance of the phase (angle about the Z axis).

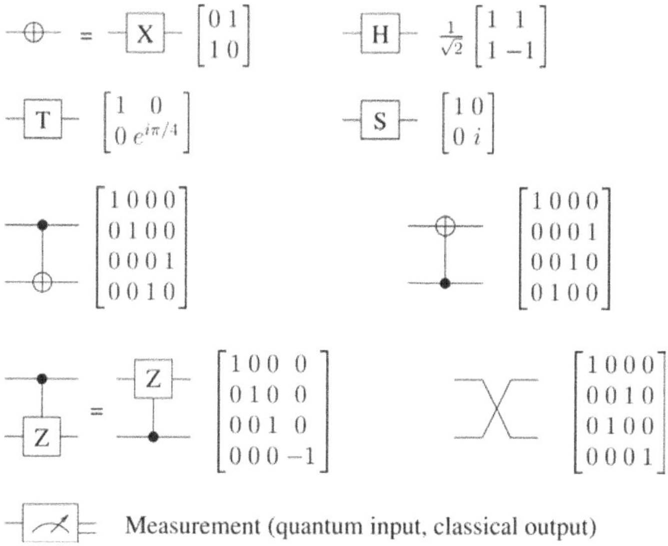

Figure 2.2. *Basic one-qubit* NOT *(X), Hadamard (H), $\pi/8$ (T), and phase (S) gates (top two rows), and two-qubit* CNOT*s, control-phase, and* SWAP *gates (bottom two rows)*

Let us look at the unitary transforms for single-qubit gates applied to two-qubit systems so we can see the form that multi-qubit operator matrices take. For operations on multi-qubit registers, we will let U_i be the single-qubit unitary operation U on the ith qubit in the register. We will number qubits from zero, with qubit zero being the "low order" qubit in the system. Qubit i then corresponds to the value 2^i in the binary expansion (note that this is in keeping with common computer architecture practice, but physicists usual number from qubit 1, starting at the left, or high-order, bit). In circuit diagrams, the low-order qubit will be the bottom qubit. Using the tensor product of equation [2.7], the transform for a Hadamard gate on the low-order qubit is

$$H_0 \equiv I \otimes H = \frac{1}{\sqrt{2}} \begin{bmatrix} 1 & 1 & 0 & 0 \\ 1 & -1 & 0 & 0 \\ 0 & 0 & 1 & 1 \\ 0 & 0 & 1 & -1 \end{bmatrix} \quad [2.57]$$

and for one on the high-order qubit is

$$H_1 \equiv H \otimes I = \frac{1}{\sqrt{2}} \begin{bmatrix} 1 & 0 & 1 & 0 \\ 0 & 1 & 0 & 1 \\ 1 & 0 & -1 & 0 \\ 0 & 1 & 0 & -1 \end{bmatrix} \qquad [2.58]$$

where I_i is the identity operation on qubit i and H_i is the Hadamard on qubit i. Because the two gates operate on independent qubits, the order in which we compose the larger unitary, or execute the two gates, does not matter

$$H_0 H_1 = H_1 H_0 = H \otimes H = \frac{1}{2} \begin{bmatrix} 1 & 1 & 1 & 1 \\ 1 & -1 & 1 & -1 \\ 1 & 1 & -1 & -1 \\ 1 & -1 & -1 & 1 \end{bmatrix}. \qquad [2.59]$$

2.4.3. *Global versus relative phase*

In the equations above, we have discussed the phase of a quantum state being the angle ϕ about the Z axis of the Bloch sphere. To be more complete, we need to discuss the difference between *global* phase and *relative* phase. Using a single-qubit state $|\psi\rangle = \alpha|0\rangle + \beta|1\rangle$ as an example, we can factor out a phase, giving

$$|\psi\rangle = e^{i\gamma}(\alpha'|0\rangle + \beta'|1\rangle), \qquad [2.60]$$

where γ is a real number such that α' is a non-negative, real number. In this form, γ is the *global phase*. Note that $|e^{i\gamma}| = 1$ so that $|\alpha'| = |\alpha|$ and $|\beta'| = |\beta|$. The probabilities of finding 0 or 1 upon measurement do not change. In fact, regardless of which basis we choose to measure our qubit in, changing that global phase will not affect the probabilities. *The global phase is unobservable and can be factored out or ignored.*

The phase of β' is the *relative phase*, showing up as a negative or imaginary term in the $|1\rangle$ entry in the state vector, and as the angle ϕ in our Bloch sphere diagram in Figure 2.1. Unlike the global phase, the relative phase is observable and indeed critical to quantum computation. It is the relative phase that is modified by a Z rotation, as in equation [2.55]. Effective use of the phase is key to creating the interference patterns used in quantum algorithm construction.

Note also that phase applies to the quantum register as a whole; although the examples discussed here are for a single qubit, applying a Z gate to *any* qubit changes the relative phase of the whole register, if the qubit is 1. This is most easily seen by recognizing that the phase is encoded in the state vector terms and that each represents a value for all of the qubits of the register. For example, in equation [2.14], a Z gate on the first qubit would modify the phase (or sign) of γ and δ, but not their value; a Z gate on the second qubit would change the phase of β and δ.

2.4.4. *Two-qubit gates*

Naturally, computations involve more than one qubit, so we must have two-qubit gates. First, let us look at the controlled-NOT gate, or CNOT. One variable (or input) is designated as the control line and the other, as the target. If the control bit is one, a NOT gate is performed on the target bit; if the control bit is zero, the target bit is left unchanged. The output is the exclusive OR (XOR) of the two bits and one of the input bits: $(a, b) \rightarrow (a, a \oplus b)$. Table 2.1 shows the truth table for a CNOT with A as the control bit and B as the target bit. Applying a CNOT gate twice to the same bits returns the system to its original state, $(a, b) \rightarrow (a, a \oplus b) \rightarrow (a, a \oplus b \oplus b) = (a, b)$.

input		output	
A	B	A	B
0	0	0	0
0	1	0	1
1	0	1	1
1	1	1	0

Table 2.1. CNOT *truth table*

When we write a CNOT gate, occasionally, it will be necessary to distinguish which qubit is which. In that case, the first subscript will be the control qubit and the second subscript the target qubit, e.g.

$$CNOT_{1,0} = \begin{bmatrix} 1 & 0 & 0 & 0 \\ 0 & 1 & 0 & 0 \\ 0 & 0 & 0 & 1 \\ 0 & 0 & 1 & 0 \end{bmatrix} \qquad [2.61]$$

and

$$CNOT_{0,1} = \begin{bmatrix} 1 & 0 & 0 & 0 \\ 0 & 0 & 0 & 1 \\ 0 & 0 & 1 & 0 \\ 0 & 1 & 0 & 0 \end{bmatrix}. \qquad [2.62]$$

In some physical implementations, a control-phase gate is the natural Hamiltonian. The control-phase or control-Z unitary is

$$CZ_{1,0} = \begin{bmatrix} 1 & 0 & 0 & 0 \\ 0 & 1 & 0 & 0 \\ 0 & 0 & 1 & 0 \\ 0 & 0 & 0 & -1 \end{bmatrix}, \qquad [2.63]$$

or, more generally, for an arbitrary rotation by an angle θ about the Z axis,

$$CZ_{1,0}(\theta) = \begin{bmatrix} 1 & 0 & 0 & 0 \\ 0 & 1 & 0 & 0 \\ 0 & 0 & 1 & 0 \\ 0 & 0 & 0 & e^{i\theta} \end{bmatrix}, \qquad [2.64]$$

which is not quite what we need for most logic. However, we can construct a CNOT gate from CZ easily by wrapping the CZ in a pair of Hadamards on the target qubit:

$$H_0 CZ_{1,0} H_0 = CNOT_{1,0}. \qquad [2.65]$$

DiVincenzo described other related constructions in an early paper [DIV 98]. The control-phase gate is actually symmetric; it does not matter which of the two qubits we treat as the control and which we treat as the target. The change in the system state is the same.

A final two-qubit gate of some interest is SWAP, which exchanges two variables and has a very simple transform,

$$SWAP = \begin{bmatrix} 1 & 0 & 0 & 0 \\ 0 & 0 & 1 & 0 \\ 0 & 1 & 0 & 0 \\ 0 & 0 & 0 & 1 \end{bmatrix}. \qquad [2.66]$$

2.4.5. *Quantum circuits*

A quantum computation, in the abstract, is a unitary transformation on the initial state of the system, creating a desired state, which we can then measure. The complete unitary transform on n qubits, of course, is a $2^n \times 2^n$ matrix; therefore, direct construction of the unitary to implement a complex function of more than a few qubits is difficult. The physical phenomena used for quantum computation do not, in general, lend themselves well to direct implementation of complex transforms; the physical interactions are two-party. Moreover, human beings are not good at imagining such large systems, but are very good at composing large systems from smaller components. Thus, the abstraction of a *quantum circuit* is important. A quantum circuit effects the overall transform via a series of smaller gates (generally, one- to three-qubit gates) applied in a prescribed order on the appropriate qubits.

Researchers have found several methods for decomposing a specific unitary transform into a series of small gates or operations that we know how to implement. Some methods find optimal evolution paths (not necessarily composed of discrete gates) but are highly theoretical, and it is not immediately clear how to compile a large program by employing these methods [NIE 06, CAR 06]. Using the most general method, the number of gates grows exponentially as the size of the problem increases, negating any advantage in computational complexity that quantum computing appears to offer [SCH 03]. Most of the work on quantum programming languages and tools for them essentially defers the decomposition problem to the programmer [GAY 05, ÖME 02, AHO 03, SVO 06]. Fortunately, many quantum algorithms depend on a few basic building blocks that have known efficient decompositions (such as the quantum Fourier transform) or on ideas translated directly from classical analogs (such as arithmetic).

Figure 2.3 shows a simple example of a quantum circuit with four qubits. Using the symbols shown in Figure 2.2, this circuit consists of two Hadamard gates and three CNOT gates. Gates on different qubits can be executed at the same time, as shown by the vertical alignment. A figure like this expresses the ordering and dependencies of the gates, but different gates may actually take very different amounts of time to execute.

2.5. Bell pairs

Teleportation, which forms the foundation of much quantum networking, consumes a resource known as a *Bell pair*. Bell pairs, described qualitatively above, are the canonical example of entanglement, and in fact, one form of Bell pair appeared in equation [2.15] in section 2.3.4.

The Bell pairs, or Bell states, were originally developed by John Bell in the 1960s as part of a thought experiment extending and clarifying the work of Einstein, Podolsky and Rosen (EPR). The three researchers recognized that the equation of quantum mechanics required "spooky action at a distance", and therefore felt that q.m. must be incomplete. Bell saw that this non-local nature could be tested statistically; if an experiment on a pair of entangled quanta could be repeated a large number of times, then the results could be analyzed to rule out the possibility that what we see as an entangled state is actually a pre-determined state we simply do not fully understand. A theory assuming that the two quantum subsystems do not actually affect each other over a distance is known as a "hidden variable" theory. Bell formulated the characteristics of local and non-local action in terms of an inequality. If the inequality holds, classical probability and a hidden variable theory are enough to describe the results; if the inequality is violated by the experiment, only non-local effects between the separated subsystems can explain the results. Violation is viewed as support for the non-local interpretation of quantum mechanics.

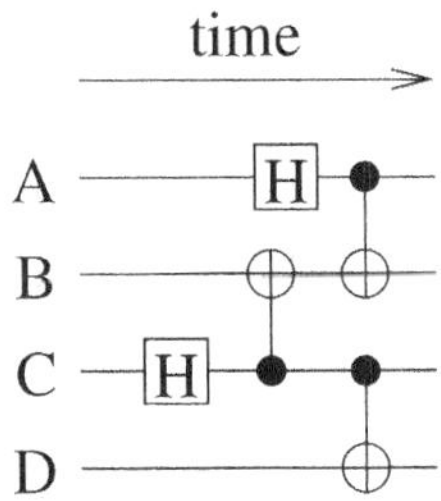

Figure 2.3. *A simple quantum circuit. Time flows left to right, each horizontal line is a qubit (a quantum variable), and each box or vertical symbol is a one- or two-qubit gate*

A Bell pair consists of two qubits whose states are not independent; they are *entangled*, as introduced in section 2.3.2. That is, when we measure one of the qubits, we get one of two possible outcomes. We cannot predict the outcome of the measurement, only the probability of the two outcomes; for Bell pairs, the probability is always 50/50. However, once we know the result of *one* measurement, we can then predict with certainty the outcome of the *other* measurement!

There are four forms of Bell pairs:

$$|\Phi^+\rangle = \frac{|00\rangle + |11\rangle}{\sqrt{2}} \qquad [2.67]$$

$$|\Phi^-\rangle = \frac{|00\rangle - |11\rangle}{\sqrt{2}} \qquad [2.68]$$

$$|\Psi^+\rangle = \frac{|01\rangle + |10\rangle}{\sqrt{2}} \qquad [2.69]$$

$$|\Psi^-\rangle = \frac{|01\rangle - |10\rangle}{\sqrt{2}} \qquad [2.70]$$

Being a little more explicit, we can write the first of those pairs in the form

$$|\Phi^+\rangle = \frac{|0\rangle_A \otimes |0\rangle_B + |1\rangle_A \otimes |1\rangle_B}{\sqrt{2}} \qquad [2.71]$$

to emphasize that the Bell pair consists of two qubits, A and B, which may be separated in space. Alice may hold A, while Bob holds B, and they can be an arbitrary distance apart without the behavior of the Bell pair changing.

In this $|\Phi^+\rangle$ Bell pair, either both qubits are zero or both qubits are one, with equal probability. When Alice measures her qubit and finds a one, she can be sure that when Bob measures his qubit, it will be a one. Likewise, if she measures zero, Bob will measure zero. In $|\Psi^+\rangle$, in contrast, we know that the qubits are anti-correlated; if Alice measures a one, Bob will measure a zero and vice-versa. Moreover, this effect does not change if Bob measures his qubit first or if they both measure their qubits simultaneously.

Even when the two parts of the Bell pair are separated, e.g. when held by Alice and Bob, it is possible to convert from one Bell state to another using only local operations. That is, Alice or Bob may operate on her or his qubit and change the state, without consuming any entangled quantum resources or requiring any long-distance quantum operations. Interconvertibility is under these constraints is known as *LOCC, local operations and classical communication.* Working out the exact operations that Alice or Bob applies to change one to another is left as an exercise for the reader.

2.5.1. *The Bell basis*

The four types of Bell pairs can also be used as a *basis set* for writing two-qubit states. The *computational basis* is the four vectors $\{|00\rangle, |01\rangle, |10\rangle, |11\rangle\}$, which can be combined in superpositions and can represent any possible, pure, two-qubit state. Alternatively, we can use the four Bell pairs $\{|\Phi^+\rangle, |\Phi^-\rangle, |\Psi^+\rangle, |\Psi^-\rangle\ \}$ as a basis set, known naturally as the *Bell basis.* In equation [2.67], we defined $|\Psi^+\rangle$ in the computational basis using terms of $|00\rangle$ and $|11\rangle$, but we can also write $|00\rangle$ in the Bell basis,

$$|00\rangle = \frac{|\Phi^+\rangle + |\Phi^-\rangle}{\sqrt{2}}. \qquad [2.72]$$

Verification of this and extension to the other cases are left as exercises for the reader. We will see full density matrices written in the Bell basis beginning in section 8.2.3 and will use them extensively in the rest of this book.

2.5.2. *Measurement in the Bell basis*

An operation we will need is *Bell measurement.* Rather than determining whether a register was in the state $|00\rangle$, $|01\rangle$, $|10\rangle$ or $|11\rangle$, Bell measurement determines if it was in $|\Phi^+\rangle$, $|\Phi^-\rangle$, $|\Psi^+\rangle$ or $|\Psi^-\rangle$. As with computational basis measurement, the pre-measurement state may have included a superposition of Bell basis terms (e.g. $(|\Psi^+\rangle + |\Psi^-\rangle)/\sqrt{2}$), but measuring in the Bell basis destroys any superposition and leaves the register in exactly one state.

Some measurement operations may be non-destructive, leaving the original qubits intact but possibly destroying some of the terms in the superposition. This *projects* the state of the qubits into a state and can be used to create entanglement; we will see such operations in Chapter 8. For our purposes in this chapter, destructive measurement of the two qubits is adequate, leaving behind only classical data without entanglement.

Figure 2.4 shows a simple circuit for determining which of the four Bell states a pair of qubits (A and B) held. The first control-NOT operation changes B to hold the *parity* of the two qubits so that a measurement of B in the computational (Z) basis will tell us whether the pair was $|\Phi\rangle$ or $|\Psi\rangle$. Measuring A in the Hadamard ($+/-$, or X) basis then tells us whether the pair was $+$ or $-$. Table 2.2 shows which state was present depending on the measurement outcomes.

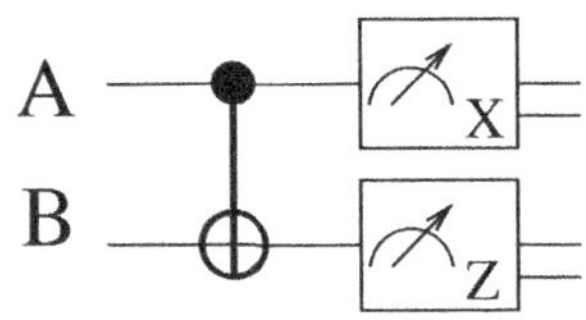

Figure 2.4. *A simple circuit showing one way to execute Bell measurement*

A	B	Bell state
$-$	1	$\|\Psi^-\rangle$
$+$	1	$\|\Psi^+\rangle$
$-$	0	$\|\Phi^-\rangle$
$+$	0	$\|\Phi^+\rangle$

Table 2.2. *Correlating the measured values of A and B after the circuit in Figure 2.4 tells us which Bell state the register was in*

2.5.3. *The Bell inequalities and non-locality*

In the 1930s, Einstein, Podolsky and Rosen (EPR) proposed a thought experiment intended to demonstrate the incompleteness of quantum theory, showing that the theory required *non-local* (or "spooky") action, that two parts of an entangled quantum system would affect each other even if physically separated and even if independently measured outside of the *light cone* of the other, that is, faster than any signal limited by the speed of light could travel from one subsystem to the other.

The modern form of Bell's inequality is due to Clauser, Horne, Shimony and Holt [CLA 69], and is called the CHSH inequality. It involves an initial Bell pair of the form $|\Psi^-\rangle$ (also known as a "singlet state", which we will discuss further in Chapter 7), and then having Alice and Bob each measure their quantum subsystem along one of two possible axes, or detector settings, chosen at random. To maximize the inequality, the two pairs of axes are different at Alice and Bob.

In our notation, we will assume measurement of A (by Alice) either in the Z basis, which will give us one of $\{|0\rangle, |1\rangle\}$ as a result, or in the X $\{|+\rangle, |-\rangle\}$ basis. If Alice chooses to measure Z, she gets the bit value (0/1) of her half, but no information about the phase, whereas if she chooses to measure X, she gains the phase information but not the bit value. Each of these measurements produces either the $+1$ eigenvalue or the -1 eigenvalue; e.g. $|0\rangle$ is the $+1$ eigenvalue of the Z measurement and $|1\rangle$ is the -1 eigenvalue.

Measurement of B is then conducted using bases set at multiples of angles halfway between those two. Recall that the angle between the Z and X axes on the Bloch sphere is $\pi/2$. We will set the basis vectors for Bob to use, which we will call ϕ and ϕ', at angles $\pi/4$ and $3\pi/4$ from Z, giving $\angle(Z,\phi) = \angle(X,\phi) = \angle(X,\phi') = \pi/4$ and $\angle(Z,\phi') = 3\pi/4$. Recalling the relationship between angles on the Bloch sphere and the elements of a single-qubit state vector, the first detector setting for Bob (the ϕ axis) becomes $\{(\frac{1}{2}\sqrt{2+\sqrt{2}}\,|0\rangle + \frac{1}{2}\sqrt{2-\sqrt{2}}\,|1\rangle), (\frac{1}{2}\sqrt{2-\sqrt{2}}\,|0\rangle - \frac{1}{2}\sqrt{2+\sqrt{2}}\,|1\rangle)\}$ and the other detector setting (the ϕ' axis) becomes $\{(\frac{1}{2}\sqrt{2-\sqrt{2}}\,|0\rangle + \frac{1}{2}\sqrt{2+\sqrt{2}}\,|1\rangle), (\frac{1}{2}\sqrt{2+\sqrt{2}}\,|0\rangle - \frac{1}{2}\sqrt{2-\sqrt{2}}\,|1\rangle)\}$.

In running an experiment, Alice and Bob will each choose one of their measurement bases at random so that there are four possible settings: (Z_A, ϕ_B), (Z_A, ϕ'_B), (X_A, ϕ_B) and (X_A, ϕ'_B). We will use N_{++} to represent finding the $+1$ eigenstate for both A and B, regardless of the choice of detector setting. Then, we define the *quantum correlation* as

$$E = \frac{N_{++} + N_{--} - N_{+-} - N_{-+}}{N_{\text{total}}}, \qquad [2.73]$$

that is, the fraction of time the detectors find the same eigenvalue minus the fraction that they differ. The CHSH inequality is then expressed in terms of E for the various detector setting combinations,

$$S = E(Z_A, \phi_B) - E(Z_A, \phi'_B) + E(X_A, \phi_B) + E(X_A, \phi'_B). \qquad [2.74]$$

Classically,

$$-2 \leq S \leq 2, \qquad [2.75]$$

while if the non-local interpretation of quantum mechanics holds, each of the four terms in S can reach an absolute value of $1/\sqrt{2}$, and the total may be up to $\pm 2\sqrt{2}$.

2.5.4. *Experimental demonstration of violation of Bell's inequality*

John Bell formulated his inequality in 1964, and over the ensuing four decades, experimental demonstrations of increasing sophistication have been performed using a variety of qubit types and apparatuses. The most famous experiments, widely acknowledged as definitively demonstrating violation of the inequality, are those by Aspect, Grangier and Roger in 1981 and 1982 [ASP 81, ASP 82, ASP 99]. The story of these experiments is told well in a lightly fictionalized form in Gilder's *The Age of Entanglement* [GIL 08].

As their qubits, they used pairs of photons entangled in the polarization degree of freedom. (These experiments predate the term "qubit", as well as the form of notation used in this book, but we will describe them in current terminology.) Recall that the Bloch sphere is a mathematical abstraction, which in general does not correspond to the physical orientation of the physical carrier of the qubit. In the case of photon polarization, $|0\rangle$ may be a vertical ($0°$) polarization of the photon, sometimes written $|V\rangle$, while the orthogonal state $|1\rangle$ may be horizontal ($90°$) polarization ($|H\rangle$). We will use horizontal and vertical polarization as Alice's first basis and $45°$ and $135°$ rotated polarization as her second basis. Alice's two choices of basis are sometimes written as the $+$ and $\times$ bases. Bob's bases are set at a $22.5°$ angle to those.

In Figure 2.5, the photon pairs are generated by the source marked **S**, with one being given to Alice and one to Bob. Alice and Bob each select a polarization basis, and the two detectors marked $+$ and $-$ then detect the $+1$ or -1 eigenstate, respectively. In the most straightforward setup, the polarizers are physically rotated to the correct angle. The box marked **CM** is the coincidence monitor, which counts the statistics used in equation [2.73] for each pair of settings.

Experimental proofs such as this involve a lot of work to eliminate possible loopholes. One such is that somehow the entangled pair generate "detects" the settings of the polarizers and generates photons accordingly so that closing the loophole requires that the setting be determined and performed *after* the creation of the photons, which is physically difficult to accomplish. Another is that detection fails frequently, due to failure to generate a pair, loss of the photons in the channel, absorption or reflection at the polarizers or failure to trigger the detector. We must, in general, depend on the assumption that such failures are not statistically biased in a way that would affect the outcome of the experiment. Under such an assumption, it is common to replace N_{total} in equation [2.73] with $N_{++} + N_{--} + N_{+-} + N_{-+}$, eliminating the cases with only zero or one photon detection.

Long-distance tests of Bell's inequality were performed in Geneva in 1997 and 1998, over a distance of more than 10 km [TIT 99]. Other measures of entanglement have also been demonstrated [ALT 05].

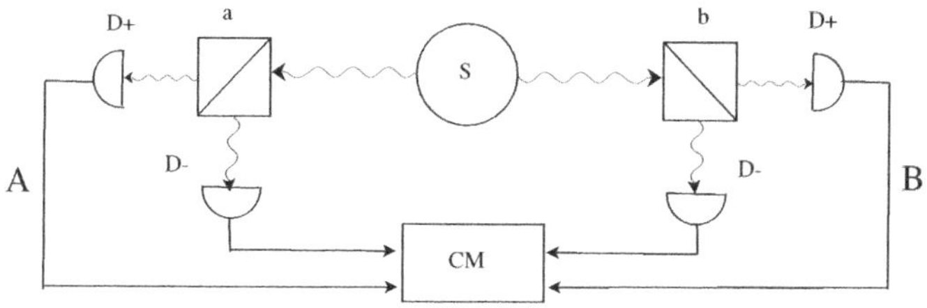

Figure 2.5. *A generic representation of a Bell inequality test setup, similar to that of Aspect et al.*

2.6. The no-cloning theorem

Quantum data is subject to a restriction known as the "no-cloning theorem". Standard, classical, methods of controlling error cannot be implemented [WOO 82]. Quantum data cannot be "backed-up" or copied for simple repetition code processing to detect and corrected for errors.

Using a CNOT appears at first to copy data; input a qubit $|\psi\rangle$ as the control and a $|0\rangle$ as the target and out come two qubits much like $|\psi\rangle$. In fact, this technique is commonly used in a "fanout" operation, when we need access to copies of the same variable in different places in a system or algorithm. However, what actually happens is not the creation of two independent copies of $|\psi\rangle$, but instead the entanglement of the two qubits into a state something like a $|\Phi^+\rangle$ Bell state. If $|\psi\rangle = \alpha|0\rangle + \beta|1\rangle$, then

$$|\psi\rangle|0\rangle \rightarrow \alpha|00\rangle + \beta|11\rangle. \quad [2.76]$$

If $|\psi\rangle = |+\rangle$, then the resulting state is exactly $|\Phi^+\rangle$; other input states will result in different outputs, of course. In fact, some input states will result in no change in the amount of entanglement, but in general, this approach alters the amount of entanglement.

Measuring one qubit of this state, we can see, results in the collapse of the superposition and both qubits assuming the same state and entanglement being destroyed. Thus, the two qubits are *not* independent copies of the original $|\psi\rangle$ and cannot be used as such.

The theorem describing this behavior was developed by Wootters and Zurek back in 1982 and is a fundamental principle affecting the behavior of quantum networks including both quantum key distribution and repeaters [WOO 82].

2.7. Conclusion

The material in this chapter should provide a solid foundation in quantum information. We have covered all of the critical concepts necessary to understand the material in this book. In fact, this foundation should be enough to allow a dedicated reader to read almost any of the referenced papers and much of the new research that is published on a continuing basis.

In addition to this chapter, we will see fundamental concepts introduced later in the book: some additional, valuable distributed entangled states in section 6.1, the stabilizer representation in section 6.1.1 and quantum error correction in section 11.1.

Chapter 3

Networking Background

This chapter gives a brief introduction to key networking problems. The special challenges posed by Internet-scale networks and internetworks are covered. This is not intended to be a comprehensive introduction to networking; for a basic introduction, see, e.g., Kurose and Ross, Peterson and Davie, or Tanenbaum [KUR 12, PET 11, TAN 10]. Moreover, much of what is in this chapter is specific to Internet-related networking; the terminology and relationship among entities will differ in other types of networks. The focus here will be on laying the groundwork to adapt Internet design principles to the development of quantum repeater networks.

Communication is either the exchange of messages or the sharing of state between two or more parties (the two models are known to be equivalent). This allows the parties to compute or make decisions based on data that someone or some other node previously held, or to synchronize some activity. Quantum communication will be either the transfer of a quantum state (entangled or not), the creation of an entangled state, or the use of a previously established entangled state.

Communication can be based on one or more *messages*, or it can be a continuous *stream*. Communication sessions may guarantee reliable delivery, in-order delivery, both or neither. Sessions may also be *unicast*, sent from one sender to one receiver, *multicast*, sent from one sender to a group of receivers, or *broadcast*, where all nodes within hearing range of the message are expected to accept it.

Communication, of course, depends on our ability to get our messages from the *source* to the *destination*. If there are only two participants and they are in the same room, this is a rare problem. With more participants, regulating conversations becomes more complex. Moreover, with multiple participants, it is normal for them to be spread out, rather than in a single room. When communication over a distance, with multiple participants, is required, we turn to a network architecture.

Metcalfe's Law states that the utility of a network grows as the square of the number of participants. This value is easily seen in the adoption of the Internet; the more people who were communicating via the Internet, the more desirable it became for those who were not yet connected to join. However, creating a network architecture that will remain robust as it grows is an enormous technical challenge. The Internet has encountered this challenge admirably, so we have used it as our primary model in this book.

In the popular vernacular, the term "the Internet" is used as if there were a single global data network. More correctly, the Internet is a *network of networks*, or an *internetwork*. In fact, the Internet consists of more than 40,000 separate networks that are capable of exchanging small messages known as *data packets*. The underlying technology of each of those networks varies dramatically in physical implementation, age, speed, stability, size and many other characteristics, and individual networks are owned and operated by many different entities, from individuals to universities to corporations to governments.

Quantum networks will need to integrate their services into the Internet. The quantum Internet will run alongside the classical Internet to provide new or more capable services to the users.

In this chapter, we have first discussed some key network concepts, then the major challenges in building planet-wide systems at scale. We have presented some important design patterns that are used in creating networks, then very briefly review the Internet itself.

3.1. Concepts

3.1.1. *Multihop communication: networks as graphs*

A *network* is a group of interacting parties, where each of the members potentially wants to interact with any of the other members. A computer network consists of *nodes*, connected via *links*, which carry messages from node to node. A *network node* can be a laptop computer used by a human, a large server computer, or a node "internal" to the network, such as an Internet *router*, which *forwards* messages toward their destination based on an address. In classical networking, nodes acting as message sources or sinks are called *hosts*. In the Internet, these messages are called *packets*, and they carry metadata information for addressing and for managing a communication session, and of course often the data that a sender (called a *source*) wishes to transmit to one or more receivers (*destinations*).

We call the collective set of messages on the network the *traffic load*. Until we can measure the traffic on a production quantum network directly, a gravity model might be a reasonable first guess [MED 02] because the human patterns driving quantum communication are expected to be similar to those for classical communication.

A node uses a *physical interface* to connect to the link. The links may be bidirectional or unidirectional, and the bidirectional links can be further divided into full- and half-duplex, depending on whether they support concurrent bidirectional transfer or must be time multiplexed. The links are also described as two-party or multiparty. Two-party links involve two known, fixed nodes at opposite ends of the link; multiparty links involve multiple nodes.

Two-party links, or *channels*, are the basis of Shannon's model of communication, but $\approx N^2/2$ links are required to directly connect everyone in an N-node network to everyone else (see Figure 3.1, left side). More commonly, such full connectivity is supported with a smaller set of links using *multihop forwarding* (see Figure 3.1, right side). Forwarding is the fundamental concept that enables scalability in physical systems, both in distance and number of nodes: individual nodes have multiple physical interfaces, receive messages and make decisions about how best to send the messages on to their respective destinations.

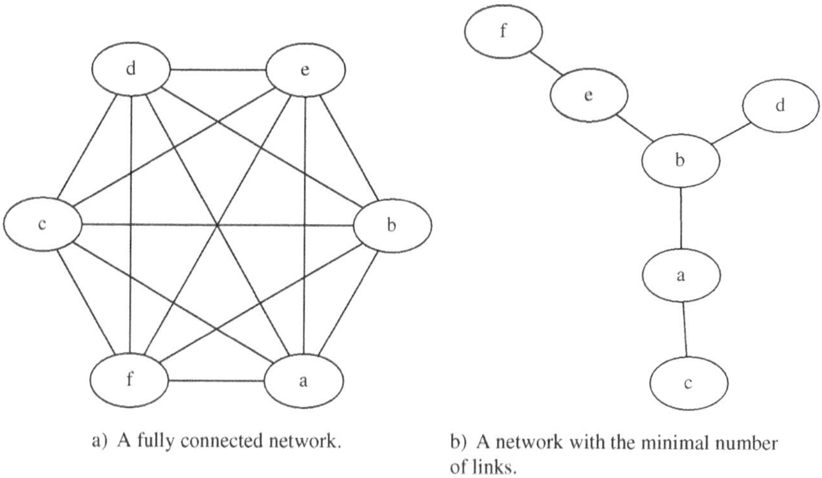

a) A fully connected network.

b) A network with the minimal number of links.

Figure 3.1. *Multiparty communication without message forwarding requires $O(N^2)$ links (left), but with message forwarding can use as few as $O(N)$ links (right)*

Given two nodes that want to communicate, a *path* must be found along the existing links, and communication along that path cascaded. *Routing* is the continuous background process of determining the topology of the network, including any recent changes, and calculating the preferred path for packets to take[1]. *Packet forwarding* is the basis of the Internet, and in contrast to routing is the real-time action taken to move a packet out of the node, using the paths pre-determined by the routing process.

Routers are specialized nodes in the "interior" of the network, responsible for accepting messages arriving on links, determining which neighboring node offers the best path toward the destination, and forwarding the messages along outbound links to the appropriate neighbor (known as the *next hop*). They also occasionally apply some policy-determined operation, such as modifying the packet (rewriting portions of it) or even dropping packets that violate some condition. They may store the packet for later processing or forwarding, depending on the availability of the outbound link, or to enforce some constraint such as message ordering or *quality of service* guarantees. Routers are also allowed to drop messages without delivering them, when overloaded or otherwise unable to forward them.

Many concepts from graph theory play an important role in the discussion of networks. One of the important concepts is the *diameter* of a network. In general, two nodes will want to communicate using the shortest path between them. The diameter is the longest such shortest path between any two nodes on the network. The diameter of the left side of Figure 3.1 is simply 1, whereas the diameter of the right side is 4. The diameter of the Internet is difficult to determine precisely, but the Cooperative Association for Internet Data Analysis (CAIDA) shows a maximum number of hops of 36 from one of its measurement points to any of its measurement destinations, with a median value of 18 hops, as of January 2014[2].

The *node degree* is the number of links that a given node has. The distribution of this number (how many nodes have only one link, how many have two, three, etc.) across the entire network graph is an important characteristic of a network, helping us to evaluate the traffic the network can carry, the latency from place to place, the robustness of the network to failures of either nodes or links. However, we must be careful not to assign too much meaning to the distribution alone. The type and bandwidth of the links matter, and each link that we can see may represent a single fiber or cable, or an entire, resilient path through a lower-layer network; Internet backbone links, in particular, are often redundantly provisioned at a level not visible to the routing protocols. Moreover, the performance and fault tolerance of the network varies depending on whether high-degree nodes are presumed to be near the center of the network or near the edges; in practice, nodes near the edge of the

---

1 In some wireless networks, this probing of the network and path calculation is done *on demand*, but here we have assumed the more common background operation.

2 http://www.caida.org/projects/ark/statistics/san-us.html.

Internet (close to the home users) have many low-bandwidth links, whereas those in the middle have fewer higher-bandwidth links [DOY 05, MAH 06].

It is worth pointing out that some researchers have been concerned about the vulnerability of the Internet to failures of, or attacks on, specific nodes. However, the network itself is actually remarkably robust, and heals itself as nodes and links come and go. A larger worry is that a previously-unknown bug in the software from a common router vendor might surface, bringing down, or allowing attacks upon, a large fraction of the routers on the network. Because the Internet is fundamentally founded on trust and cooperation among the participants, it is also vulnerable to, and occasionally suffers from, deliberate or accidental misconfiguration of the routing system that determines how packets get where they are going.

3.1.2. *Resources*

Execution of any communication or computation consumes *resources*. Resources may include memory, bandwidth or access to a network link, or CPU time. Effective management of those resources, especially in the context of heterogeneous technology, distributed operation and possibly autonomous management, is a challenging problem. Over-commitment of resources or an offered traffic load exceeding the network capacity can result in *congestion*; occurring even in only a few places, it can result in failure of communication sessions or even collapse of the entire network's ability to make forward progress. In contrast, under-utilization of the resources is economically inefficient.

Resources are managed to further some specific system design goal. Our choice of *multiplexing discipline* determines how a resource may be shared by multiple active tasks. The multiplexing discipline may be designed to maximize throughput, minimize idle time or share the resources as broadly as possible.

The simplest abstract network with a shared link for separate connections, giving distinct resource contention on different links, is the dumbbell network shown in Figure 3.2. A conversation or Internet connection between New York and Los Angeles may use a link between Washington and Denver, and a second conversation between Baltimore and Phoenix may also choose to use (or be assigned to use) that same Washington-Denver link.

Jain's fairness measure for resource allocation gives us a precise mathematical means of assessing the balance of a system with competing users [JAI 91]. If the performance of user i is x_i, with n competing users, the fairness is

$$\mathcal{J}(x_1, x_2, \ldots, x_n) = \frac{(\sum_{i=1}^{n} x_i)^2}{n \sum_{i=1}^{n} x_i^2}. \quad [3.1]$$

A fairness of $\mathcal{J} = 1.0$ indicates perfectly even distribution of resources among the users, whereas $\mathcal{J} = 1/n$ indicates that one user acquired all of the resources, shutting out all other users. We will use this measure in Chapter 13 when we discuss multiplexing of channel use and buffer memory.

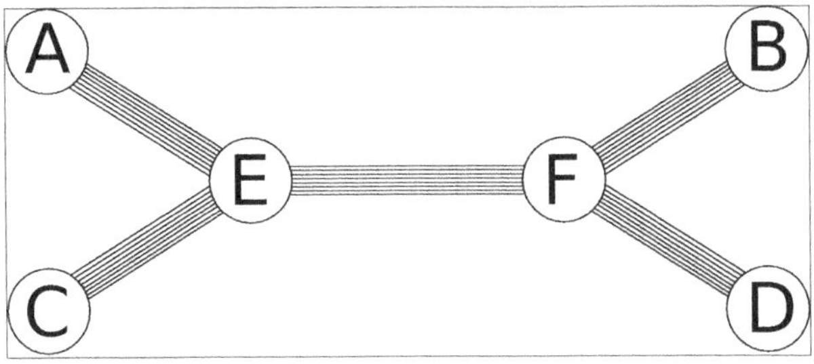

Figure 3.2. *The dumbbell network, the simplest network with a shared link (EF). When A wishes to communicate with B and C wishes to communicate with D, the resources (represented by multiple lines) must be allocated to the two competing sessions*

3.1.3. *Protocols*

A *network protocol* is the set of rules that two parties use to communicate. It is the *language* that they share, and the rules for what they should say and when. A human telephone protocol, for example, includes the shared understanding that the caller creates the connection, but the callee initiates the conversation by saying, "Hello?" once he has answered the phone and brought it to his ear. A protocol specifies the format, the contents and the semantics of the messages that are exchanged. Typically, the order and timing of messages are included.

The role of a protocol is carefully circumscribed. It is common for several protocols to be composed to build a complete system. One protocol, for example, may provide only a means to transmit data across a single link, whereas another protocol may use the service provided by the first in order to create a network that carries data across multiple hops. We see the common means of composing protocols when we discuss *layering* later.

The protocols are implemented at the nodes using hardware, software or a combination of the two. Other software accesses the services provided by the protocol through an *interface*. On systems with a UNIX heritage, some of the protocols are implemented inside the operating system itself, whereas others are implemented in user software. The internal and external interfaces may be different. The interface known as *sockets* provides access to a variety of protocols.

Note that this software interface is independent of the protocol specification; a protocol specifies only the externally visible behavior – what messages are exchanged across the wire, and what behavior is expected of the nodes.

3.1.4. *Naming and addressing*

Delivery of messages depends on *naming*, *addressing* and *routing*. A name is a label attached to an entity, and intended for use in identifying that entity; any given entity may have many names. An address is a form of name intended to guide delivery of a message, and as such often exhibits some sort of structure, such as hierarchy. This distinction is much like that between a person's name and the street address at which the person lives. It is often said that naming is *who* you want to talk to, addressing is *where* they are and routing is *how* a message can find its way to them.

Names, and many forms of address, are not directly actionable when we need to deliver a message. First, we must translate the name we have into an address appropriate for our current context. In computer systems, *name resolution* is a critical issue. A name is *resolved* in a particular *context* or *namespace*, or *closure*. Names may be either *globally* or *locally* unique, depending on the namespace. Translation is typically done when an entity (e.g. a program or software module) is holding one form of the name of another entity with which it wishes to communicate, but a different form of name must be used to actually conduct the communication.

Entities that need to be named come in many different forms:

Nodes: both end hosts and the routers in the middle of the network are the recipients of messages; hence, we must be able to name them. At a minimum, a node must have at least one address, and it may also have other forms of name.

Networks: individual networks or collections of networks also need to be named, for a variety of purposes. In Internet terminology, a large organization typically has one or more *autonomous systems* (ASes), which form the granularity of routing control at the global level. An AS is identified by its *AS number*. Each AS may, in turn, consist of a number of *subnets*, which generally consist of a power-of-two block of numeric Internet Protocol (IP) addresses.

Services and ports: communication sessions are processed by the software entities at both ends, generally called *client* and *server* processes. Applications wanting to connect to a server must have a name by which they can identify the machine itself, and a name for the *service* which they wish to access, which is usually instantiated as a software process to handle the session. In the Internet world, those services are often identified by a *port number*, and many services (e.g. mail, Web service) use *well-known ports*, identified in a standardized table and known to the client software.

(Note that "port" here is a software construct; the word is also used to refer to the physical connection point on an Ethernet switch.)

Users: sometimes, we need to identify the users themselves.

Memory locations: the memory location at which a specific data element is stored inside a computer is generally considered a private matter. However, a few protocols do share such addresses outside of the computer itself, and a means for doing this may come in handy in quantum networks.

Inside computer systems, another form of naming is translation of the *virtual memory addresses* used by application programs for data to the *physical memory addresses* that ultimately are used to store and retrieve the data. This translation process is handled so that the application program is not aware that it occurs; the application never sees or handles the physical addresses. Both of these concepts – client-visible name translation and client-transparent name translation – can be used in the design of quantum repeater networks.

Requests: within a communication session, it is often necessary to refer to a specific request. A *request ID* that can be unambiguously shared is valuable.

3.1.5. *Security*

A key motivation for quantum networking is to enhance the security of software systems deployed on the Internet. We have already introduced QKD; other security-related uses of quantum networks include secret sharing and secure leader election. We will not explicitly discuss the entire field of computer security in this book, but having in mind a few critical concepts will help in understanding these applications, as well as issues in the security of the quantum networks themselves.

Authentication is the process of confirming the identity of an actor: are you who you claim you are? For confirming the identity of an individual, the problem may be solved by something you *are* (biometrics), something you *have* (a physical door key) or something you *know* (a password). With public-key cryptography, the last of these can be restated as something you *can calculate*.

Authorization is the right to use a resource: the right to access a file or connect to a network, for example. In computer systems, authorization is often expressed as a *right* or *permission* and may be carried in a cryptographic form called a *capability*.

A system that exhibits *data integrity* allows the user to confirm that the data have not been modified by an unauthorized party. This is often achieved by applying a *digital signature* to the data. A digital signature is often a cryptographically secure hash of the data, allowing the verifier to confirm the integrity with high probability.

Data privacy is what most people think of when they think of encrypting data: the data are kept secret from anyone not having the correct decryption key.

In systems carrying important transactions, such as financial commitments, another useful factor is *non-repudiation*, in which an actor that makes an agreement, evidenced by e.g. digitally signing a document, cannot later claim *not* to have done so.

Finally, an issue we see in later chapters in the context of both quantum key distribution and leader election is that it must be impossible for either a participant or an attacker to *bias* an outcome, modifying the probability of each state in order to achieve a desired result.

3.2. Challenges in scaling up networks

The difficulty of every item in the design of a network grows as the scale of the network increases. In this section, we highlight a few interesting areas, emphasizing the solutions that have evolved in the Internet research and development community over several decades.

3.2.1. *Heterogeneity*

A key reason for the success of the Internet is its resilience in the face of heterogeneity. IP runs over link layers using copper wire, optical fiber, packet radio, short-range wireless LAN, and more, with varying characteristics including loss rates. IP has been run over links with bandwidths that range from a few hundred bits per second to over 40 gigabits per second, some eight decimal orders of magnitude, with no signs of showstopper problems as bandwidths continue to grow.

IP has also proven to be robust as the software protocols running on the Internet change, with different routing protocols coexisting peacefully, and even seamless interoperability of different versions of TCP, which guarantees delivery of messages in order and helps to quell network congestion.

This flexibility is both a cause and an effect of the continuously evolving nature of the Internet, which has been likened to rebuilding the airplane while in flight. Doubtless a few of the earliest systems connected to the Internet are still running, alongside brand-new supercomputers and small embedded devices such as Arduino boards.

3.2.2. *Scale*

One of the largest challenges to the operation of the Internet is its sheer scale: by some estimates, the number of devices connected to the Internet exceeded the human population sometime around 2009, and is predicted to grow to 50 billion devices by 2020 [EVA 11]. It is, of course, impossible to know for certain because there is no global registry of nodes on the network. This lack of a registry, in fact, contributed to the ease of growth, and hence the Internet's popularity.

This enormous size means that new solutions are required for naming and addressing, routing, security, obsolescence and practically every design issue. Critically for quantum networks, it is difficult for nodes near the edge to understand anything about the interior of the network, such as the details of the path a given communication session will take through the net.

3.2.3. *Dealing with out-of-date information*

One reason that networking is hard is simply dealing with out-of-date information. Because information (even in quantum networks) does not flow faster than the speed of light, both the transfer of data and information about the state of the network itself are inevitably delayed.

This may include dynamic connection state. For example, which of my messages have arrived? How can I tell? In particular, how can both ends of the session create a shared understanding of the connection state?

For routers, our problem includes out-of-date information about the status of the network: routing protocols take time to *converge* (reach a new, functional steady state after a change to the topology of the network) after an event such as a link going down. Information about congestion in one part of the network also cannot be expected to propagate in a timely fashion to all parts of the global network; indeed, Internet engineers initially chose not to create an explicit system for sharing such information, and direct detection and notification remain rare.

3.2.4. *Organizational needs*

The organizations that want to take advantage of the network often have disparate needs that affect their willingness or ability to join the network. They may prefer to have control over their share of the network, deciding what kind of technology to use and even how many nodes to connect, as opposed to having to call the telephone company each time they need a new extension. That information may be considered proprietary, something they are not willing to share beyond the bare minimum necessary to make the network work.

In modern network design, it is considered prudent to make the basic platform flexible enough to allow organizations to develop and deploy new services with limited coordination with any sort of authority. In fact, companies will use the network to supply services to their customers, and they may even want to become a network provider and compete with the existing network providers themselves.

3.2.5. *Misbehaving nodes*

In order for a network to be robust, it must not be possible for a single node or small group of nodes to disrupt operation of the entire network. The Internet, as we just noted, is founded on the autonomy of nodes, networks and services, which in turn means that it is fundamentally impossible to predict all of the behaviors that will arise. A node or group of nodes actively preventing others from completing their own work is said to be executing a *denial of service* (DoS) attack. DoS is the flip side of the authorization discussed earlier.

Even small groups of nodes must be able to detect misbehavior by members. Reasoning about the faults in a distributed system is aided by the state machine approach introduced later [SCH 90].

Many problems of detecting misbehavior and creating an actionable shared understanding of a state can be expressed in terms of *Byzantine agreement*, or the *Byzantine generals problem* [LAM 82]. In the Byzantine generals problem, several cooperating armies surround a city. In order for an attack on the city to succeed, all the armies must attack simultaneously. The armies can communicate only via messenger, and unfortunately the messengers may be delayed or prevented from delivering their messages altogether. Leslie Lamport, Robert Shostak and Marshall Pease articulated this problem and proposed solutions under two sets of conditions: when the authenticity of a message can be confirmed (signed, written messages), and when it cannot (oral relay of messages). Others have since improved upon the practicality of solutions, such that this is widely considered a solved problem in classical systems, though new twists continue to arise [CAS 99, KOT 10]. We discuss both the classical problem and quantum solution to a variant in more detail in section 6.5.

3.3. Design patterns

3.3.1. *Hierarchy*

Perhaps the first design pattern discovered in networks is simply the application of *hierarchy* to the organization. The earliest telephone network showed this characteristic, with local telephone exchanges and trunk lines that connected smaller cities to larger ones. This is seen even in the telephone numbers themselves, which were originally allocated strictly according to geographical hierarchy.

The *backbone* of a network is the higher-speed, more centralized links and nodes that connect distant parts of the network, and whose costs are amortized over many users. The major city-to-major city telephone backbone was a fairly richly connected topology even early on, but routing was largely based on a *tree* structure, with calls first working their way up the hierarchy until they reach a level where their address (phone number) is shared with the destination, or the backbone itself, then working their way back down another part of the tree.

The ARPANET (an early predecessor of the Internet) adopted a two-level hierarchical system even in the earliest days, when the network consisted of only a few tens of nodes. Within a building or campus, a *local area network* (LAN) was used, and between organizations the *wide area network (WAN)* was used. The challenge was connecting the two types of networks. This role was filled by a *gateway*, a special-purpose machine that translated protocols as necessary. We can see hierarchy appearing again and again in this issue of network topology.

3.3.2. *Layering*

In section 3.1.3, we presented the idea of a protocol and hinted that protocols are often designed to be composed with one another. In modern communication architectures, the functionality is divided into a set of *protocol layers*, as shown in Figure 3.3. Each layer has a different role in supporting the end-to-end communication requested by the originating application. Each layer uses services provided by its lower layers to communicate with its peer layer at the remote node. In practice, the software implementing these layers may be integrated into a single module, but the layers are commonly described as if they were implemented separately.

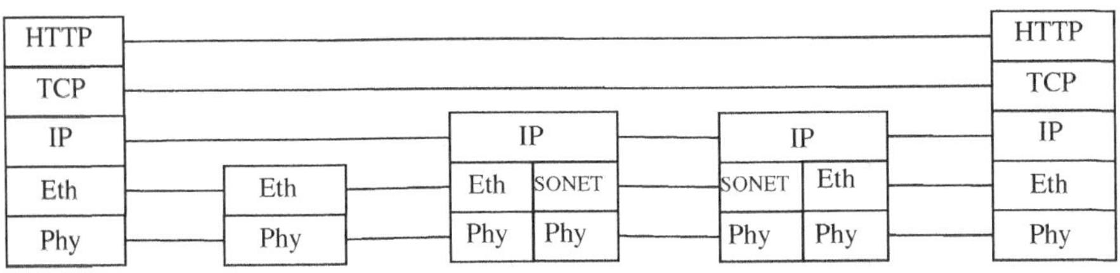

Figure 3.3. *Classical multihop/multilayer architecture, including forwarding steps as part of a multihop path using the Internet protocol*

An important feature of this approach is that it enhances scalability by limiting the required functionality at nodes in the middle of the network. A single router in the middle of the network may carry traffic that is part of millions of ongoing connections; if every router in a path had to be involved in the setup of each connection, the resources required (time, processing power and memory) would be substantial. Moreover, deployment of new services becomes difficult, if the new service conflicts with the actions of the router or requires new functionality to be implemented in the router. (As an alternative architecture, we are compelled to mention the original analog and digital instantiations of the telephone network, which in fact performs such an end-to-end setup operation, including allocating resources along the path to guarantee a certain level of performance. However, the early network was dedicated to delivery of voice traffic, and the difficulty of adapting it to other uses proves our point.)

3.3.3. *Narrow waist*

The brilliant insight by Vint Cerf and Bob Kahn in the early 1970s to allow messages to transit multiple networks by using an additional layer of addresses, now called *Internet protocol addresses* or *IP addresses*, made the internetworking architecture possible [CER 74].

This concept is encapsulated in IP itself. IP serves as a common meeting point, a platform on which services are built. IP runs over many different kinds of link layers, and in turn, the higher-layer protocols use the basic services provided by IP. IP itself makes minimal assumptions about the capabilities and promises of the underlying link or network, and promises only that it will endeavor to deliver the packets it is given, with no guarantees of ordering of messages or even their successful delivery. A view of this architecture as a set of layers leads to the notion of the *hourglass* or *narrow waist*: "IP over everything, everything over IP", as depicted in Figure 3.4 [DEE 01].

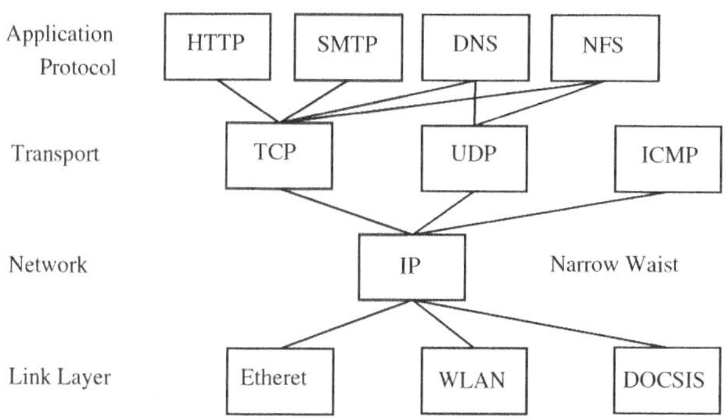

Figure 3.4. *The "hourglass" protocol layering architecture of the Internet, in idealized form*

As quantum networks are used around the world, a similar internetworking architecture will be critical. One proposed architecture is covered in Chapter 15.

3.3.4. *Multiplexing resources*

Perhaps the simplest approach to managing network resources is *circuit switching*, in which a path is selected when a session is initiated, and all of the resources along that path are dedicated to the sole use of that session until it ends, after which they can be reallocated. This approach arises from the original telephone network, in which analog circuits were physically connected. Naturally, this results in other sessions being *blocked* if the usage of the network is even moderately high. In Figure 3.5, if the AB session is initiated first, the EF link will be dedicated to that session. When the end nodes C and D attempt to initiate the CD session, they will be forced to wait until the AB session completes. In the meantime, the CE and DF links are left idle. In a larger, more complex network, depending on how such requests for the network are queued, the overall efficiency of the network may be poor or connections may be required to wait a very long time. In early telephone networks, this was resolved through the use of *trunk lines* with multiple wires, allowing multiple sessions to run concurrently. This approach is not discussed here.

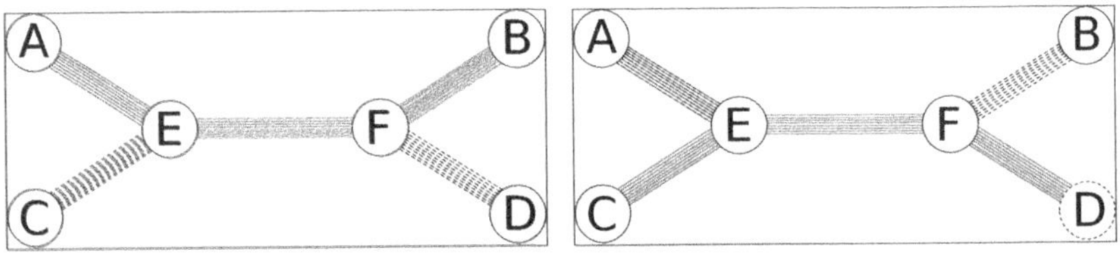

a) Circuit switching with the EF link allocated to AB, as marked with solid lines.

b) With the EF link allocated to CD.

Figure 3.5. *Simple circuit switching*

In time division multiplexing (TDM) or time division multiple access (TDMA), we establish a set of *time slots* that are allocated to the various communication sessions. A session can only use the link during its assigned slot, and if that session is not ready to do so, the link goes unused. A session typically wants to transfer more data than it can send in one time slot, so it will use many time slots. Application software at both ends must patiently await its turn to use the network. TDM by its nature is easy to implement and analyze for a single channel, and is guaranteed to be fair. With independent allocations on each hop of a multi-hop path, the analysis is more complex. TDM was used in early digital telephone networks, asynchronous transfer mode (ATM) networks and some second-generation cellular networks.

The Internet is built around uncoordinated use of the network; end nodes and intermediate nodes send traffic independently and asynchronously when they have data ready to transmit. The immediate availability of a link thus becomes a probabilistic process, dependent on aggregate statistics of a large number of traffic sources, leading to the name *statistical multiplexing*. When a link is not immediately available, traffic is commonly *buffered* to be sent later.

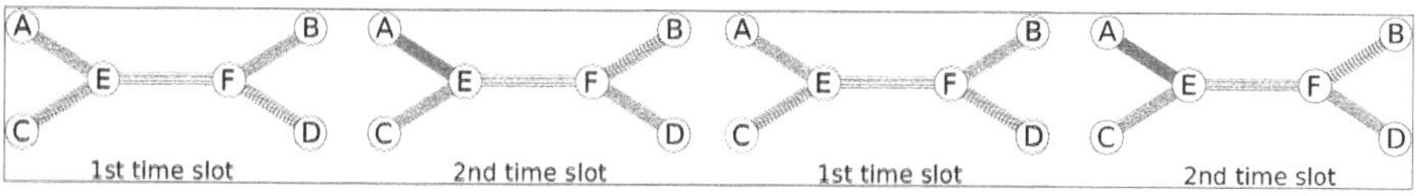

Figure 3.6. *Time division multiplexing of two communication flows. The solid lines indicate links enabled for the active flow*

Quantum repeater networks, as we will see in Chapter 13, are very dependent upon the availability of buffer memory at the repeater nodes. An approach with such a limited resource would be to split the memory into n equal parts and dedicate the use of each part to a separate communication session. We can refer to this as *buffer memory division multiplexing*. In Figure 3.7, the AB session and the CD session are each given half of the memory resources at nodes E and F. This fixed allocation scheme allows for predictable, if slower, behavior for each connection. This approach is not common in classical networks, except implicitly as part of circuit switching operation.

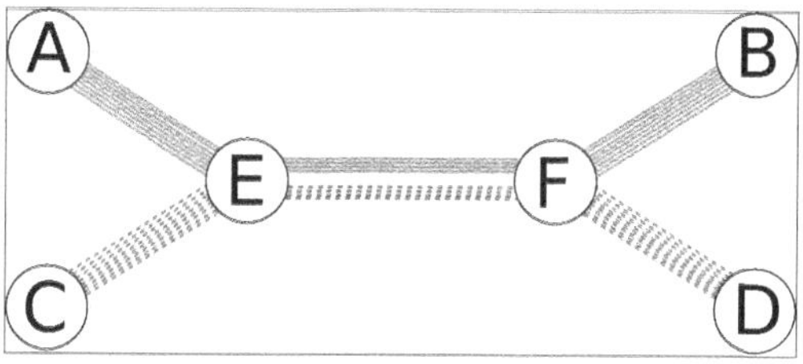

Figure 3.7. *Buffer space multiplexing*

When a session is blocked from communicating end to end, the most natural approach may be to simply wait until the end-to-end path becomes available. An alternative approach is to aggressively use the portions of the path that are available, anticipating the chance to use the other portions. This occurs, in effect, in classical networks that send data partway with the assumption that it will be buffered at intermediate nodes, pending the availability of transmission bandwidth. In quantum repeater networks, as can be seen in Chapter 13, this can have a significant positive effect on the overall network throughput.

3.3.5. *Smart versus dumb networks*

The routers in the middle of the Internet are intended to provide basic minimal functionality. In the Internet, delivery of messages is unreliable and ordering is not guaranteed, leaving the enforcement of such needed semantics to the end nodes. In fact, in a file transfer, determining that the data in a file were correctly transferred all the way from the disk at one end to the disk at the other can *only* be done by the end points themselves; confirming correct transfer on a hop-by-hop basis *cannot* guarantee that the data were correctly handled by the receiving host. There *must* be a final check before the complete operation is declared a success.

This idea of deferring functionality to those who know the needs best, the end nodes, is known as the *end-to-end argument*, as articulated by Jerry Saltzer, David Reed and Dave Clark [SAL 84, CLA 90]. They summarized the argument as follows:

> The function in question can completely and correctly be implemented only with the knowledge and help of the application standing at the endpoints of the communication system. Therefore, providing that questioned function as a feature of the communication system itself is not possible. (Sometimes an incomplete version of the function provided by the communication system is useful as a performance enhancement.)

David Isenberg later reformulated this as the "rise of the stupid network", arguing forcefully for limiting the actions of interior nodes, for fear of accidentally precluding desirable new, unanticipated functionality [ISE 97]. This approach also helps to keep the functionality of nodes simple, making implementation of high-performance systems easier and cheaper.

As mentioned in section 3.3.2, the original telephone network involved a number of architectural decisions that placed intelligence in the switches in the network, rather than close to the customer, and the choices had the unintended effect of tying the network to a specific data type – human voice – which impeded the creation of new services.

3.3.6. *Distributed management and autonomy*

The Internet is fundamentally founded on the notion that nodes, networks and even the software services built on top are independent and autonomous. No permission from a central authority is required to deploy a new service. Centralized coordination is applied only at the level of widely used service names, such as email and the Web, which are assigned port numbers by the Internet Assigned Numbers

Authority (IANA). Although, in principle, any node can reach any other, in practice, business agreements and legal considerations sometimes result in traffic from one network to another being blocked; there is no *requirement* that all networks connect, though the value of the network to some users is diminished by the degraded connectivity.

Individual networks can do pretty much as they wish, provided that they conform to a few basic guidelines when connecting to the other networks, as we will see later when we discuss routing protocols. The choice of underlying link technologies is not only completely up to the individual networks, it is completely invisible to those outside the network.

One of the few areas where any centralized control is exerted is distribution of IP addresses; to be globally reachable on the Internet, each network must acquire a set of addresses for its exclusive use. These addresses are distributed by the IANA to regional registries that in turn allocate them to individual networks. However, once an organization has received those addresses, it may do pretty much with them as it will, without further consulting any authority.

3.3.7. *State machines*

We need a way to reason rigorously about the behavior of parties in a distributed communication session, in order to prove that operation is robust and efficient [YUA 88]. A critical element in such reasoning is to determine *who* knows *what, when,* so that we can show that each node has enough information to make the decisions we expect of it, and that the error model is understood. Robust operation also involves showing that *deadlock* does not occur, in which two nodes are each waiting on the other to send them a message. We can describe the node operation as being controlled by a *protocol state machine*. The use of state machines in distributed systems design is covered well in the classic book by Nancy Lynch [LYN 96].

We can describe the behavior at each end of a communication session in terms of the conditions that have been encountered, and what happens when an event occurs that changes that status. A *state* generally has a name describing either the condition that has been met, or the next event upon which it is most commonly waiting. The state has one or more allowable events, typically messages that are received or local occurrences (most commonly the expiration of timers). An event always results in a *state transition*, although it may be a self-transition, back to the same state, after executing some operation. Operations caused by transitions are called *side effects*. For our purposes, side effects are either local quantum operations, the sending of messages to one or more nodes, or both.

Figure 3.8 shows a simple example. If we connect two traffic signals together, the goal is for a light to turn green after receiving a message that its partner has turned red.

A specified interval after turning green, it will turn yellow, then red and send out its own "red light" message. (We ignore the fact that we must figure out how to start the system in the state where one is red and one is green.) In this book, we have used the notation that states are circles with text, and arrows are state transitions. Text next to an arrow represents an internal event such as the expiration of a timer. Text labeled (in) or (out) represents messages, which are marked as (in) if they are received, triggering the transition, or as (out) if they are transmitted when the transition occurs. Most of the state machines in this book are designed to be composed with other state machines, so we ignore the initial states; instead, we will usually mark an incoming transition as an event that invokes the state machine, and another transition out as terminating the operation of the machine.

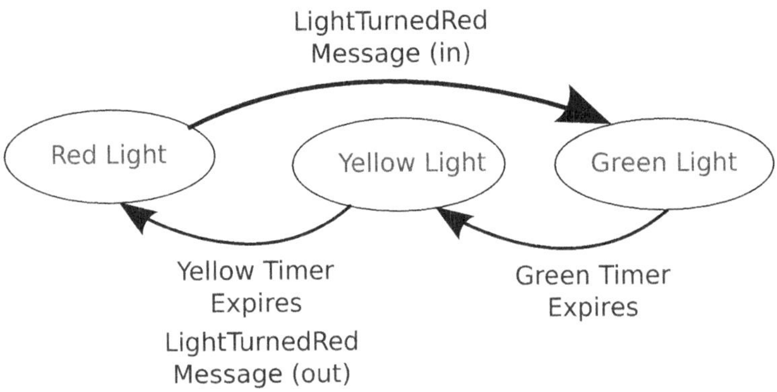

Figure 3.8. *State machine for a traffic signal, assuming two connected signals and ignoring initialization*

In this book, *we will assume that messages are always delivered reliably, in order and in a timely fashion.* The first two requirements can be encountered by using TCP, but TCP does not guarantee the timely delivery. Working out the details, including the variance in latency allowed, is beyond the scope of this book.

3.3.8. *Weak consistency and soft failure*

A design choice that allows the Internet to scale is the use of *weak consistency* in key protocols. The domain name system (DNS), which translates a human-readable name to an IP address, for example, allows nodes to keep a copy of, or *cache*, a translation [MOC 88]. Changes to that translation may not be seen immediately by all nodes across the network. Thus, for a period of time nodes may be attempting to use incorrect IP addresses and may fail; however, the system will gradually correct itself, and the number of users affected should be small. This eventual consistency allows for tremendous scalability in the system through the use of caches holding possibly outdated data and non-authoritative answers. More recently, demands on services have increased, and the desire to have greater consistency in DNS has grown.

Likewise, the routing system cannot guarantee that all nodes will immediately see any changes to the topology of the network as nodes and links go up and down. The routing system is guaranteed to converge eventually, so any outages of reachability are expected to be temporary.

3.3.9. *Distributed routing protocols*

A routing algorithm chooses a path on a graph, and consists of two parts: a definition for the cost of a single link and a function for calculating the cost of a path based on those link costs, allowing us to extend a single point-to-point channel to a richer network. Dijkstra's Shortest Path First algorithm, for example, takes a simple scalar cost for each link and treats the sum of link costs as a cost for a candidate path [DIJ 59].

More formally,

$$C_{\text{path}} = \sum_i c_i, i \in \{P\}, \qquad [3.2]$$

where $\{P\}$ is the set of links in a path and c_i is the cost for link i. The challenge is to extend an abstract algorithm such as this to one that operates in a distributed fashion, allowing all of the networks in the network to reach consistent conclusions that result in packets reaching their destination whenever possible, without being discarded or traveling in loops. This robustness is the most important feature of a routing protocol; finding an optimal or near-optimal path for some metric of interest is actually a secondary goal.

As the Internet grew from its predecessors, various routing protocols were devised, and a two-layer hierarchical structure created (using what are known as *interior gateway protocols* (IGPs) and *exterior gateway protocols* (EGPs)) to hide internal topologies and provide scalability, management autonomy and privacy. This hierarchy has other layers, but until recently the number and structure of layers has been fixed.

The Open Shortest Path First (OSPF) protocol built on a distributed form of Dijkstra's algorithm is typically deployed in networks of up to a thousand nodes or so, with a diameter typically less than 20 (often less than 5 in modern practice) and an average path length of 4–7 even in the largest networks [BAS 01, GOV 02, MOY 97]. In OSPF, each node learns the topology of the entire network and calculates the best path from itself to each other node. OSPF is used as the IGP in many autonomous systems, but each AS is free to choose its IGP from a set of common ones, to create its own, or even to maintain internal routing entirely by manual configuration. The topologies of two Internet ASes, as measured by the Rocketfuel tool, are shown in Figure 3.9. These figures include only the routers, not the end nodes. The number of end nodes is orders of magnitude larger.

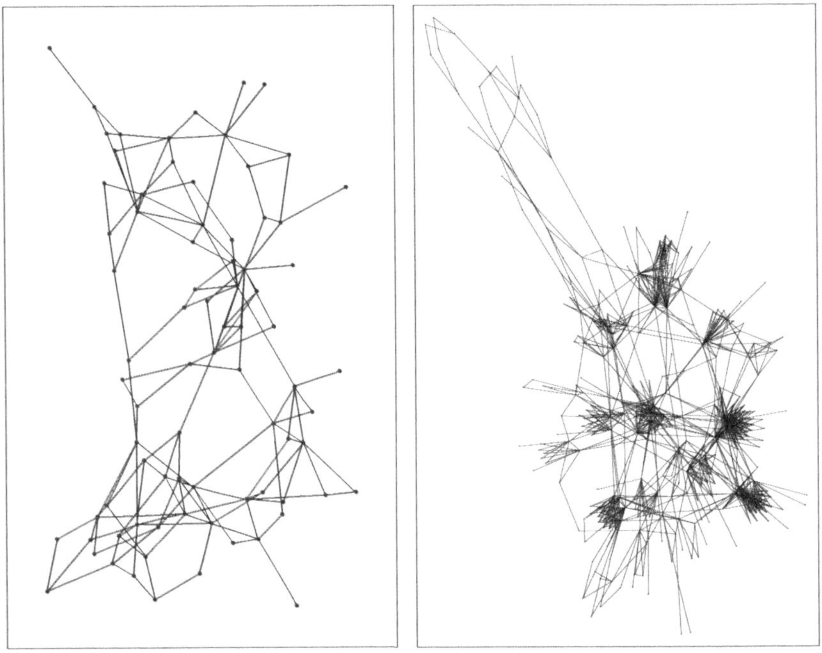

a) An intermediate-size AS. b) A large Internet AS.

Figure 3.9. *Topologies of two Internet ASes*

Routers in one AS *never learn about the interior topology of other ASes*, preserving the autonomy and privacy of other networks, and enhancing scalability; learning about the exact configuration of millions of other nodes and links is impractical. They learn of the existence of a path from one side of the AS to the other, allowing a network to support *transit* traffic received from one neighboring AS and forward it to another neighboring AS, in accordance with the local policies determined by business practices.

3.3.10. *Overlays, virtualization and recursion*

Over time, it became desirable to connect two networks *through* the network but as if they were directly connected, to build one network on top of another, or to translate addresses as data packets cross network boundaries. This can be accomplished using a *tunnel*, which *encapsulates* a message inside another message and sends it from one *tunnel end point* to a partner tunnel end point, which then *decapsulates* the packet and forwards it on. This allows other elements in the network to behave as if the two end points were connected by a physical link. Tunnels are desirable because they can effectively combine two networks for management purposes, such as connecting two distant offices as if they are in the same building. The tunnel may also encrypt the packets, guaranteeing the privacy of packets as they cross the larger network, although not the two networks at each end. This is an example of a *virtual private network (VPN)*. We see this approach when we discuss IPsec in section 5.4.1. Using more than one tunnel, we can create a *overlay network*, one that creates a virtual topology on top of another network.

Virtual networks, tunnels and overlays, as well as mobile IP and network address translation (NAT), are used to achieve various technical and operational goals. However, they all interfere with the original uniform scheme for addressing and routing, and they have resulted in much duplicated engineering effort as functionality is re-implemented at various levels [DAY 08a]. Recently, these layers have been understood as instances of a more general and flexible *recursive architecture* that supports arbitrary layering [DAY 08b, TOU 08, TOU 10].

A subset of a network can be embedded in the overall topology; such embeddings are useful in classical networks to hide complex subnet structure, as in Figure 3.10. A recursive network represents the embedded subnet as a router at the higher level, where the ingress and egress nodes of that subnet act as hosts inside the subnet (see Figure 3.10, detail). Such embedding hides the network topology, simplifying routing and enforcing autonomy. It can happen many times, sometimes on top of existing embeddings, which is why we use the term *recursive*. Note the similarity of Figure 3.10 to Figure 15.1. When a given layer in the protocol stack provides an interface to its clients that is identical to the interface of the services it uses (e.g. packet forwarding to nodes in a particular namespace), implementation of recursion is straightforward.

Recursive networking was developed in 2000 [TOU 01], and has since evolved as a possible architecture for the future Internet [DAY 08a, DAY 08b, TOU 01, TOU 06, TOU 08, TOU 10]. It has been used to unify the layering of protocol software, message forwarding and topology embedding. Quantum networking is described well by the emerging concept of recursive networks, as we see in Chapter 15.

3.4. The Internet

Beyond the numerous hints provided in this chapter, a few points are helpful to solidify an image of the Internet, far from complete but good enough for now. The history of the Internet is often traced to the early ARPANET experiments, begun in 1969, but we can argue that the birth of the Internet is more properly dated to January 1, 1983. On this date, IP itself became mandatory on the nodes on the nascent network, replacing the older NCP protocol. NCP and IP had coexisted within the ARPANET backbone for several years, but NCP was unable to reach directly to the end nodes (the computers themselves), stopping instead at the entrance to the LAN, known as an IMP. The introduction of the EGP protocol formalized the two-tier routing structure, with division of management responsibility including routing and autonomous systems (ASes).

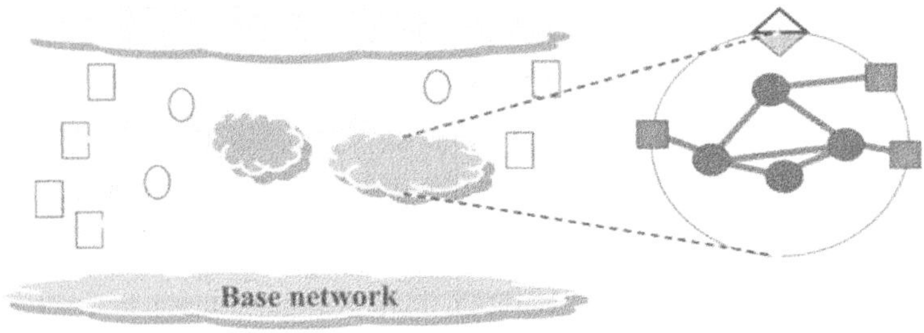

Figure 3.10. *Classical network recursion, showing a subnet (cloud) that acts as a router in the overlay, where the recursive router (inset circle) consists of ingress and egress virtual hosts (shaded squares on the inset) and interior routers (gray circles)*

It has been known since the ARPANET days that a protocol that operates properly when all nodes behave correctly is not necessarily robust against misbehavior such as hardware problems. In 1980, a single stuck bit in a single node resulted in the collapse of the entire network because the network drowned in a flood of its own routing messages [ROS 81].

In fact, even a network of properly functioning nodes can show an emergent behavior, such that the entire network fails. Although the concept of *congestion* was known earlier, it came in full force in 1986, when *congestive collapse* of the Internet occurred. *Useful* traffic stopped flowing almost entirely, as nodes tried repeatedly to push their own messages through a network that was already overflowing with data; the repeated attempts exacerbated the problem, reducing effective throughput for most connections down to a few tens of bits per second. Problems continued until Van Jacobson's congestion control algorithm was more or less fully deployed [JAC 88]. Research in the area continues, with many congestion control variants with names such as Vegas, Reno and QBIC, and Steven Low's control theory-based approach [LOW 99]. Quantum networks, with a strong need for addressing the real-time decay of fidelity, appear to face congestion as an especially urgent problem.

The last incident we mention here is the 1988 Internet Worm, which is sometimes called the "loss of innocence" of the Internet, making it clear that not all of the actors (nodes, software, human beings, etc.) can be trusted to behave cooperatively. The primary effect of the worm was a massive, and apparently unintended, DoS attack on most of the Internet.

Although the history of the Internet is a complex story and involves many landmark events, we use these events from the 1980s as demarking the end of the Internet's childhood and its arrival in something like its present form. For our purposes in this book, understanding the implications of these four events carries us a long way toward making good engineering choices for quantum networks.

3.5. Conclusion

In this chapter, we have covered a lot of ground, perhaps in an eclectic manner. It is, of course, impossible to cover all of the work in classical networking in so few pages, but the material here should suffice to allow the reader to grasp the key issues in designing a network from scratch. We have used the Internet as an example because our engineering approach to designing quantum networks adopts many concepts from the experience of the Internet's designers, builders and operators. Equally important, the services that are enabled by quantum repeater networks will inevitably be required to integrate smoothly with the classical Internet. We see the most important example of this when we discuss quantum key distribution in Chapter 5.

Chapter 4

Teleportation

Much of quantum networking depends on teleportation. Chapter 2 included an extended discussion of *Bell pairs* and violations of the Bell inequality in its Clauser-Horner-Shimony-Holt (CHSH) form. With this foundation, we move into teleportation proper, first explaining the procedure, then briefly introducing experimental demonstrations, and closing with discussion of the classical state machine and communication necessary to support teleportation in a simple fashion, without worrying about issues such as correct identification (naming) of repeater memories.

4.1. The basic teleportation operation

Charles Bennett and several collaborators described *quantum teleportation* in the famous paper in 1993 [BEN 93]. Teleportation begins with a single unknown data qubit $|\psi_D\rangle = \alpha|0\rangle + \beta|1\rangle$, held by Alice, and a Bell pair shared between Alice and Bob. Any of the four Bell states are acceptable, but in this chapter, we have assumed that they share a $|\Psi^-\rangle$ state.

During the operation, the entanglement in the Bell pair and Alice's original $|\psi_D\rangle$ are destroyed. Alice must send Bob classical messages that direct Bob as to what operations he must perform. At the end of the operation, Bob is left holding a quantum state that is identical to the original $|\psi_D\rangle$, even if neither he nor Alice knew anything at all about the original state.

The procedure runs as follows:

1) Alice and Bob prepare a $|\Psi^-\rangle$ Bell state, with each of them holding one qubit of the pair. We will refer to the qubit Alice holds as $|\Psi^-\rangle_A$, and likewise for Bob $|\Psi^-\rangle_B$.

2) Alice prepares (or acquires) the state she wishes to teleport, $|\psi_D\rangle$. This may happen concurrently with the first step.

3) Alice measures $|\psi_D\rangle\,|\Psi^-\rangle_A$ *as a pair, using the Bell basis*, as described earlier. This gives Alice two classical bits as results. Bob is left holding a single qubit with some relationship to the original $|\psi_D\rangle$, but he does not yet know the exact relationship.

4) Alice transmits her two classical bits to Bob.

5) Bob uses the two received classical bits to determine whether to apply X and Z corrections to the qubit he holds.

6) After applying any necessary corrections, Bob's qubit now contains Alice's original state $|\psi_D\rangle$.

Let us look in detail at the evolution of the entire three-qubit state. The initial state is

$$|\psi_D\rangle\,|\Psi^-\rangle = (\alpha\,|0\rangle + \beta\,|1\rangle)_D(|01\rangle - |10\rangle)_{AB}/2, \qquad [4.1]$$

which can be rewritten as

$$\begin{aligned}|\psi_D\rangle\,|\Psi^-\rangle &= \frac{1}{2}[\,|\Psi^-\rangle_{DA}(-\alpha\,|0\rangle - \beta\,|1\rangle)_B \\ &+ |\Psi^+\rangle_{DA}(-\alpha\,|0\rangle + \beta\,|1\rangle)_B \\ &+ |\Phi^-\rangle_{DA}(\beta\,|0\rangle + \alpha\,|1\rangle)_B \\ &+ |\Phi^+\rangle_{DA}(\beta\,|0\rangle - \alpha\,|1\rangle)_B] \end{aligned} \qquad [4.2]$$

Proving this is left as an exercise for the reader. If we can perform the Bell measurement described above on the two qubits D and A, we will pick exactly one of the four terms in equation [4.2]. It is easy to see that the corresponding state of B has the same coefficients α and β in its superposition as were in the original $|\psi\rangle_D$.

For example, if DA is found in $|\Psi^+\rangle$, then the corresponding state of B is $(-\alpha\,|0\rangle + \beta\,|1\rangle)_B$. Correcting this state to the original $|\psi_D\rangle$ requires only the application of a Z gate (phase gate), which will flip the sign of the $|1\rangle$ term. Table 4.1 lists the corrections that are necessary for each of the states that may occur.

Note that in this case and in the $|\Psi^-\rangle$ case, straightforward factoring will show a minus sign on the $|0\rangle$ term and a plus sign on the $|1\rangle$ term. However, as we saw in section 2.4.3, we are unable to observe the global phase of a quantum state, and we are free to multiply any equation by -1, provided that all the terms receive the factor. Thus, $(-\alpha\,|0\rangle + \beta\,|1\rangle)_B$ is functionally equivalent to $(\alpha\,|0\rangle - \beta\,|1\rangle)_B$ but not to $(-\alpha\,|0\rangle - \beta\,|1\rangle)_B$.

D meas.	A meas.	Bell state of DA	X	Z
1	1	$\|\Psi^-\rangle$	N	N
0	1	$\|\Psi^+\rangle$	N	Y
1	0	$\|\Phi^-\rangle$	Y	N
0	0	$\|\Phi^+\rangle$	Y	Y

Table 4.1. *X and Z corrections that must be applied to qubit B by Bob, depending on which Bell state Alice finds when she measures DA, assuming Alice and Bob started with a shared $|\Psi^-\rangle$ Bell pair. The relationship among the columns will differ, depending on which Bell state was originally shared*

Interestingly, the four Bell states are all equally probable, independent of the state $|\psi\rangle$. No one can predict which of the four states Alice will find, and therefore which corrections Bob will have to apply in order to recover $|\psi\rangle$. The two bits indicating the Bell state found must be transmitted via a classical channel from Alice to Bob. Without this correction information, Bob cannot measure B and recover any useful information. It is the latency induced by this operation that prevents teleportation from violating Einstein's theory of special relativity.

The basic concept is illustrated in circuit form in Figure 4.1. Five of the phases are marked on the figure: first, the Bell pair creation and distribution, then the Bell state measurement (BSM); classical communication of the Bell state results, and a classical decision of which correction operations, if any, need to be applied. Both the Bell state creation and the measurement can be executed using a variety of physical mechanisms, but all will have the logical effect of the circuits shown (as long as we are considering only pure states).

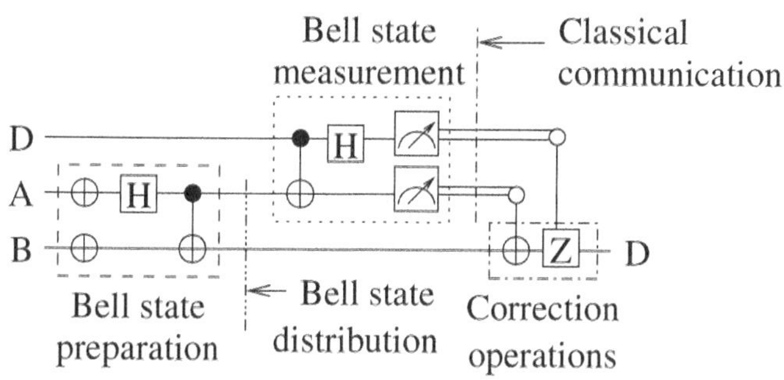

Figure 4.1. *The basic teleportation circuit. The gates in the dashed box create the $|\Psi^-\rangle$ state, assuming both A and B start in $|0\rangle$. Some physical systems create this state directly, rather than via gates as drawn here. After the state creation, A is given to Alice, and B to Bob. The final X and Z gates are not controlled quantum operations, but rather a classical decision by Bob on whether or not to execute them, based on the results of the BSM*

Note that the Bell pair creation may take place anywhere – Alice may do it, Bob may do it or a third party may do it. Once created, the Bell pair is split into two, with one member retained by (or sent to) Alice and the other sent to (or retained by) Bob. In Chapter 8, we can see various link arrangements that result in different timings happening at Alice and Bob.

Network resources. Teleportation consumes exactly one Bell pair. Two classical bits of measurement result must be communicated (ignoring the extensive management protocols described later), resulting in the operation time equaling the one-way latency. Careful engineering may allow pipelining of this operation with others.

4.2. Experimental demonstration of teleportation

Anton Zeilinger's group in Vienna performed one of the first important teleportation experiments, 4 years after teleportation was initially proposed [BOU 97]. While the initial image of teleportation may be that of a stationary qubit being teleported using an entangled state of light, in fact their scheme teleported the state of one *photon* from Alice's location to Bob's. Operating entirely with photons, although difficult, is a substantially easier task than interacting photons with matter.

This experiment used a pair of photons in the $|\Psi^-\rangle$ state, generated directly using a physical mechanism known as *parametric down conversion* (PDC), which is described in Chapter 8. The BSM device in their setup was designed to unambiguously detect only the $|\Psi^-\rangle$ state between Alice's data qubit and her member of the initial, distributed Bell pair. Because, as we saw, Alice is equally likely to measure any of the four Bell states, the experiment succeeded only one quarter of the time, even when the photons were not lost. An overview of the experimental setup is shown in Figure 4.2.

The following year, Jeff Kimble's group at Caltech conducted a substantially different teleportation experiment using continuous quantum variables (CV) encoded in states of light, rather than two-level binary qubits. Mapping points in the continuous variable space to the Bell states, their experiment worked for all four possible state detections and included the necessary correction operations [FUR 98]. In the process, they also closed several pending logical loopholes to prove that the teleportation actually occurred. Akira Furusawa, who was a researcher in Kimble's group at the time, has continued experimental work on continuous variable systems in his own group. Figure 4.3 shows a photo of a recent teleportation experimental setup in Professor Furusawa's laboratory at the University of Tokyo.

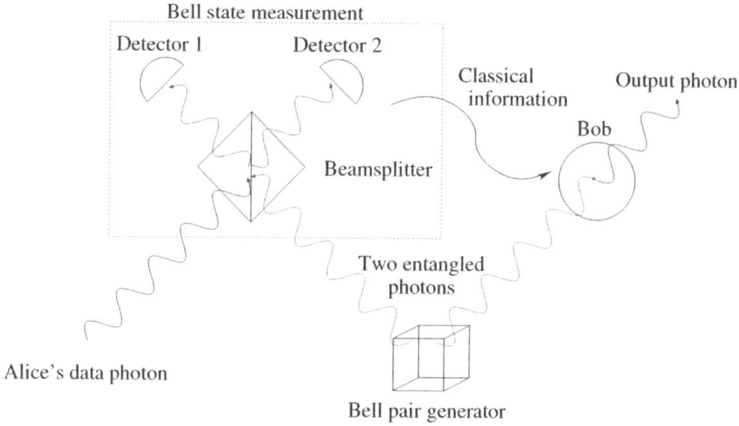

Figure 4.2. *Overview of the Vienna setup for quantum teleportation. On receiving the BSM results, Bob knows whether or not his photon contains the same state Alice teleported. In this setup, correct teleportation is detected 25% of the time, and correction operations do not need to be applied. Other states are discarded*

Figure 4.3. *Photo of a recent teleportation experiment in the lab of Prof. Akira Furusawa, University of Tokyo. Courtesy of Akira Furusawa*

Note that the definition of teleportation involves only abstract quantum states. Those states can be represented in a variety of physical (or logical) phenomena. It is possible, in fact, to teleport from a photon to an atom, or vice versa. The task of coupling photons to stationary memories has been tackled using teleportation in this manner. Without the ability to interchange quantum states between photons and some form of stationary memory, we will neither be able to build quantum computing systems with elements placed at a substantial distance from one another, nor many forms of quantum repeater. Eugene Polzik's group at the Niels Bohr Institute achieved this important milestone in 2006 [SHE 06]. In their experiment, Sherson *et al.* coupled a strong light pulse to the state of an ensemble of atoms, then transmitted the pulse over 50 cm. At the far end, the "data" state was a continuous variable state of a small number of photons. The data state was jointly measured with the strong light pulse, resulting in the teleportation of the data state to the atomic ensemble at the origin.

Teleportation between two atoms has been achieved although the original Bell pair was created by interacting the atoms directly and then separating them, rather than coupling them using light [RIE 04, BAR 04]. Using light as a coupling medium, teleportation has been achieved between two atomic memories held a meter apart [OLM 09], and more recently, using photons transmitted through 150 m of optical fiber [BAO 12]. One recent experiment achieved 88% fidelity over 21 m between two atomic memories, with a success probability of 0.1%, several orders of magnitude higher than previous experiments [NÖL 13]. Teleporting the state of one photon to another, distant photon has also been done under increasingly real-world conditions. Two stations that are 550 m apart, connected via 800 m of fiber deployed as a part of Swisscom's installed telecommunications network, served as the physical plant [LAN 07]. In this last experiment, portions of the control traffic were sent via a TCP/IP network. All three of those experiments produced fidelities of approximately 90%. In 2012, Zeilinger's group achieved the startling task of photon-to-photon teleportation over 143 km of free space, using two telescopes in the Canary Islands [MA 12]. The fidelity was solidly more than 80% and the achieved rate was some one hundred teleportations per hour.

4.3. State machines for teleportation

Figure 4.4 shows the protocol state machine (PSM) used to manage a qubit at the teleportation destination. This simple PSM is coupled into a larger software stack. Control of the qubit is handed to this machine (the bottom arrow) once two conditions are confirmed: (1) the existence of a Bell pair of adequate fidelity, shared with the prospective teleportation source, and (2) confirmation that this qubit is the teleportation target.

The central (indeed, only) state in this protocol is TELEPORTLIMBO: Bob knows only that Alice is highly likely to teleport a qubit onto this one. Until he receives a message telling him so, Bob does not know when, or even if, Alice has done so. In general, Bob cannot use this qubit for further operations until he has applied the correction operations and the teleportation is complete. Thus, we say that the qubit is in *teleportation limbo*.

Alice may send either a TELEPORTCORRECTION message or an *Abort* message. The TELEPORTCORRECTION message tells Bob which correction operations, if any, need to be performed in order to recover Alice's original $|\psi\rangle$. Once these are applied, the teleportation is complete; this PSM's work is done, and control of the qubit is passed to a higher layer in the protocol stack, or to the application awaiting the arrival of the teleported data.

Robust operation of any network requires that nodes be able to recover from the failure of other nodes in the network, or of the communication channels that connect them. At a minimum, timers are set to recover in-use resources if a failure results in a partner being unable to communication for an excessive period of time. In general, the recovered resources will be returned to a pool of available memory, where they must be reinitialized and allocated to some operation before being brought back into play. The TIMEOUT transition in the figure represents this event.

Technically the ABORT is unnecessary; once the Bell pair is created (a precondition of entering this PSM), the circuit in Figure 4.4 has no obvious failure mode in which Alice would wish to abort the teleportation. Moreover, TIMEOUT is both necessary and sufficient; any abort operation could simply be left to be recovered by the timeout mechanisms, rather than explicitly signaled. However, some BSMs use physical mechanisms that may result in ambiguous results, or Alice may be conducting some operations in a speculative fashion that allows errors to occur. Adding an ABORT message can substantially improve the system efficiency.

In this formulation, the teleportation operation is divorced from the process of Bell pair creation, which may include such complex operations as purification (Chapter 9) or even entanglement swapping and full repeater operation (Chapter 10). It may even be conducted at the logical level, on error correction-encoded rather than physical qubits (Chapter 11).

Integration of the Bell pair creation directly into this state machine is possible, at the cost of software modularity and hence flexibility. However, if done carefully, this may result in more asynchronous operation, potentially improving performance and reducing waiting times, which in systems with poor memory lifetimes can be critical. Increasing the asynchrony involves guaranteeing that conditions (1) and (2) *will* be met at some point in the future. Condition (1) generally requires a reliable link layer; one end receives its half of the Bell pair substantially in advance of the other and proceeds to operate upon it. Condition (2) is a tricky problem in classical distributed systems, akin to ensuring that two processors with very high latency access to a shared memory will use a given memory location for the same purpose. This may be achievable by implementing a set of rules at both ends that ensure the correct allocation of memory.

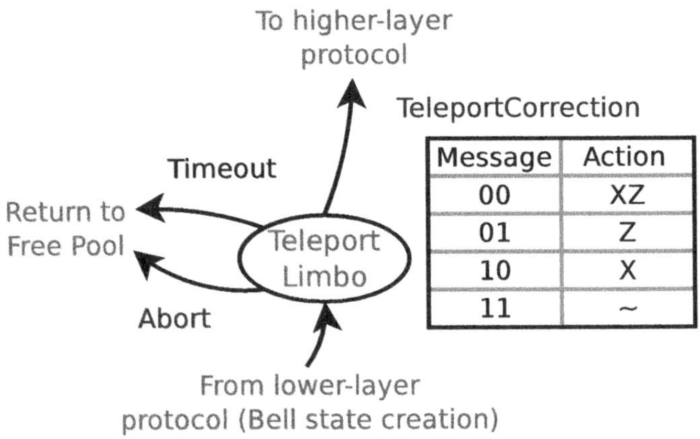

Figure 4.4. *Simple state machine for the teleportation receiver qubit. The* TELEPORTLIMBO *oval is the only state; text alongside arrows represents events. The table on the right lists the gates to applied as side effects of receiving a* TELEPORTCORRECTION *message*

There is an important class of exceptions to the rule preventing Bob from using the qubit while in limbo, known as *Clifford group operations*, which allows reordering of operations followed by batched correction operators at the end. Modifying the state machine to support out-of-order execution here requires meeting the same guarantees just described, as well as adding a mechanism for tracking the batch of correction operations across multiple memory locations, and potentially across several other nodes. We can see this in Chapter 12.

The first repeater communication session approaches assumed synchronous operations. Recently, both of these approaches to increased asynchrony have been explored, although the necessary changes to the state machines have not been described in detail in the literature. We can see these architectures in Chapter 12.

4.4. Teleporting gates

A moment's thought shows that the generic teleportation operation is equivalent to a gate on the data qubit as it moves from one site to the other although the exact gate is probabilistic: if the measurement returns 11, nothing has happened to the state, that is, we have done the identity gate I. If the measurement returns 01, we have already performed a Z on the state. If that was a desirable outcome, such as if the next operation planned was to apply a Z anyway, then we can simply leave the state alone, and the teleportation operation has performed the gate for us. Similar statements apply to the 10 and 00 results.

But those are relatively simple, single-qubit gates, and probabilistic at that; to get what we always want, we may still have to conduct fix-up operations. To have a complete gate set for universal computation, we require arbitrary single-qubit rotations and one form of entangling operation. In fact, if we can do the entangling operation over a distance, we can build distributed quantum computers (quantum multicomputers) out of building blocks consisting of only a few qubits per node [JIA 07b, LIM 05, OI 06, VAN 06].

In 1999, Gottesman and Chuang showed how a CNOT gate can be executed remotely between two qubits held far apart, by using a special four-qubit entangled state [GOT 99]. This approach benefits some physical implementations, such as photonic computers, as well as potentially being useful when managing variable placement and algorithm execution in multicomputers [VAN 08].

The required four-qubit state is

$$|\chi\rangle = \frac{(|00\rangle + |11\rangle)\,|00\rangle + (|01\rangle + |10\rangle)\,|11\rangle}{2}. \qquad [4.3]$$

The first two qubits of $|\chi\rangle$ are held by Alice and the latter two by Bob. Each holds a data qubit as well; Bob's qubit will be the control and Alice's the target on the CNOT as executed.

Gottesman and Chuang provided two constructions for $|\chi\rangle$, one using two $|\Phi^+\rangle$ Bell pairs and a CNOT, the other using two three-qubit $(|000\rangle + |111\rangle)/\sqrt{2}$ GHZ states and a BSM. (GHZ states are described in more detail in section 6.1.2.) If implemented exactly as described, Gottesman's first protocol requires a teleported CNOT gate itself. In the left half of Figure 4.5, we provide a different construction that requires only a single qubit to be teleported.

Bob begins with the qubits B, C and D and creates the state $(|000\rangle + |111\rangle)/\sqrt{2}$. The box labeled Tp teleports B to Alice, who then applies a CNOT gate using A, which is in $|+\rangle$. This results in the state $|\chi\rangle$, with Bob holding the two qubits used with the control qubit and Alice holding the two qubits used with the target. The BSM is done at each end, corrections are applied (single-qubit rotations based on the BSM outcomes), and the teleported gate is then complete.

Network resources. As with teleportation, exactly one Bell pair is consumed, if the $|\chi\rangle$ construction of Figure 4.5 is used. The required classical communication is bidirectional rather than unidirectional, two bits in each direction. The results are independent, and transmission may be concurrent.

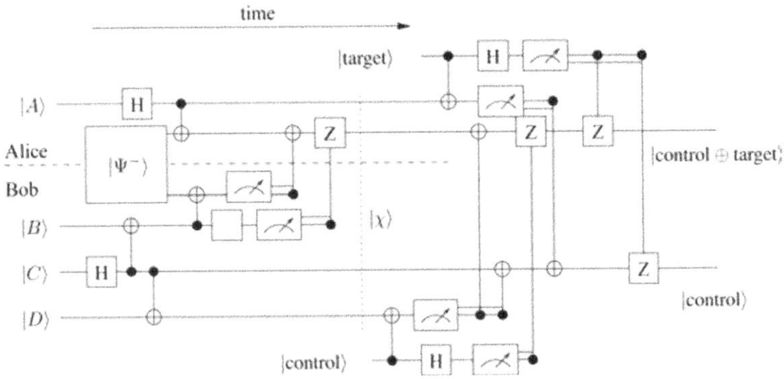

Figure 4.5. *Complete teleported gate circuit for* CNOT. *The box labeled* $|\Psi^-\rangle$ *is generation of a Bell pair*

4.5. Conclusion

Even as this book is being written, the experimental state of the art in teleportation continues to advance [KRA 13, TAK 13, STE 13]. Improvements in the probability of success have taken the process from a rare but detectable event to an effectively deterministic operation. The ability to couple light to matter qubits, output fidelity and conversion to telecommunication wavelengths all are improving more rapidly than a book can capture. Readers are encouraged to check the literature for advances since the publication of this book.

Teleportation depends on the ability to create entangled Bell pairs over some distance. Naturally, many of the experimental groups involved in teleportation have also pushed the boundaries of what is possible to create larger, longer-distance, higher-fidelity or longer-lived entangled states. We have discussed some of these physical mechanisms in more detail in Chapter 8.

In this chapter, we have discussed teleportation as a standalone operation. In the following chapters, we can see how teleportation is integrated into complete repeater networks and distributed applications. An important factor in more distributed systems is sharing of the classical information used to make the corrections shown in Table 4.1. In practice, the reception of this information is of course limited by the speed of light and classical networking factors, but under some circumstances, this does not prevent us from using the teleported state for further operations.

In fact, we can perform teleportation using not only any of the four Bell states but even a *currently unidentified* Bell state. Classical information telling us which Bell state we are holding, or how to modify the state we are holding to turn it into a specific Bell state, can be delayed either intentionally or as a result of the inevitable transmission delays. The corrections can even be deferred to be combined later with corrections from other operations, into a collective process called *Pauli frame correction*. We can see this in practice in Chapters 10 and 11.

PART 2
Applications

Chapter 5

Quantum Key Distribution

In this chapter, we cover the most important, commercial application of quantum communication technology, quantum key distribution (QKD) [GIS 02, LO 08]. QKD creates shared, secret random numbers between two parties. These numbers are typically used as cryptographic session keys to support secure communication across insecure networks.

The core idea in QKD is the use of quantum mechanics to detect the presence or absence of an eavesdropper. Alice and Bob exchange quantum states during the protocol. An eavesdropper (Eve), if present, will measure the states being exchanged, in an attempt to learn the data values. This measurement, as we have seen, results in collapse of the quantum state. Sensitive statistical methods can reveal the changes caused by Eve's measurement, even when she tries her best to limit her impact on the state. Fundamentally, the more she learns, the greater the chances of her presence being detected. Over time, the odds of her going undetected are vanishingly small. Rather than a distributed numeric computation, then, we can view QKD as the canonical example of a *quantum sensor network*. In Chapter 7, we can see another form of sensor network or cyber-physical system, when we study the use of shared quantum states as reference frames.

We first briefly present the concept behind QKD, then study its rationale. Because QKD is the current leading motivation for the deployment of quantum repeater networks, we go into some detail on the current status of cryptographic communication. Only once the rationale is established do we turn the QKD protocols themselves. Both the original BB84 protocol [BEN 84] and Ekert's entanglement-based protocol [EKE 91] are discussed. Two major QKD testbeds, the DARPA network and the Tokyo network, will be outlined, then the European SECOQC testbed will be used as an example to discuss the engineering necessary to go from link-based QKD to a full-fledged network. Recent attacks and responses and theoretical advances in security proofs are mentioned, but as the focus of this book is not strictly QKD, they are not examined in detail.

5.1. QKD and the purpose of cryptography

QKD systems generate shared, secret random numbers between two distant parties: nothing more, nothing less. This function alone is important enough to warrant large investments in research projects, product development and startup companies, not to mention the time (and even careers) of many researchers. Researchers may be attracted for the simple reason that a new, and initially unexpected, physical phenomenon is involved, and exploring it is inherently worthwhile. But product investment is another matter. Why are shared, secret random numbers valuable?

The answer lies in the critical field of computer systems and network security, specifically *key generation* or *key agreement* for cryptographic sessions. Shared random numbers, if provably secret, can be used as *cryptographic keys*, allowing secure communication across physically insecure networks such as the Internet. ("Secure" in this context can mean having any or all of privacy, integrity, authentication, and non-repudiation, as discussed in section 3.1.5. The details of these characteristics are beyond the scope of this book; the interested reader can learn more from a security or applied cryptography book [SCH 96].)

Ideally, such keys are used exactly once and discarded. XORing the data with a key that is never reused is known as a *one-time pad* (OTP), or *Vernon cipher*, and is provably the only perfectly secure cryptographic mechanism. However, because OTP consumes exactly as much key material as the amount of data transferred, this becomes impractical in most cases. Worse, in general it requires the keys to be transferred securely, well in advance, between each pair of parties that may want to communicate. In a general network where any person may decide to communicate with any other, pre-sharing keys for every possible conversation would mean that $O(n^2)$ OTP key streams need to be shared, a decidedly impractical approach.

Instead, practical cryptographic methods such as the *Advanced Encryption Standard* (AES) are used to encrypt the bulk data to be transferred. AES is a symmetric, secret-key system. It and other such methods depend upon a key that must be shared and secret, and so we are dependent on a key agreement mechanism. Key agreement is most often done using *Diffie-Hellman key exchange* or *Diffie-Hellman-Merkle key exchange*, sometimes also called *key agreement* because the key is not actually exchanged across the network [DIF 76]. D-H allows two parties to create a shared secret by exchanging the unencrypted messages over a public network. Using a publicly agreed-upon prime number p and base g, Alice selects a secret a and Bob selects a secret b. Alice can send in public the value $A = g^a \bmod p$ and Bob the public value $B = g^b \bmod p$. Alice now calculates $B^a \bmod p$, and Bob calculates $A^b \bmod p$. Because $(g^a)^b \bmod p = (g^b)^a \bmod p$, $B^a \bmod p = A^b \bmod p$, and Alice and Bob have now calculated the same number, without directly exchanging that number across the network. Although the messages are sent in public, it is difficult for a passive observer to discern the key decided upon. Alice and Bob can agree to use that number as the secret key. Note that the senders of the messages must be *authenticated* to be Bob and Alice, in order to prevent a *man-in-the-middle attack*.

The security of such mechanisms is, in general, dependent on hypothesized but unproven facts about the computational difficulty of certain problems [SCH 96]. The three functions, or phases of a session, necessary to support an encrypted communication session, and possible attacks on the functions, are summarized in Table 5.1. The computational difficulty of factoring and discrete logarithm are related, and are not known to be outside of the computational complexity class P. The best-known algorithm for factoring is the *number field sieve*, commonly used in successful assaults on factoring challenges [KLE 10]. For symmetric key cryptography, the goal in a design is to create a system where each additional bit of key doubles the number of cases that an attacker must examine; he or she should gain no information that allows shortcuts. Of course, it remains possible that the cryptographic algorithm has a latent bug that will surface later.

Table 5.1 allows us to reason about the conditions under which deploying QKD makes sense. QKD is an obvious candidate to replace the Diffie-Hellman key agreement, and in fact, that is its most common use. Ideally, we would like to replace AES as the bulk data encryption method with OTP using QKD-generated keys. However, that requires enough key material to keep up with the bulk data encryption rate. Bulk encryption operates at in excess of a gigabit per second, whereas even the fastest QKD systems under ideal conditions remain in the megabit per second range [DIX 08] and repeater-based systems will be quite a bit slower for the foreseeable future.

With current and near-term technology, then, we have only one practical usage scenario: standard classical authentication, key agreement based on QKD and bulk data encryption using standard classical encryption. (Augmenting the systems using multilayer encryption is possible and perhaps desirable, if it forces an attacker to break *both* layers, but here we will assume only a single-layer system.) Who are the potential users of such an arrangement?

Cryptography is used by governments, by corporations both for intra-company communications and for secure e-commerce transactions with their customers, and by individuals both acting as those customers and when communicating with family, friends, and especially their lawyers, financial advisers, doctors and presumably spiritual advisers. All of these purposes today are served satisfactorily by classical encryption. Under what conditions will they find the just-described scenario compelling?

The answer lies in the *information sensitivity lifetime*. When an attack has the goal of altering a financial transaction (altering the integrity of the data directly, or conducting a man-in-the-middle attack), it must be conducted in real time; likewise, cracking some encryption scheme now might allow emptying someone's bank account. Other secure systems include real-time control of systems such as power plants and the electric grid. Cracking that information is of limited, if any, use 50 years from now. In contrast, the contents of diplomatic cables may be of more than mere historical interest even 50 years from now, and governments do their very best to ensure their privacy. This concept is illustrated in Figure 5.1, for the four cases of AES encrypted keyed using Diffie-Hellman, AES keyed using QKD, some advanced symmetric encryption labeled AES++ keyed using QKD and the ideal case of an OTP with QKD-generated keys. Richard Hughes of LANL refers to decryption of old stored communications as a "retroactive vulnerability".

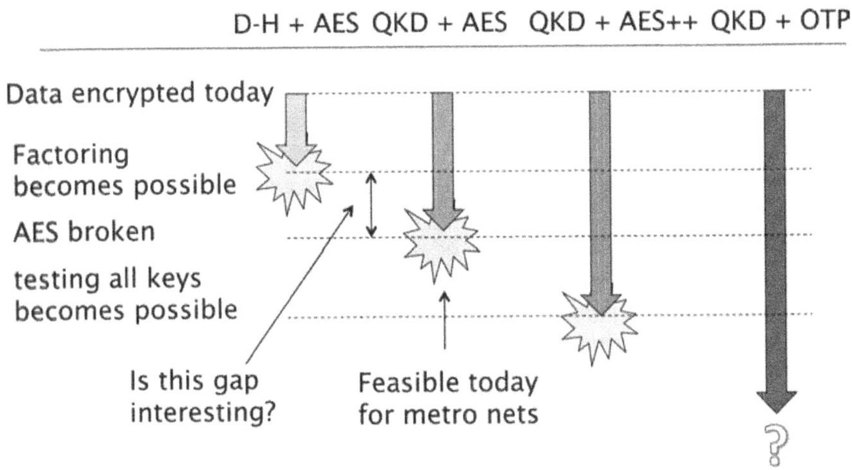

Figure 5.1. *The concept of information sensitivity lifetime allows us to reason about when encrypted data become retroactively vulnerable*

Which brings us back to the relative difficulty of cracking the factoring problem, and of either trying all AES encryption keys or finding an exploitable shortcut, as summarized in Table 5.1. A plausible, perhaps widespread, belief is that factoring will in the long run turn out to be an "easier" problem than cracking a well-designed symmetric-key system such as AES. We might, for example, posit that factoring of 1,000-bit numbers is currently out of reach but will be achieved within 20 years – perhaps using a quantum computer [SHO 97], perhaps via a classical algorithmic breakthrough – and simultaneously posit that AES remains effectively out of reach for 50 years. Under such a scenario, replacing Diffie-Hellman with QKD today gains us 30 years of reassurance that our data remain secret.

Function / communication phase	Classical implementation	Attacker's action	Cracking timeframe	Required capability
Authentication	RSA or pre-shared secret	Man-in-the-middle	Session initiation time only	Real-time factoring (RSA); testing all possible keys (pre-shared secret)
Key agreement	Diffie-Hellman	Passive recording	Indefinite future	Discrete logarithm
Bulk data encryption	e.g., AES; ideally, OTP	Passive recording	Indefinite future	Testing all possible keys (or encryption flaw)

Table 5.1. *Possibility of attacks on different session phases for discovering data in encrypted communication sessions*

With this understanding of our goals in mind, let us next turn to the two primary QKD algorithms before returning to the issue of how to use QKD in a real network.

5.2. BB84: single-photon QKD

Alice and Bob's approach to discovering the presence of Eve is to detect her influence on the exchanged quantum states. This is most easily done by forcing her to commit to a quantum measurement that will reveal her presence. In the Bennett-Brassard protocol, published in 1984 by Charles Bennett and Giles Brassard and known as BB84, Alice sends a series of individual photons to Bob, who attempts to measure them as they arrive. Many of the photons never arrive or are not detected. Of the resulting bits, some are kept and used as key material, the rest are used to search for Eve.

The fundamental insight that makes BB84 possible is that Alice can choose to encode a bit in a qubit using any basis. Measurement in the original basis will produce Alice's original classical bit, but measurement in a different basis will alter the state of the qubit such that a subsequent measurement may produce a different bit. Thus, if Eve measures the qubit using the *wrong* basis as it passes her station, then lets it go on to Bob, who measures it using the *right* basis, he will be able to detect that the qubit has been altered in flight.

Alice has several choices of photonic qubit that can be used, but the simplest explanation involves polarization. Alice can choose the horizontal/vertical basis (with e.g. $|V\rangle \equiv |\uparrow\rangle \equiv |0\rangle$ and $|H\rangle \equiv |\rightarrow\rangle \equiv |1\rangle$), for which we will write $+$, or the diagonal basis (e.g., $|\nearrow\rangle \equiv |0\rangle$ and $|\searrow\rangle \equiv |1\rangle$), for which we will write $\times$.

The protocol runs in several phases:

1) Alice sends quantum states to Bob, using two random classical bits for each qubit she sends to select (a) the encoding basis she uses ($+$ or $\times$), and (b) the bit she sends (0 or 1).

2) Bob receives and measures the qubits, using a random classical bit or quantum effect to select his measurement basis. This is the last quantum operation in the protocol.

3) Bob tells Alice via a public channel which qubits he received.

4) Bob and Alice exchange via a public channel the bases they chose. (Bob's messages can be combined with the previous step.) When Alice and Bob chose the same basis, the measured bit is kept; when they chose different bases, the bit is discarded. This is known as *sifting*.

At this point, Alice and Bob each hold a set of bits. In an ideal world and with no eavesdropper, those two sets of bits will be identical. In practice, before they can be used as a cryptographic key, two tasks remain: performing the actual eavesdropping detection and conducting error correction on the bits that remain. To check for the presence of Eve, Alice randomly chooses to reveal some of the bits she encoded. If Bob received the same bits, Eve has not been interfering with the quantum states, and the remaining (still secret) bits can be used as a cryptographic key. However, errors are likely due to many natural processes, and determining Eve's presence actually becomes a sophisticated exercise in statistics. Standard classical error correction is applied to the remaining key material. Finally, a technique based on the classical information theory known as *privacy amplification* is used to squeeze out the last bit of doubt about the amount of information Eve has gained. Many of these steps depend on continuous streams of bits flowing between Alice and Bob to generate statistical confidence.

An example of the core principle, excluding the error correction and privacy amplification, is shown in Table 5.2. In this example, Eve is not present, so Alice and Bob find the same values when they disclose a fraction of their bits as part of the eavesdropper detection. This assures them that Eve is not measuring the qubits anywhere in the path.

Figure 5.2 shows the user interface for an interactive demonstration of QKD principles, created for a Keio University public exhibition in fall 2012. Alice's choices of bit value and encoding basis are shown before execution of the transmission phase. Equivalent interfaces were created for Bob and Eve, and a hardware device using a bright polarized laser rather than single photons was built and showed.

Alice's data		1	0	0	0	1	1	1	0	1	0	1	1
Alice's basis		+	×	+	×	×	+	×	+	+	+	+	×
Alice sends	→	→	↗	↑	↗	↘	→	↗	→	→	→	→	↘
Bob's basis		×	+	+	×	×	+	×	×	×	+	+	+
Photon detected at Bob			◇	◇	◇		◇			◇	◇		◇
Bob measures			↑	↑	↗		→			↘	→		→
Bob reports reception & basis	←		+	+	×		+			×	+		+
Alice discloses basis	→		×	+	×		+			+	+		×
Same basis used				◇	◇		◇				◇		
Alice discloses some bits	→			0			1						
Bob discloses same bits	←			0			1						
Remaining key data					0						1		

Table 5.2. *An example of the core BB84 sequence. The middle column indicates the direction of messaging*

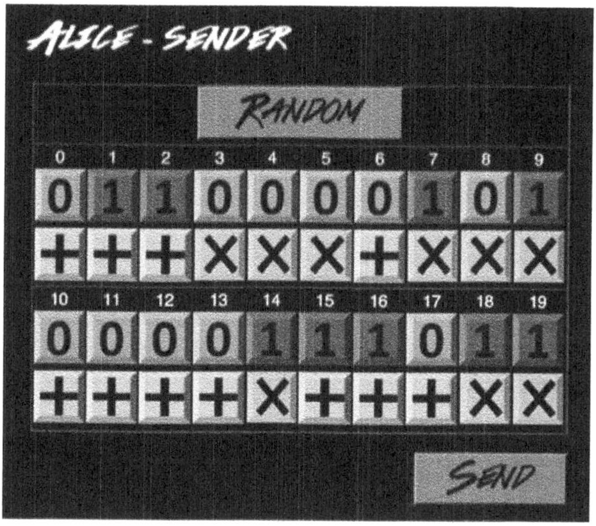

Figure 5.2. *iPad-based user interface for Alice in an interactive, educational demonstration of the principles of QKD*

Eve's simplest attack is to choose a measurement basis for the photon as it passes her, measure the photon and allow it to continue on to Bob. She will be right half the time and wrong half the time, if she and Alice both choose randomly. Detecting her presence is then very simple: half of the time that Bob and Alice announce that they agree on the choice of basis, Bob's measured value will be wrong. If p is the number of bits they disclose in eavesdropping detection, then, Eve lucks out and goes undetected at a probability of 2^{-p}.

Eve actually has an array of attacks at her disposal. A true man-in-the-middle attack is foiled by (and explains the need for) authentication and data integrity checks on the classical channel. More subtle ones involving the quantum portion of the process are numerous, most attacking characteristics of specific implementations. The proofs that all can be foiled have consumed much effort over the past 20 years, but as of this writing the proofs are considered complete; no attack by Eve against a properly implemented QKD system can succeed. The security proofs are beyond the scope of this book; readers are referred to surveys on QKD for the details [GIS 02, LO 08].

BB84's biggest shortcoming is its limited distance. It can cover only a few hundred kilometers in fiber or through the atmosphere, though remarkably it can probably be used with a satellite in orbit [FED 09, ASP 03, VIL 08, WAN 13]. The reach can be extended through the use of relays; all of the existing QKD networks use this approach. However, those relays must be trusted, which is considered a severe handicap by security specialists. One approach to mitigating this concern is routing QKD generation through two disjoint paths in the network at the same time [SAL 10]. Repeater-based QKD, which we address next, does not have this drawback.

5.3. E91: entanglement-based protocol

In 1991, Artur Ekert proposed a form of QKD using entangled pairs of photons, in a protocol now called E91 [EKE 91]. A Bell pair generator (which may be owned by Alice or by a third party) is used to make photon pairs, and Alice and Bob each receive one photon. Alice and Bob can each measure their photon and use the resulting bit as part of the cryptographic key.

Eve's best attack on E91 is to entangle a third qubit with the photon flying down the channel. We saw in section 2.5.3 a procedure for testing for the presence of entanglement, when we discussed Bell's inequalities. Alice and Bob are attempting to create pure bipartite entanglement. When Eve is listening in, from Alice and Bob's point of view, their Bell pair has become a mixed state, entangled with some unknown environmental subsystem, so a test of a Bell inequality becomes a test for Eve.

Using a stream of Bell pairs, we can easily test for the existence of bipartite entanglement, ruling out the possibility that a third photon (or other qubit) is entangled with Alice and Bob's. We use S as defined in equation 2.74, and if $|S| = 2\sqrt{2}$, then our Bell pairs are pure, and any Bell pairs left over from the eavesdropper detection we can measure and use directly to get our key bits.

Although Ekert initially phrased his protocol in terms of directly generated photon singlets ($|\Psi^-\rangle$), the protocol works equally well for any Bell pairs regardless of physical form or how they were generated. In fact, this protocol serves as one of the key motivations for the long-distance multi-hop repeater networks that are the focus of this book.

Network resources. Bell pairs are the only form of resource required; desired data rates (post-eavesdropper and privacy amplification) are discussed later. As the infidelity corresponds to the amount of information that leaks out of a Bell pair, with low fidelity, the eavesdropper detection becomes more difficult and consumes a larger fraction of the end-to-end Bell pairs. A recent paper addresses in detail the relationship between Bell pair fidelity, local gate operation fidelity and the end-to-end key distribution rate for a range of parameters and two purification scheduling schemes [BRA 13]. More specific repeater hardware proposals have also examined the key generation question [LOO 06, LAD 06].

5.4. Using QKD

In this section, we present two scenarios for using QKD over quantum repeater networks, then discuss the resilience of networks dependent on QKD. Many other approaches are, of course, possible, but these two applications mesh smoothly with existing IP-based network infrastructure, demonstrate point-to-point and hub-and-spoke communication, can be implemented in the near term, and allow us to examine performance demands more concretely than anywhere else in this book.

5.4.1. *Campus-to-campus virtual private network*

Figure 5.4 shows a simple arrangement for incorporating QKD into a production classical network. The goal is to securely connect two networks in two locations, perhaps belonging to the same organization, such as the Keio University Mita Campus in the middle of Tokyo, and the Shonan Fujisawa Campus (SFC) some 50 km away. Data between the two networks are to be encrypted. The network protocol chosen here is IPsec, which is a mechanism standardized by the Internet Engineering Task Force (IETF) for this exact purpose [FRA 11, KEN 05]. Data traffic originating from a computer at SFC and addressed to Mita is routed to the IPsec gateway, where it is encrypted and placed into a *tunnel* between the two gateways. This involves placing the data packet in an "envelope" addressed to the Mita gateway, and sending the encapsulated packet via the ordinary Internet. The packet is then removed from the envelope and decrypted when it arrives at the Mita gateway, and forwarded on toward the final destination. Note that this means the data are unencrypted when flowing over the *internal* network at SFC or at Mita. This approach is known as a *virtual private network*, or VPN. As we saw in section 3.3.10, a VPN is a type of overlay network, in which the network interconnection topology seen by the nodes does not correspond to the physical arrangement of links and nodes.

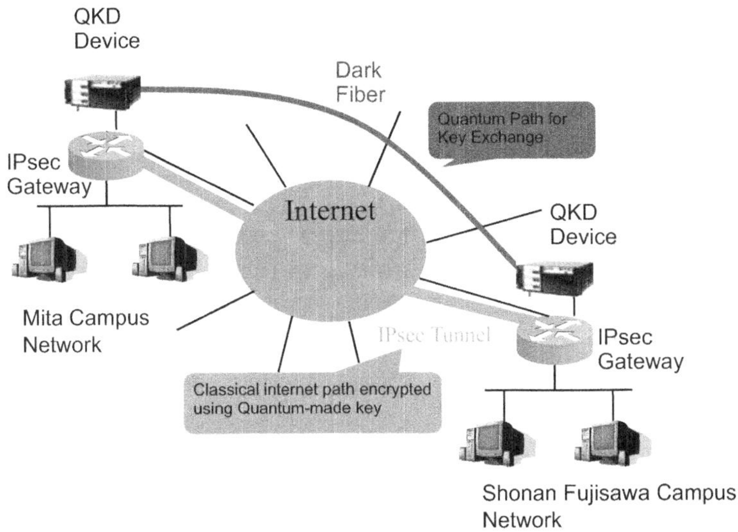

Figure 5.3. *Virtual private network setup for using QKD with IPsec. The dark gray line is dark fiber or QKD signals multiplexed through amplifier-free fibers carrying other traffic. The light gray line is an IPsec tunnel, acting as a single link in an overlay network*

IPsec is split into two parts, the bulk encryption of packets from the computers on the two networks, and the key generation and management, known as Internet Key Exchange (IKE) [RFC 05]. The QKD devices are connected via some physical path, such as dark fiber, and are continuously generating key material [NAG 09, ALL 09]. The classical channel used for sifting, basis announcements, etc. can be an authenticated connection transiting the Internet.

For the long-distance systems, we will endeavor to replace the dedicated systems with a network of repeaters.

Network resources. In this configuration, the Mita-SFC tunnel is long-lived. Ideally, as noted, we would like to be able to use QKD-generated bit strings as OTP, which would dictate Bell pair rates in the gigahertz range. Failing that, we need only enough to *rekey* the tunnel at the designated *key rollover* limit, which may be either after using the classical key for a specified amount of time, or a specified amount of data. A time-based limit is known as the *cryptoperiod*. Key lifetimes today are often set to be on the order of 24 h, but substantially quicker rollover gives greater security. A data rate of 10 bits of key material per second would allow rollover every 30 s; the exact relationship between the number of Bell pairs and key bits is complex, but a rate of even a few Bell pairs per second will permit rollover at a rapid rate. *This is perhaps the most feasible and attractive use of quantum repeater networks in the near term.*

5.4.2. *Transport-layer security (TLS)*

Transport-Layer Security (TLS), which evolved from the secure sockets layer (SSL), is used to encrypt a single communication session, e.g. between a Web client and a web server. Mink *et al.* have proposed an approach to basing TLS on QKD [MIN 09].

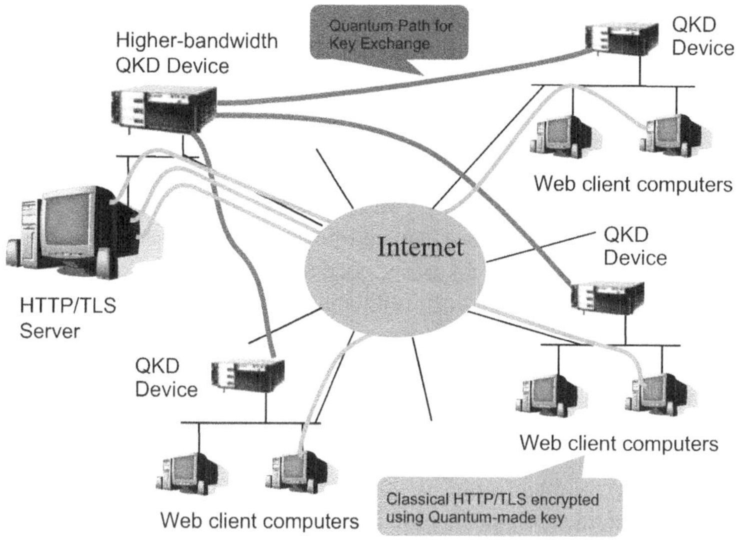

Figure 5.4. *Network setup for using TLS with QKD. Dark gray lines are dark fiber or QKD signals multiplexed through amplifier-free fibers carrying other traffic. Light gray lines are TLS-encrypted connections. Every client must either co-reside with a QKD device it trusts, or include a QKD interface itself. Each computer must have a secure connection to its QKD device*

In the VPN scenario, we discussed an essentially static network configuration with only a few participating networks. With TLS, in contrast, clients are very dynamic, with sessions typically last seconds to minutes. Moreover, being a client-server architecture, the topology of the connections is typically hub-and-spoke. Individual clients may need only a few connections an hour, but a server or collection of servers (a *farm*) will require thousands to millions of times as much capacity to serve its collective customer base.

Network resources (client). Website customers typically demand a latency of 3 s or less [NYG 10]. In TLS, as in IPsec, the bulk encryption is done using e.g. AES with a 256-bit key. To generate this key in 3 s, a client must be able to generate approximately 100 bits of key material per second, when all network overhead is taken into account.

Network resources (server). A server or farm supporting 10,000 customers arriving per second will require a megabit per second of post-privacy amplification key material. Presumably, initially few clients will be able to take advantage of the availability of QKD-secured communication, so it should be possible for a service to begin with a low-throughput server implementation and gradually upgrade to a faster one.

Arguably, this is the second most plausible usage scenario for repeater networks in general use. Data rates are higher, but hopefully within the range of evolution a few years after the IPsec scenario becomes feasible.

5.4.3. *Resilience of networks dependent on QKD*

Discussions of QKD tend to focus on its positive side: it provides detection of eavesdroppers using a physical mechanism (quantum mechanics), providing a novel means of guaranteeing the privacy of an operation (key generation). However, this capability comes with a downside: although an attacker cannot learn the key generated, he or she can prevent parties on the QKD network from creating keys, via the simple act of listening in. This becomes a form of DoS attack. If clients depend on the QKD service, such a DoS attack can have potentially severe operational consequences.

Of course, any communication is vulnerable to disruption of the channel, and neither common Internet communication nor QKD is immune to fiber or cable breaks or other physical problems. As we saw in section 3.2.5, the openness of the Internet makes it particularly vulnerable. The architecture of the Internet, however, gives it the means to work around many such problems.

QKD networks, at least initially, will not have as rich a topology as the Internet, and hubs or the links out of major service providers will become attractive targets to vandals bent on disrupting commerce or other communication.

Nagayama and Van Meter have proposed that, at least in the IPsec use case, network managers be given the capability to define *fallback* behavior when eavesdroppers are detected [NAG 09]. Encrypted communication through the IPsec tunnel and the QKD key generation are inherently somewhat decoupled. When key rollover time arrives, new key material may be unavailable. This may occur due to the eavesdropper detection halting key generation, degraded links under-performing planned generation rates, or congestion in larger networks, especially once entanglement-based QKD over repeater networks is deployed. What should a gateway do?

Three options immediately present themselves:

– Halt. This is safest, but least robust, operation.

– Continue using the existing key. Assuming an IPsec tunnel had already been established using QKD-generated keys, and a short key rollover time is in effect as described earlier, it may be acceptable to allow the key to remain in use in the hope that any outage or delay is temporary. Care must be taken with this choice.

– Fall back to Diffie-Hellman or another classical technique. Assuming the vulnerability of D-H was one of the driving reasons for the deployment of QKD in the first place, this option is likely to be unattractive to most network managers. However, it may be useful for limited temporary operation, such as work by the network managers themselves, while the outage is investigated and corrected.

The TLS scenario, which uses temporary, on-demand keys, will present different pragmatic considerations.

5.5. Existing QKD networks

Implementations of QKD are well beyond the experimental phase. QKD has been demonstrated in many labs throughout the world, with early work at government laboratories, such as NIST and LANL in the United States and NICT in Japan, and at corporate laboratories, such as HP, IBM, NEC, NTT and Toshiba. Two companies, MagiQ Technologies and id Quantique, have had QKD gateways on the market for several years. Both products incorporate classical encryption for the bulk data with the QKD itself. The enormous breadth of work over nearly 30 years is beyond the scope of this book; here we wish to focus on network-level issues, rather than link-level issues.

Metropolitan-area testbed networks have been built in Boston, Vienna, Geneva, Barcelona, Durban, Tokyo, several sites in China and elsewhere throughout the world [CHE 10, ELL 03, ELL 05a, DOD 09, PEE 09, STU 11, URS 07]. Here, we discuss the Boston-area network supported by DARPA, which was the world's first deployed QKD network, and the Tokyo QKD network. In the next section, we can also see some of the contributions of the European SECOQC in the area of QKD network engineering.

The DARPA QKD network, illustrated in Figure 5.5, consists of 10 nodes running several different QKD implementations. The first four nodes, Alice, Bob, Anna and Boris, use attenuated, weak coherent pulses from lasers to generate the single photons. In the naming scheme used, "A" nodes contain the transmitters and "B" nodes contain the receivers. Anna resides at Harvard University and Boris at Boston University, whereas the other two nodes are at BBN, which led the DARPA network development. An optical switch (at BBN) allows Anna to talk to either Boris or Bob, and Alice to talk to either Boris or Bob. Because the switch is purely optical, its presence does not affect the security proofs of the system. The inter-site communications are done through fibers installed specifically for the purpose in existing underground conduits. The photos of these first four nodes in Figure 5.6 show the volume of equipment used in early QKD implementations; more recent, production-oriented units are often four rack units high (4U).

In addition to the optically switched nodes, the network consists of three QKD links implemented by three different organizations and incorporated in the DARPA network, using BBN's software. Two of the links are free-space links built using optical telescopes at each end. The last, built by Boston University, uses entangled pairs of photons, created using PDC (discussed in section 8.1). One subsystem creates photonic Bell pairs, one of which is measured (consumed) at Alex, the other at Barb. Although the Bell pair source could sit halfway between two receivers, in this implementation Alex incorporates the pair generator and measures his photon without transmitting it.

Of course, the "A" nodes are not directly connected to all of the "B" nodes, and some nodes are multiple hops away. BBN developed a quantum key *relay* protocol to allow those nodes to share secret keys [ELL 05b]. Relaying requires an end-to-end path of quantum links, which means that some of the nodes are actually colocated and share a private network connection. In this implementation, the nodes in the middle of a path must be *trusted*, although the non-overlapping path technique can force an adversary to subvert multiple nodes [SAL 10].

QKD implementations require substantial error detection and correction; one of the links in the DARPA network, for example, is reported to have a typical error rate of approximately 5%. The most common protocol for managing these errors is the Cascade protocol developed by Brassard and Salvail [BRA 94]. BBN implemented both Cascade and their own Niagara forward error correction (FEC) protocol [PEA 04]. For a bit error rate of 3%, Niagara reduces the amount of classical communication by a factor of 40, the number of communication round trips from 68 to 1, and the CPU consumption in the error correction phase by a factor of 16.

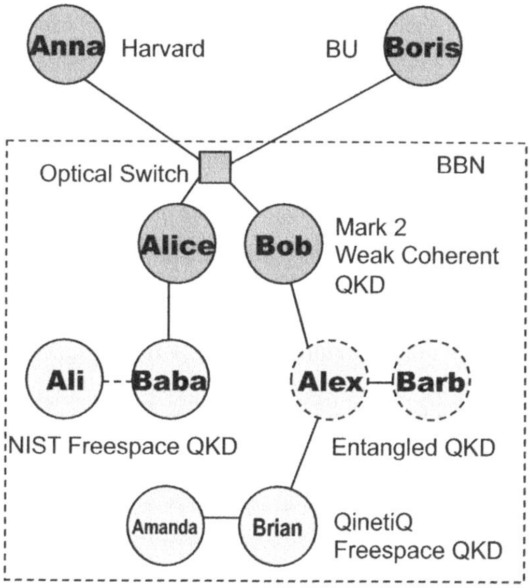

Figure 5.5. *The nodes and links in the Boston QKD network. Image courtesy of Chip Elliott (BBN)*

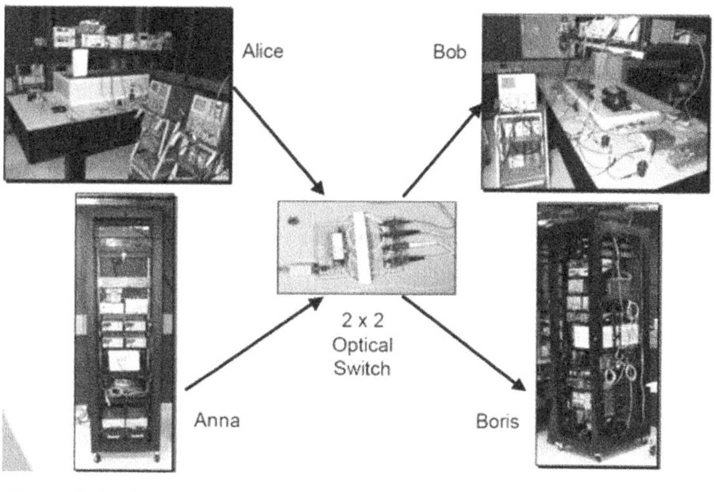

Figure 5.6. *Photos of the first four nodes and the optical switch in the Boston QKD network. Image courtesy of Chip Elliott (BBN)*

These protocols, or similar ones, will also be used in repeater-based implementations of QKD using the E91 protocol. The existing Alex-Barb link, although it uses entangled photon pairs, runs the BB84 protocol; substantial additional software development would be required to complete the CHSH inequality test required to confirm the fidelity of entanglement required for E91.

Figure 5.7 illustrates the software stack created by BBN for this network. The software runs on a Unix-based system, with the QKD protocols all running in a user-level process. The QKD-generated random numbers are ultimately used as keys for an IPsec tunnel, as described in section 5.4.1. This is achieved by modifications to IKE. A standard software implementation of IKE + IPsec places the key management in a user-level process implementing IKE, and the actual data packet encryption and forwarding in the kernel. The IKE module maintains the keys and installs them in the kernel as necessary. The IP kernel module responsible for forwarding packets applies a security policy as appropriate to choose whether or not to encrypt and which key to use.

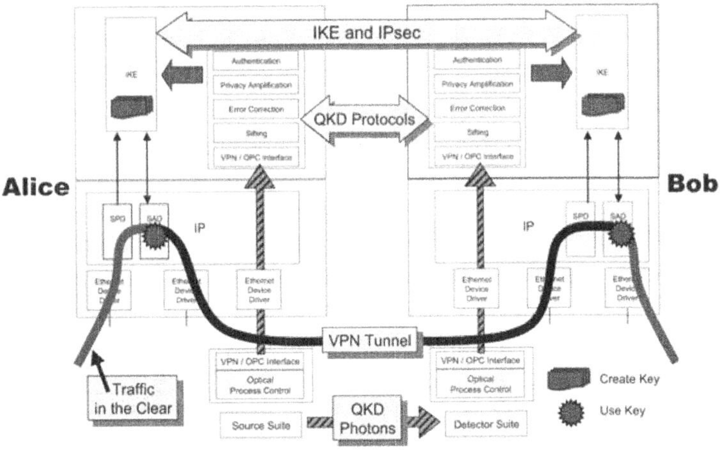

Figure 5.7. *The software components and protocol stack implemented by BBN. Image courtesy of Chip Elliott (BBN)*

The Tokyo QKD network uses six links spanning three locations, Koganei, Otemachi and Hongo, with some links emulating longer distances by doubling back through the same fiber. This network also uses several different QKD implementations.

5.6. Classical control protocols

One of the most important QKD testbeds is the *Secure Communication based on Quantum Cryptography*, or SECOQC, network in Europe [ALL 07, PEE 09]. We have reserved discussion of this network to this point in order to use it as an example of the many complex classical control protocols needed to move from point-to-point connections or small testbeds to larger, more robust networks. Much of the work here deliberately abstracts the work of the network into services without focus on the physical implementation, allowing it to apply equally well to both BB84 and E91-based QKD.

The general information architecture has been codified into three network "planes", or overlays: the quantum plane, the (classical) secrets plane, and the data plane, where actual data encryption is performed, as in Figure 5.9 [DIA 08, WEI 11]. The quantum plane includes the physical network, whereas the secrets plane takes care of network management and operations such as ongoing creation of link-level key material. It provides software services such as multi-hop key creation in topologically complex networks.

Quantum backbone (QBB) nodes are those directly connected to the quantum network. QBB nodes accept requests for connections from applications running on client computers on networks behind those QBB nodes. These clients are assumed to have a secure connection to their QBB node. Networks without a QBB node can also be integrated into the overall secure network scheme via a quantum access node (QAN). A QAN serves as a classical encryption gateway only, but provides the same software services to its clients as a QBB node on the secrets plane. The top-level plane in this network, sometimes called the "data plane" and sometimes the "application plane," is where software clients of the network construct encrypted connections, potentially over multiple hops, resulting in an overlay topology potentially very different from the underlying quantum network.

Members of the SECOQC project have detailed the packet contents for the exchange of QKD control information [DIA 07]. An important facet of the entire project is standardization. Some of this work is taking place in the context of an industry specification group (ISG) inside of the European Telecommunication Standards Institute (ETSI) [1].

---

1 Most of this work has nothing at all to do with IP-based networks, hence the standardization effort is outside the scope of the IETF. The interface between QKD and IPsec, however, falls within the scope of the IETF [NAG 09].

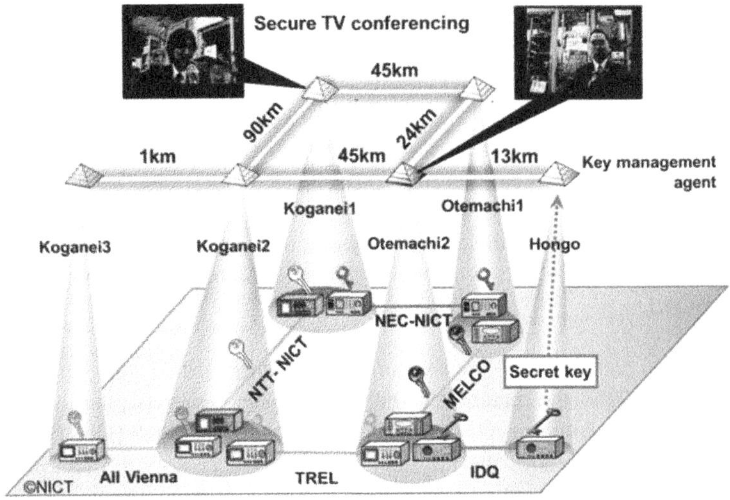

Figure 5.8. *The Tokyo QKD network. Image courtesy of Masahide Sasaki (NICT)*

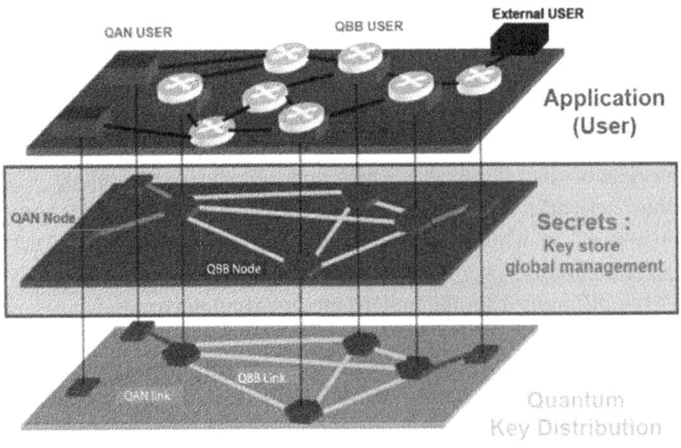

Figure 5.9. *The overlay network configuration of the SECOQC QKD architecture. Note that the topology of the planes may be different. Image courtesy of Romain Alléaume*

The SECOQC project also defined routing protocols for multi-hop QKD networks, based on a form of Dijkstra's algorithm [DIA 08]. Standard OSPF is used, usually with hand-set link costs. The authors propose running OSPF independently for each interface, in order to facilitate multipath transmission, and modifying link cost dynamically depending on the traffic load.

5.7. Conclusion

QKD is the most completely developed commercial use of quantum information technology. While the BB84 protocol is only indirectly related to quantum repeater networks, which are the focus of this book, it represents a key test case for the ability quantum technology to capture attention in the marketplace, and most of the network and business issues translate directly as repeater-based QKD becomes feasible.

We must integrate the services provided by a quantum network with those provided by a classical network such as the Internet. This integration actually works in both directions, as QKD is dependent on classical communication, as well. Much work is still required, and timely standardization will advance the business.

The question looming over all efforts here is when the service provided becomes compelling enough that a large market for devices will spontaneously develop. Implementation costs for QKD devices will decline as the technology matures, and radical new approaches have even been tested that would allow simple implementation in a handheld device such as a cell phone, for short-distance operations such as transactions with an ATM. For wide area networks, the ability to multiplex quantum signals with classical ones is a critical capability, potentially reducing deployment costs by a tremendous amount.

Although QKD is perhaps the most compelling used case for the quantum repeater networks, other nascent applications are arising, which are covered in the next two chapters.

Chapter 6

Distributed Digital Computation and Communication

This chapter may be valuable for those interested in learning what quantum networks can *do* as opposed to how they will *work*. Applications of long-distance entanglement for more efficient distributed computation will be presented. In terms of how they are used, these can be divided into two categories, distributed agreement protocols and distributed computation, though the underlying theory is essentially the same. The goal of distributed agreement protocols is for multiple parties on a network to converge upon a shared decision in fewer rounds of communication than can be achieved classically under similar circumstances. Closely related are topics such as multi-party secure communication, which we will cover below.

Note that most of the protocols presented in this chapter are building blocks rather than applications themselves; the most polished actual application remains QKD. However, the building blocks may be useful for distributed decision algorithms, cryptographic purposes and such critical issues as development of quantum money [WIE 83, AAR 12].

A separate topic is the use of distributed quantum computers for numeric computation. Such computations might be conducted in a single system, if one were available with the right capabilities. However, it may be necessary to operate in a client-server fashion due to the availability of computational resources or data. We may also need to yoke together multiple quantum computers to have a large enough multicomputer arrangement to solve the problem we are attacking [VAN 06], although that is perhaps more likely to be done in system-area networks rather than the wide-area networks that are the primary focus of this book.

The field of *communication complexity* assesses the question of efficient distributed computation, focusing on the number of bits that must be exchanged and the number of communication rounds that must be executed to calculate some function of state held by the parties to the computation. Over a decade ago, enough was already known about quantum communication that Brassard could produce an extensive review [BRA 03], covering the early work of Buhrman, Cleve, Watrous, van Dam and others. Earlier papers on distributed quantum computing, including quantum games and distributed versions of basic quantum algorithms, are also helpful historical references [CLE 97, BUH 03, DE 02, MEY 04]. More recent formal models for some of these can be found in d'Hondt's Ph.D. thesis [D'HO 05a], Cavoille *et al.* [GAV 09], Kerenidis [KER 09] and Elkin *et al.* [ELK 13], focusing on the number of communication rounds necessary to complete a given task. Broadbent's perspective on the relationship between classical and quantum distributed computing is valuable [BRO 08]. The reader interested in theoretical bounds is encouraged to read these papers.

In this chapter, we will not focus directly on the underlying communication complexity theory. Instead, we will assess several distributed quantum algorithms for their real-world applicability, their theoretical advantages over equivalent classical algorithms and provide an initial assessment of the demands made on repeater network resources in order to actually execute the algorithms.

6.1. Useful distributed quantum states

In section 2.5, we saw two-qubit entangled states known as Bell pairs, such as $|\Phi^+\rangle = \frac{|00\rangle+|11\rangle}{\sqrt{2}}$. Bell pairs are the most commonly-discussed form of distributed, entangled state, and the most straightforward quantum repeater networks will focus on generating high-fidelity end-to-end Bell pairs, creating and consuming shorter, lower-fidelity Bell pairs in the process.

There are other, larger multi-party entangled states that are useful for a variety of tasks. Here, we will briefly introduce GHZ, W and graph states.

6.1.1. *The stabilizer representation*

So far in this book, we have used the state vector and density matrix representations when discussing quantum states. When discussing entangled states and quantum error correction, we will find another representation to be useful. We can write down the set of *stabilizers* for some pure states. A stabilizer S of a state $|\psi\rangle$ obeys the condition $S|\psi\rangle = |\psi\rangle$; that is, $|\psi\rangle$ is an eigenstate of the operator S.

For a single qubit, for example, $Z|0\rangle = |0\rangle$, we can say that Z stabilizes $|0\rangle$ and $-Z|1\rangle = |1\rangle$; therefore, the stabilizer of $|1\rangle$ is $-Z$. Similarly, the stabilizers for $|+\rangle$

and $|-\rangle$ are X and $-X$, respectively. The stabilizers for the Bell states are shown in Table 6.1.

Bell state	stabilizers
$\|\Psi^-\rangle$	$-XX, -ZZ$
$\|\Psi^+\rangle$	$XX, -ZZ$
$\|\Phi^-\rangle$	$-XX, ZZ$
$\|\Phi^+\rangle$	XX, ZZ

Table 6.1. *Stabilizers for the four Bell states*

With n qubits and n such stabilizers, the state is fully constrained; only one quantum state (potentially consisting of a large number of superposition terms) can fulfill all of the conditions. With only $n - k$ stabilizers, the state has k degrees of freedom that can be used to encode qubits. This feature is used in quantum error correction.

The stabilizer representation for a quantum state requires only a few terms, linear in the number of qubits, whereas the full state vector or density matrix grows exponentially with the number of qubits. This compact representation allows for easy manipulation and is a quick, friendly means of discussing states. The tradeoff is that the stabilizer representation can express only a small portion of the huge range of the possible quantum states.

6.1.2. *GHZ and W states*

The Greenberger-Horne-Zeilinger state, or GHZ state, is a generalization of $|\Phi^+\rangle$ [BRA 06, GRE 89]. Either all of the qubits are in the zero state or all of them are in the one state

$$|\psi\rangle_{\text{GHZ}} = \frac{|0\rangle^{\otimes N} + |1\rangle^{\otimes N}}{\sqrt{2}} = \frac{|000...\rangle + |111...\rangle}{\sqrt{2}}. \qquad [6.1]$$

In what they called "open destination teleportation", Jian-Wei Pan's group demonstrated creating a four-party GHZ state and then using it to teleport the state of a fifth qubit from one location to any of the other three [ZHA 04]. This requires the cooperation of the two parties who are not recipients of the state; they measure their qubits in the $\{|+\rangle, |-\rangle\}$ basis and send the results to the receiver, who uses them to correct the remaining qubit, if necessary.

A state such as this is also useful for sharing the same information among multiple parties. In an extension to teleportation, we can use it to "fan out" copies of variables to multiple destinations. We can also use it to make a decision, equivalent to a coin flip, in a delayed fashion, by first distributing the state and then measuring it. When the state is measured, all parties will of course find the same value, either 0 or 1. It should be obvious that a GHZ state is very fragile. Losing one qubit out of the state results in a mixed state that is then of no use in our application. Premature measurement of any of the qubits in the state of course collapses the superposition.

An extension of the $|\Psi^+\rangle = (|01\rangle + |10\rangle)/\sqrt{2}$ state is the W state [DÜR 00]. The W state is the superposition of all Hamming weight one states. The three-qubit form is

$$|\psi\rangle_W = \frac{|001\rangle + |010\rangle + |100\rangle}{\sqrt{3}}. \qquad [6.2]$$

This state is more robust than the GHZ state against the loss or accidental measurement of one of its qubits. The loss of a qubit here leaves us with a mixed but still entangled state, while accidental measurement may leave us with either an entangled or unentangled state depending on the result.

One Bell state can be converted easily to another, even when the two qubits are held in separate locations. For example, $|\Phi^+\rangle$ can be converted to $|\Psi^+\rangle$ by simply applying an X gate to either qubit, without the use of any additional entanglement and without changing the amount of entanglement present. Two states that can be converted using only such local operations, possibly augmented with classical communication, are said to be *LOCC equivalent*.

Interestingly, although the two-qubit $|\Phi^+\rangle$ and $|\Psi^+\rangle$ are LOCC equivalent, their larger multi-party states are not. A three-party (or more) GHZ state cannot be converted to an equal-size W state.

6.1.3. *Graph states*

Graph states are a class of entangled states of some interest in distributed quantum systems. Graph states are created by defining a graph and executing an entangling operation corresponding to each edge of the graph. We begin by placing all of the qubits (vertices of the graph) in the $|+\rangle$ state, then executing a control-Z gate between each pair of qubits connected by an edge in the graph. If the graph $G = (V, E)$, where V is the set of vertices and E is the set of edges, we can write down the creation process as

$$|0\rangle^{\otimes V} \rightarrow |+\rangle^{\otimes V} \rightarrow \prod_{(a,b)\in E} CZ_{(a,b)}\, |+\rangle^{\otimes V} = |G\rangle\,. \qquad [6.3]$$

The first step in this transformation is trivial, involving only initialization of each qubit and application of a Hadamard gate. If the qubits are held in different locations, the second step involves an entangling operation over a distance. This can be done by teleporting a control-Z gate onto the qubits [GOT 99] or via direct physical mechanisms coupling the light to the stationary qubit, which may operate probabilistically [CAM 07, HEI 06].

The stabilizers of a graph state can be written as

$$K_G^a = X^a \prod_{b \in \mathrm{ngbr}(a)} Z^b \qquad [6.4]$$

where ngbr(a) is the set of neighbors of a, or those connected to a via an edge in E. This stabilizer expresses the condition that if you flip the value of a qubit from 0 to 1 or vice versa, you must flip the phase of all of its neighbors, and you will recover the initial state. There is one such stabilizer for each vertex in the graph.

GHZ states are known to be LOCC equivalent to some graphs. Interestingly, the two known GHZ-equivalent graphs are very different: a star graph, with one qubit acting as a hub and others all pair-wise entangled only with the hub, and the fully-connected graph, in which all $n(n-1)/2$ pairs are entangled [HEI 06].

The stabilizers for a four-qubit star graph with node A as the hub and B, C and D as leaf nodes are

$$X_A Z_B Z_C Z_D \qquad [6.5]$$

$$Z_A X_B I_C I_D \qquad [6.6]$$

$$Z_A I_B X_C I_D \qquad [6.7]$$

$$Z_A I_B I_C X_D. \qquad [6.8]$$

Writing down this state in the ket notation would require expanding out the full sixteen terms, showing the value of the compact stabilizer notation.

Graph states on a regular lattice, especially the two-dimensional square lattice, are called *cluster states*. Cluster states and some other graph states can be used for *measurement-based quantum computation* (MBQC), developed by Robert Raussendorf, Daniel Browne and Hans Briegel [RAU 03]. In MBQC, the generic cluster state is created first; then, measurements are performed on the individual qubits. Those measurements can be in the Z, X or Y basis. Naturally, each measurement removes the qubit from the entangled state. The pattern of measurements and choice of bases result in changes to the remaining entangled state that are identical to executing gates, when Pauli frame corrections are applied, as in gate teleportation (section 4.4). In the basic form of MBQC, a cluster state in the shape of a square lattice is created, where one edge of the lattice is $O(n)$ lattice cells and the other edge is $O(k)$ lattice cells, where n is the number of qubits in the computation and k is the number of gates to be executed.

While creating then immediately destroying entanglement may seem wasteful, this approach is easier in some physical systems, such as photonic systems, where direct execution of gates such as CNOT are hard. In photonic systems, the basic entangling operations and measurements are technically difficult to achieve but are well understood. However, they operate only probabilistically even in the ideal case [KNI 01] and are further limited in scalability due to loss in the channels and imperfect detectors. Because the graph state itself is generic, operations that grow it only probabilistically are also allowed even if they damage part of the existing cluster when they fail, provided that on average the size of the cluster grows. Thus, MBQC using a probabilistically-grown graph appears to offer an attractive path to scalability in systems built with physically probabilistic basic operations.

Graph states can also be used to transfer quantum data from place to place. The Pauli frame corrections naturally require the transmission of classical information, limiting the effective use of the modified state to speed-of-light propagation.

6.2. Coin flipping

6.2.1. *The simplest multi-party distributed quantum protocol*

Choosing a random bit and sharing that value is a common computational task. In quantum systems, this can be done using a single qubit in the $|+\rangle = (|0\rangle + |1\rangle)/\sqrt{2}$ state, which can be used as-is in a quantum algorithm or measured to derive a classical random value. In a reliable, trusted network environment, we can accomplish the same action in a distributed fashion, simply by creating a GHZ state among all the parties. Unreliable networks or untrusted partners can be dealt with using cryptography-related protocols, discussed below.

Network resources. If the network supports direct creation of GHZ states, only a single n-party state is required. If the network does not support direct GHZ state creation, the simplest protocol is for a leader to create the GHZ state and pair-wise teleport each component qubit to one of the other $n - 1$ parties.

6.2.2. *QKD-Based protocols*

When two parties trust each other, but not the repeater network, and wish to flip a coin together, a simple approach is to use Ekert's version of quantum key distribution [EKE 91].

Network resources. As in QKD, a number of Bell pairs are created and measured in order to establish the fidelity and security of the connection between the end points.

6.2.3. *Practical, optimal quantum strong coin flipping*

The more interesting case is that in which the two parties do not trust each other. Is it possible for them to flip a coin over the telephone? Blum proposed both the problem and a solution in 1983 [BLU 83]. Blum's solution depends on computing a secure one-way function, whose security is dependent on the assumption that the partner (or opponent) does not have access to unlimited amounts of computational power. If one party, who wishes to cheat, has access to unlimited computational power, he or she can easily break one-way functions, public key cryptography and related concepts. In this case, no known solution holds. A prospective large-scale, high-speed quantum computer could potentially render the classical solution insecure. Could a quantum network, in turn, restore that balance by making quantum coin tossing practical?

In the quantum context, the problem has been divided into two cases, *strong coin flipping* and *weak coin flipping*. In strong flipping, neither party is associated with a particular outcome. In weak flipping, each result is associated with a given player, e.g., Heads is Alice, Tails is Bob. In weak flipping, the players may attempt to bias the outcome toward their own state, but do not attempt to bias the process in order to lose; bias is single-sided.

With a dishonest player having unlimited (quantum) computational power, except only the ability to directly affect the qubits held by the honest player, it has been proven that no scheme for totally unbiased strong flipping exists; it is always possible for a player to influence the outcome, though not always perfectly. Mochon developed a method for weak flipping that reduces the cheating success probability to $1/2 + \epsilon$, for $\epsilon > 0$ [MOC 07]. Chailloux and Kerenidis extended this protocol to one for strong flipping that limits the cheater's success probability to $1/\sqrt{2}$, about 71% [CHA 09], and their protocol has been analyzed in the context of direct physical photon transmission [PAP 11].

Network resources. Mochon's weak coin flipping protocol operates in a small number of rounds in which the players pass a quantum state back and forth. Chailloux and Kerenidis's strong flipping protocol requires about the same number of rounds. Hence, each use of the protocol will require only a small number of Bell pairs, used for teleportation.

6.3. Leader election

In leader election, the goal is to identify a random single actor among a group of peers, with no previously-established priority. Solving this problem supports more complex distributed algorithms, as well, including finding the maximum value among a set and real-world problems such as distributed spanning tree.

6.3.1. *The second simplest multi-party distributed quantum protocol*

As with coin flipping, in the simple case of trusted partipants and a trusted network infrastructure, a straightforward protocol exists. We could call this the *New Year's Cabbage Protocol*: a tradition in some families in southern West Virginia is to place a dime in a pot of cooked cabbage on New Year's Day, and the person who finds the dime will have luck during the year. However, naming rights belong to d'Hondt and Panangaden, who described this protocol in 2005 [D'HO 05b].

The protocol works as follows: build an n-party W state with each party holding one qubit; recall that the W state is Hamming weight one, having a single non-zero qubit in every term in the superposition. Measure the state, and the party that finds the non-zero qubit is declared the leader. As with other quantum algorithms, the measurement operation can be delayed until a classical value is actually needed. Of course, the quantum protocol uses a superposition where the one bit's position is truly indeterminate until measured, in constract to the cabbage protocol, where the dime's position is not uncertain, merely unknown.

We refer to this protocol as the second-simplest distributed protocol because the algorithms for creating W states are more complex than those for the GHZ states used in the trusted coin flip protocol.

Network resources. If the network supports direct creation of W states, only a single n-party state is required. If the network does not support direct W state creation, the simplest protocol is for a leader to create the W state and pair-wise teleport each component qubit to one of the other $n - 1$ parties.

6.3.2. *Tani et al.'s quantum protocol*

In 2005, Tani, Kobayashi and Matsumoto proposed two leader election protocols [TAN 05, TAN 12]. These protocols, almost uniquely among quantum algorithms, pay careful attention to the connectivity among parties, representing the network as a graph and detailing the interaction between neighbors. They are, however, rather heavyweight protocols, in terms of communication volume and the number of rounds. We will look only at the first algorithm.

The important characteristics of the algorithm are that it solves the leader election problem in a fixed, rather than probabilistic, number of rounds, that it operates on anonymous networks (those in which nodes do not have pre-assigned identifiers) and that it works reliably regardless of the network topology and does not depend upon any source of randomness. No classical algorithm is known to work reliably under this set of conditions, though algorithms that work well under current real-world constraints are in constant use in real networks. This algorithm, however, may be useful in future networks and tells us something about the relative power of quantum and classical computation and communication.

Network resources. Tani's first algorithm runs in $n - 1$ full *phases*. Each phase uses a subroutine which itself requires creating and then exchanging $4(n - 1)$ qubits with every neighbor on the graph in a series of rounds. The amount of work for this subroutine varies at each node according to its degree d. The execution time is $O(n^2)$ for the quantum portion assuming all links can be used in parallel or $O(Dn^2)$ if the bottleneck is work by the node's processor, where D is the maximum degree of any node in the graph. The total quantum network load is $O(n^2|E|)$ two-party teleportations, where E is the set of edges in the graph. In the absence of further information about the graph, we know that at best $|E| \sim n$ and worst $|E| \sim n^2/2$; therefore, we can say that the network load is $\Omega(n^3)$ and $O(n^4)$.

6.4. Quantum secret sharing

6.4.1. *Semi-classical, multi-party secret creation*

Below, we will discuss protocols for sharing a secret quantum state $|\psi\rangle$, with the goal of being able to reconstruct that state. Quantum protocols for multi-party sharing of classical secrets also exist; one has been demonstrated by Wolfgang Tittel, H. Zbinden and Nicolas Gisin using a GHZ-like state where the reconstructors must share measurement outomes in order to find the secret value [TIT 01]. In this algorithm, rather than sharing a pre-determined secret known to the originating party, the sender creates a generic GHZ state and sends one qubit to each receiver, and the receivers manipulate their qubit such that a final classical result is created, or calculated, by their actions.

The original experiment was expressed in terms of optical operations, but we can describe it roughly as follows: the sender distributes a GHZ state to two or more parties; here, we will assume three, Alice, Bob and Clare. The receivers (reconstructors) each choose at random either 0 or $\pi/2$, and apply a phase shift, giving the total state

$$|\psi\rangle = \frac{1}{\sqrt{2}}(|000\rangle + e^{i(\alpha+\beta+\gamma)}|111\rangle), \qquad [6.9]$$

where α, β and γ are the phase shifts chosen by Alice, Bob and Clare. Each then measures her or his qubit in the $\{|+\rangle, |-\rangle\}$ basis, obtaining the eigenvalue $+1$ or -1 for their respective measured values a, b and c.

The "secret" is then the value of one of a, b or c in the equation $abcd = 1$, where $d = \cos(\alpha + \beta + \gamma) = \pm 1$. To find this value, each participant announces her or his choice of phase. Only cases where the phases sum to 0 or π are kept. Each participant now knows his or her value and the value of d. Up to this point, all participants in the process must be honest and cooperative. The first phase of the protocol ends here.

Later, any two of the three participants can collaboratively continue and reconstruct the third participant's measured value, with or without her or his cooperation. If, for example, Alice and Bob decide that Clare's value is to be the secret, they exchange their measured values, after which Alice and Bob each have a, b and d, from which they can calculate c. (Note that this requires that Alice and Bob already have the ability to privately share their measured values.) This value can then be used as a secret classical bit shared by all three participants.

With a large number of such bits, a large random secret can be created. To guarantee the security of the secret, a fraction of the bits are used as check bits, as in Ekert's version of quantum key distribution (section 5.3), using the three-party equivalent of the CHSH inequality (section 2.5.3), which will show if an eavesdropper has entangled with or otherwise modified the state. The algorithm thus becomes a multi-party form of quantum key distribution.

Using this particular protocol, with n participants, any $n - 1$ of them can recreate the value held by the nth participant and declare that to be the shared secret. The classical communication (including required secret channels) scales poorly with increasing n, however.

Network resources. This protocol requires a supply of n-party GHZ states either created by the network or by a participant who then teleports the qubits using bipartite Bell pairs.

6.4.2. *The basic quantum secret sharing protocol*

Secret sharing is an important (classical) cryptographic capability. In secret sharing, one party (in some contexts, referred to as the *dealer*) creates a secret he or she wishes to share with others, under certain circumstances. This might be, for example, the key that decrypts a document which is supposed to be revealed at a particular time, such as a will to be read after the author's death. For safekeeping, rather than keeping the secret in one place, the dealer wants to give the key to multiple people, such as her children. Naturally, she does not want any one child to be able to decrypt and read the will, but a group of the children can get together and read it. (Presumably they will agree to do so only after her death.)

To accomplish this for n children, the secret is divided into n pieces, and each piece is given to one child (referred to as a *player*). The pieces are not simply subsequences of the original secret (which would naturally reveal a portion of the secret, making it easier for subgroups to guess the original), but instead are given cryptographically encoded pieces. Secret sharing systems are usually designed so that there is a *threshold* number of shares which can be used to reconstruct the original secret. It might be possible, for example, for a majority of the children to get together and agree to read the will. Any subgroup of players below the threshold are guaranteed to gain no information at all about the orignal secret. If t is the number of cheaters (or holdouts) that the encoding can tolerate, then we can call such a scheme an $(n-t,n)$ threshold secret sharing protocol.[1]

Cleve, Gottesman and Lo developed a quantum equivalent based on quantum error correcting codes [CLE 99, GOT 00]. Let us use their example to demonstrate the principle. We will use *qutrits*, which can have three basis states $|0\rangle$, $|1\rangle$, $|2\rangle$, rather than the usual qubits. The dealer creates her secret qutrit, then encodes it in three qutrits

$$\begin{aligned} \alpha|0\rangle+\beta|1\rangle+\gamma|2\rangle &\rightarrow \alpha(|000\rangle+|111\rangle+|222\rangle) \\ &+ \beta(|012\rangle+|120\rangle+|201\rangle) \\ &+ \gamma(|021\rangle+|102\rangle+|210\rangle), \end{aligned} \qquad [6.10]$$

and gives one qutrit to each of her children, Alice, Bob and Clare. It is relatively easy to see that any individual qutrit contains no information; treated as a standalone state, it is in a totally mixed state. Looking down the terms as vertically aligned in the equation, we can see that the value $|0\rangle$ for the leftmost qutrit is part of a term with coefficient α, one with coefficient β and one with coefficient γ, and likewise for the states $|1\rangle$ and $|2\rangle$. If Alice receives the first qutrit as her share of the secret, then she has no ability to reconstruct the dealer's original state or even to extract any information about it.

Two of the three children working together, however, can reconstruct the dealer's original state. To do so, they need an operation that does addition modulo three on the qutrits. Alice's qutrit is first added to Bob's, then the resulting value is added to Alice's qutrit. This creates the state

$$(\alpha|0\rangle+\beta|1\rangle+\gamma|2\rangle)(|00\rangle+|12\rangle+|21\rangle)/\sqrt{3}. \qquad [6.11]$$

---

1 Some papers focus on the number of shares necessary to reconstruct the state, others on the number of cheaters that can be tolerated; so, the reader is encouraged to pay careful attention to the notation in each paper.

Alice now holds the original secret qutrit, and Bob and Clare are left with a generic state of no particular interest. Needless to say, Bob should trust Alice before agreeing to this reconstruction protocol!

More elaborate versions of this basic idea work with qubits and use error correcting codes directly. A quantum code that corrects for t erasures (lost qubits) can be used as a threshold t secret sharing scheme.

Network resources. The dealer creates the entire multi-party state herself and distributes it to the players. In our qutrit example, if Alice, Bob and Clare no longer live with their mother, then she must use a network to teleport the individual shares to her children during the sharing phase. During the later reconstruction phase, either Bob teleports his share to Alice, who performs the reconstruction, or they execute gates remotely using teleported gates. The former is less demanding of network resources, using only one teleportation. In the more general scheme, the dealer must teleport n messages, and the reconstructor must gather in at least $n - t - 1$ more messages. Each message will be m qubits, where m is at least the size of the original secret that the dealer wishes to share. Because each of these are bipartite operations, the network must create $(2n - t - 1)m$ end-to-end, high-fidelity Bell pairs for both phases. In theory, each phase can be done in parallel, for $O(1)$ depth, but in practice will likely be limited by the bandwidth at the dealer and reconstructor. Beyond some basic coordination concerning timing of operations and selection of the reconstructor, which can all be done ahead of time, the protocols require no round-trip communications; they are not *interactive* protocols.

6.4.3. *Verifiable quantum secret sharing and secure multi-party quantum computation*

The protocol just presented has a shortcoming: it is not possible for the players to determine whether or not the dealer has upheld her end of the bargain. The dealer must do a substantial amount of work, taking the initial secret and encoding it in an error correction code before sending it out. For many reasons, the state she created and distributed may not actually encode a valid word in the error correction code. The players would like to be able to confirm that the state is valid, that at a later time they will be able to recover the original secret. (Of course, this does not guarantee that the recovered secret will be anything meaningful.)

Crépeau, Gottesman and Smith extended Cleve, Gottesman and Lo's original protocol to a two-level system that allows such verification, called *verifiable quantum secret sharing* (VQSS) [CRÉ 02, SMI 01]. First, a security parameter k is chosen, where the goal is for an honest reconstructor to be able to recreate an honest dealer's original state $|\psi\rangle$ with fidelity $F = 1 - 2^{-\Omega(k)}$, provided that the number of cheaters is $t < n/4$. Here, we will give an overview of the network activities required for this protocol without examining the details of the quantum operations or security proofs. This protocol serves as the basis for a simple form of distributed quantum computation and as the basis for quantum Byzantine agreement, which we will discuss later in this chapter.

The dealer creates $(k+1)^2$ quantum registers consisting of at least n qubits each. One of the registers contains an encoded copy $|\psi\rangle_L$ of the original state $|\psi\rangle$, k of the registers contain locally transformed values based on the input state and the rest of the registers contain copies of an encoded logical zero, $|0\rangle_L$. Each register is divided into n parts, and the parts are distributed to the n players. In this round, each player receives $O(k^2)$ qubits from the dealer.

Next, each player encodes each qubit it has received into n qubits and distributes a share to each of the other $n-1$ players, receiving in turn a share of each of their stacks of qubits. In this round, each player sends and receives $O(k^2 n)$ qubits. Most of the qubits are used for verification. Local operations are performed; then, the qubits are measured, and the results broadcast on a public, authenticated, classical, broadcast channel. Each player can then verify, with high probability, that the dealer has indeed sent a legitimately-encoded state and that there are t or fewer cheaters among them.

In the reconstruction phase, $n-1$ of the players send their remaining qubits to the reconstructor, who receives $k(n-1)$ qubits and performs the reconstruction.

To extend this scheme to distributed computation, each player begins with a qubit and runs as the dealer to distribute his qubit. Each player then holds a small share of each of n qubits. With an appropriate choice of error correction scheme, it is then possible for the players to collectively execute an agreed-upon circuit (application algorithm) on the shared n-qubit logical state, simply by running local operations. The reconstruction phase is then run for each of the n qubits, after which their states can be measured. This measurement phase requires distributed operations, which have not been quantified in detail in the literature. The initial description of the protocol focused on describing the basic distributed computational functionality, leaving the operational details to future work. (Determining the resource requirements and expected operational parameters of most quantum algorithms has only just begun as an organized, community effort, even for algorithms for single systems; for distributed algorithms this area has not yet received the attention of the engineers who will perform this analysis.)

At the end of this total sequence, the classical result of the entire computation is calculated using the results of the measurement operations. None of the players learn anything in particular about the input data, assuming the collectively-executed algorithm does not allow such inferences from the results. Note, however, that the players are all aware of the *program* that is executed, a condition that will be removed when we discuss universal blind quantum computation.

Network resources. It is apparent that the VQSS protocol requires both substantial computational capabilities at every node, especially the dealer, and substantial numbers of Bell pairs for teleportation. Every pair-wise path between nodes is used, which may place a substantial load on the core of a network. Each of the n players sends and receives $O(k^2n)$ qubits, for a total network load of $O((kn)^2)$ teleportation operations.

The share of each secret that node i sends to node j is unique; there is no particular advantage in this protocol to having a network that can provide GHZ, W or graph states. Only basic point-to-point Bell pairs are required.

The distributed computation protocol requires n times as many operations even without counting the unspecified distributed measurement operation.

Open problems. The protocols described in this section are defined without reference to a network and only using pure states. While the principles are sound and no doubt will translate into real-world operation, the engineering tradeoffs need to be examined. The exact relationship between the security parameter k and the Bell pair fidelity needs to be articulated. In a general network, we cannot assume that a particular communication session either does, or does not, pass through a given node. If a malicious repeater can affect the fidelity of all of the Bell pairs, for example, it will be able to acquire small amounts of information about each of the shares rather than the total information about some number of shares controlled by up to t malicious players. How do these two sources of information leakage combine? The exact network cost of the distributed measurement scheme necessary in the distributed computing protocol has not been articulated.

6.5. Byzantine agreement

6.5.1. *The original problem*

In computer systems, agreement among various parts is critical for correct operation, but parts may be faulty or even misbehave in malicious fashion. Lamport, Shostak and Pease articulated the problem as the *Byzantine generals problem* [PEA 80, LAM 82]. The abstract of their 1982 paper captures the issue and core result nicely:

> Reliable computer systems must handle malfunctioning components that give conflicting information to different parts of the system. This situation can be expressed abstractly in terms of a group of generals of the Byzantine army camped with their troops around an enemy city. Communicating only by messenger, the generals must agree upon a

> common battle plan. However, one or more of them may be traitors who will try to confuse the others. The problem is to find an algorithm to ensure that the loyal generals will reach agreement. It is shown that, using only oral messages, this problem is solvable if and only if more than two-thirds of the generals are loyal; therefore, a single traitor can confound two loyal generals. With unforgeable written messages, the problem is solvable for any number of generals and possible traitors.

In this formulation, the messaging system is reliable and the sender of a message is known. The solutions involve the commander sending messages to all lieutenants and the lieutenants exchanging messages such as, "I was told to attack". There are two forms of the problem, one using only oral messages and one involving signed, written messages. With oral messages, a disloyal lieutenant may receive an attack order, but instead tell his peers that he received retreat orders. A loyal general's signature cannot be forged, and anyone with a copy of a message can verify its validity.

Lamport *et al.*'s original solution for oral messages required at least $n \geq 3m + 1$ generals in order to tolerate the presence of m traitors and used a recursive algorithm with depth m that sends $O(n^m)$ messages. This approach is not practical for large n or m. Their solution using signed messages sends $O(n^2m)$ messages of size $O(m)$, for a total network load of $O((nm)^2)$, and tolerates $m \leq n - 2$ traitors. In 1999, Castro and Liskov developed an approach using replicated state machines and a three-phase commit, in which the first phase is n messages and the second and third phases are $O(n^2)$ fixed-size messages [CAS 99].

6.5.2. *Ben-Or and Hassidim's quantum Byzantine agreement*

In 2005, Ben-Or and Hassidim developed quantum algorithms that operate in constant time, even against a computationally unbounded, full-information, adaptive adversary, in a synchronous communication environment [BEN 05a]. (Note that each of these adjectives is meaningful, and distinguishes classes of difficulty for the problem.) Their two algorithms are a fail-stop algorithm, in which up to $\lfloor (n-1)/3 \rfloor$ traitors are detected but robust operation in their presence is not guaranteed, and a more complex Byzantine agreement algorithm, which also works in the $m < (n-1)/3$ case and is guaranteed to reach agreement.

In contrast to the quantum secret sharing and secure computation algorithms of section 6.4, which share and operate on quantum data that remains quantum, the focus of these algorithms is on reaching a *classical* consensus. As such, if a quantum network were easily accessible and the theoretical benefits realizable in practice, quantum Byzantine agreement could be a drop-in replacement for a classical Byzantine agreement module in classical distributed systems.

As noted, the quantum Byzantine agreement reaches consensus even in the presence of traitors. However, its correct operation does require a functioning quantum network capable of generating the required Bell pairs. Physical networks, both classical and quantum, are of course vulnerable to physical denial of service attacks. Quantum networks are perhaps more vulnerable due to the delicate nature of quantum information; a simple eavesdropper will result in purification operations failing and Bell pair creation being aborted. The likelihood of this should be balanced against the possibility of traitorous end nodes when choosing whether to adopt the fail-stop or Byzantine protocol.

Network resources. As with VQSS, every pair-wise path between nodes is used, which may place a substantial load on the core of a network. Each of the n players sends and receives $O(k^2n)$ qubits, for a total network load of $O((kn)^2)$ teleportation operations.

6.6. Client-server and blind computation

One of the most natural, and compelling, reasons for creating quantum repeater networks is the same as the rationale for creating classical networks in the first place: to share computing resources, including powerful machines and unique databases, that are geographically distant from potential users.

If the user merely wishes to log into the distant system and use the quantum server remotely, classical networking will suffice. For the foreseeable future, quantum computers will have classical front-end machines directing all operations, and a remote connection to the front-end is straightforward.

An obvious extension is the transfer of quantum data between a client and the server, either for input or output data. Of course, if the data to be transferred is purely classical, there is no need for the quantum network; therefore, here we assume it is in superposition or entanglement. Naturally, this requires a quantum-capable client and a network for building Bell pairs between the client and server, to allow teleportation of the data.

The most basic implementations of such client-server operation would leave the data unencrypted and vulnerable to snooping by eavesdroppers. With superposed or entangled data, directly measuring the data offers less than complete information about the state, but in many cases will still reveal information that the client would prefer to keep secret. The client can apply arbitrary single-qubit gates to the data before sending to the server, chosen using a classical encryption scheme or even random numbers. The (classical) description of the gates necessary to undo this encryption can be sent to the server via an encrypted classical channel. This naturally leaves the data unencrypted at the server, and the server must know what program to execute upon the data; therefore, the client must trust the server. However, we can go a step better than this and avoid trusting the server altogether.

In 2009, Craig Gentry described classical *homomorphic encryption* [GEN 09, GEN 10, GEN 11]. In homomorphic encryption (now also called Gentry encryption), the client can encrypt data in a form that allows a compute server to calculate functions of the data without learning either the input or output value. The client could ask for two encrypted numbers to be multiplied, for example. This system has two drawbacks: first, the client does have to tell the server what computation is to be performed and second, the overhead for using this scheme is enormous – many millions of times larger than the unecrypted data transfer and computation. However, the basic idea is fascinating and points researchers in interesting directions.

Blind quantum computation carries the idea a step farther and in fact has been experimentally demonstrated in a simple form [BRO 09, BAR 12]. Blind computation allows a client to ask a server to perform a computation, while keeping the input, output and even *what computation is being performed* hidden from the server, even as it is executing the computation.

Blind computation takes advantage of measurement-based quantum computation. Rather than the simplest square lattice, the client creates enough qubits to build a modified form known as the brickwork state, and the qubits are rotated by randomly chosen angles. The state can be created a little at a time and sent to the server via teleportation. The server performs the entangling operations and measurements and sends the results back to the client. Without information about the rotation angles, the server can execute these operations, effecting the computation on the client's behalf, but without any ability to discern the data values or any details of the computation except an upper bound in its size.

Chia-Hung Chien, Sy-Yen Kuo and Van Meter analyzed the resources necessary to run some algorithms in this fashion, adding in quantum error correction and fault tolerance [CHI 13]. We found that the needed cluster size grows by about a factor of a thousand relative to the unecrypted, non-fault tolerant MBQC computation, using a single round of basic quantum error correction. This factor appears to be tolerable. It does, however, require that the client still be able to perform basic quantum computing operations including storing small numbers of qubits for some time and running quantum error correction. A separate approach allows the client to work with *only* measurement of qubits, with no buffer memory or QEC required, but adding fault tolerance to this system is onerous [MOR 13].

At essentially the same time as Broadbent's development, Dorit Aharonov, Michael Ben-Or and Elad Eban created a separate, fully blind protocol [AHA 08], couched in the language of interactive proofs. This system uses some elements of secure multi-party communication and fault tolerant techniques. We will not detail this system here.

Network resources. Only Bell pairs are required. The workload of performing a significant computation in client-server fashion, using quantum data, will be enormous. This will demand high-performance quantum networks, capable of creating thousands, or perhaps millions, of end-to-end Bell pairs per second. Adding blinding and fault tolerance to the requirements will raise demand by three or more additional orders of magnitude, putting traffic loads on quantum networks that are comparable to those of classical networks today. Thus, this use of quantum repeater networks is among the most demanding and will remain far from feasible even as the first networks are deployed.

6.7. Conclusion

In this chapter, we have examined a few uses for distributed quantum states. All of these examples, except for the trivial ones, are related to security. Examples of using entanglement to accelerate more general computations seem to largely fall in the category of distributed forms of monolithic algorithms. These are more likely to be used in quantum multicomputers than wide-area networks and will not be detailed here. Likewise, extending the theoretical bounds in the papers discussed at the beginning of the chapter to practical algorithms will help to determine if quantum networks have advantages over classical ones for common tasks, *in practice*.

Ultimately, for quantum networks to justify their value, the distributed uses of quantum information must offer new capabilities, operate under different constraints (such as known or expected capabilities of adversaries) or operate more quickly or cheaply than classical counterparts. To help determine the range of attractive operational parameters, a more detailed comparison of the classical and quantum alternatives would be valuable: for a given system function, is it possible to say, e.g., "One Bell pair is worth x gigabytes of classical communication" or something similar? Especially for security-related functions, such a direct comparison will be difficult. This remains open work, but will likely figure in decisions made by venture capitalists considering supporting quantum information and ultimately by end users considering buying and deploying the equipment.

Chapter 7

Entangled States as Reference Frames

In Chapter 6, we discussed the uses of long-distance entanglement for enabling different forms of digital quantum computation. In Chapter 5, we talked about QKD. QKD can be viewed as a form of *sensor network*: the goal of the underlying quantum operation is the physical detection of eavesdropping on the quantum channel. This can be achieved either using unentangled or entangled states; in this chapter, we will look at other sensor uses of the latter. Much of the work involves the use of a quantum state as a *reference frame* [RUD 03].

Distributed entanglement is an extremely sensitive physical state, and can be used as a physical probe for other applications. We will examine two applications: improving the resolution of optical telescopes using interferometry and comparing the relative time of two clocks separated by a distance. Both of these applications are far from practical given both the current state of the technology and the very demanding nature of the existing proposals, but they serve as important signposts on the road to the merger of quantum information and real-world sensors and actuators.

7.1. Qubits in the environment

Many types of quanta, and hence qubits, are exquisitely sensitive to changes in the environment, resulting in the rapid decoherence discussed in section 2.3.5. This is an undesirable phenomenon for attempting to compute using qubits, but it has an upside: they can then be used as probes of their physical environment. With careful use of superposition and entanglement, we can measure more precisely or with asymptotically fewer operations than is possible classically.

7.1.1. *Precession*

All physical qubits oscillate in some fashion as they evolve according to Schrödinger's equation (equation [2.10]); for example, the axis of spin of an electron or atom precesses about the direction of the environmental magnetic field when it is not precisely aligned with that field. This behavior of an electron (or nuclear) spin is known as *Larmor precession*. This precession is a change to the phase of the qubit, and it is normally compensated for in the execution of quantum algorithms by carefully tracking the expected phase, using a *rotating reference frame*.

The frequency of this precession depends on the local system. For a spin-1/2 qubit (e.g. electron spin), we may write our basis states $|\uparrow\rangle$ and $|\downarrow\rangle$ instead of $|0\rangle$ and $|1\rangle$. The up arrow corresponds to a spin axis aligned with the experimental magnetic field (again by convention, the $+Z$ axis), and the down arrow corresponds to a spin axis anti-aligned with the magnetic field. The energy level of the anti-aligned state is higher. The energy difference between the two states and the Larmor precession frequency are determined by the local magnetic field strength. The exact state of a qubit in the $|\pm\rangle = (|\uparrow\rangle \pm |\downarrow\rangle)/\sqrt{2}$ state is

$$|+(t)\rangle = \frac{1}{\sqrt{2}}(e^{-i\Omega t/2}|\uparrow\rangle + e^{i\Omega t/2}|\downarrow\rangle), \qquad [7.1]$$

$$|-(t)\rangle = \frac{1}{\sqrt{2}}(e^{-i\Omega t/2}|\uparrow\rangle - e^{i\Omega t/2}|\downarrow\rangle), \qquad [7.2]$$

where

$$\Omega = \frac{E_1 - E_0}{\hbar} \qquad [7.3]$$

and E_1 and E_0 are the energies of the one and zero states, respectively, assuming that the one state is the higher-energy anti-aligned state.

For a single electron, the Larmor frequency will typically be in the tens of GHz for experimentally accessible magnetic field strengths. For nuclear spins, it will typically be in the tens of MHz. This frequency is related to the speed at which single-qubit gates can be executed.

The singlet Bell state $|\Psi^-\rangle$ does not precess in the same fashion just discussed, assuming the two qubits are in identical environments. The time invariance of the singlet state occurs because

$$(U \otimes U)|\Psi^-\rangle = (\det U)|\Psi^-\rangle \qquad [7.4]$$

for any single-qubit unitary U. This fact is used in the Jozsa, Abrams, Dowling and Williams (JADW) algorithm, in section 7.2.2. However, minor differences between the local reference frames can result in the two qubits precessing at slightly different rates, requiring complex compensation techniques, which are not detailed here.

A Bell pair does not have to be formed of two qubits of the same physical type; indeed, for long periods of time, the states we care about are photons or other quantum states of light entangled with stationary qubits such as ions, quantum dots or NV centers in diamond. We may define the Bell pairs as $|\Phi^+\rangle = (|\uparrow V\rangle + |\downarrow H\rangle)/\sqrt{2}$ and similarly, where the left-hand ($|\uparrow\rangle$/$|\downarrow\rangle$) qubit is the spin qubit and the right-hand ($|H\rangle$/$|V\rangle$) qubit is the optical qubit. In this case, the optical qubit may be oscillating at a different frequency than the spin qubit. When measuring the two qubits, the two reference frames must be independently tracked and compensated for.

Those reference frames can vary over time. In fact, it is possible to vary the magnetic field or other control parameter, and deliberately change the oscillation frequency adiabatically; superconducting flux qubits, in particular, have a very broad range over which they can be tuned.

The reference frame can also change as the location of a qubit changes. Jozsa, Abrams, Dowling and Williams (JADW) suggest that a spin qubit would have to be transported in a box that maintained the magnetic field [JOZ 00]. The Z axis is aligned with the magnetic field, leaving the X axis in principle to be freely chosen somewhere perpendicular to the field. However, its position must be stable and, in the case of distributed multi-qubit states, the same in all locations, suggesting that an external reference such as aligning with the Earth's axis of rotation should be used. Polarization planes, if defined relative to local topography, will vary with latitude and longitude, affecting the correct interpretation of photonic qubits. Moreover, time itself slows down in a stronger gravitational field; this effect requires global positioning system (GPS) clocks to compensate for the difference between Earth's surface and Earth's orbit, for example.

Thus, our basis states for most qubits are inevitably defined relative to external references: magnetic field direction and strength for spin qubits, the horizontal and vertical planes for polarization. Altering these alignment markers can result in changes to the time-dependent evolution of the qubit, or misinterpretation of its state.

7.1.2. *Quantum optical interference*

The canonical demonstration of interference of two waves is done with a plate with two narrow slits, placed a short distance away from a piece of film. When light passes through a slit, it diffracts. With a single slit, the film will show a wide fuzzy area. Passing a strong coherent light beam from a laser through a plate with two slits results in the film receiving light from both slits. The light retains the phase after passing through the slit, so that it impinges on the film with a phase specific to the distance it has traveled from the plate. The length of the path followed by the light from the plate to the film varies depending on the position on the film, so that the phase of the light arriving from the two slits varies.

For light of wavelength λ, and path lengths l_L and l_R, the aggregate wave function is $\sin(l_L/\lambda + \phi_0) + \sin(l_R/\lambda + \phi_0)$. Bright fringes are created where the phase reinforces (constructive interference), which happens when $(l_L - l_R)/\lambda = 2k\pi$ for some integer k. Dark fringes (destructive inteference) are created where $(l_L - l_R)/\lambda = (2k + 1)\pi$. Any standard optical textbook such as Hecht [HEC 02] will include a good description of the two-slit experiment, as well as a two-armed interferometer.

Quantum optical interferometry can be demonstrated in an experiment in which the strong light beam is replaced by a very weak light beam, with individual photons passing through the plate [GRA 86, TAY 09]. Confirming the quantum nature of light, the individual photons impinge on the film in locations with a probability corresponding to the interference fringes. The only conclusion is that the photon actually passes through *both* slits, *interfering with itself* before hitting the film! This phenomenon clearly demonstrates the dual wave/particle nature of light, with the photon's wave function spanning the distance between the two slits, creating the interference pattern one photon at a time. This behavior can be shown with photons, electrons, or even atoms or neutrons [SCU 91, GRE 88]. The basic experiment was replicated with electrons by Jönsson, and Tonomura *et al.* produced an enlightening short movie showing the buildup of interference patterns [JÖN 61, TON 89].

In a two-armed interferometer, the light beam is first split and routed into two paths (arms) by a beam splitter. After traveling potentially different distances through the two arms, the light beams are recombined at a second beam splitter.

We can reason about the wave function as follows: passing straight through a beam splitter causes no important change to the wave function, but each reflection modifies the phase by a factor of i, or a rotation by $\pi/2$ in our notation. Each exit port from the last beam splitter is the sum of terms from the two paths initiated by the first beam splitter. For one port, both paths have been reflected the same number of times (twice). For the other port, one path has been reflected once, whereas the other has been reflected three times, giving a relative phase shift of $i^2 = -1$ (π). With in-phase terms, the first port sees constructive interference and light is emitted, assuming the interferometer arms are the same length. With out-of-phase terms, the second port sees destructive interference and no light is emitted.

In many experiments, a single-photon detector is placed at each of the two beam splitter output ports, and the lengths of the interferometer arms are balanced to give $(l_L - l_R)/\lambda = k\pi + \pi/2$, resulting in a 50/50 probability of detection at each detector.

It is crucial that we be unable to distinguish which of the two interferometer arms the photon passed through. If something in one of the arms detects the presence or absence of the photon in that arm, then the superposition collapses and interference will not be seen. The single-photon detectors have a natural *timing window*; photons arriving within that window are indistinguishable. We must set up the interferometer so that the lengths of the arms are close enough that we will not be able to tell which arm the photon traveled along based on when the detector clicks.

For our repeater networks, we will often need to interfere two photons, as in optical entanglement swapping. It is necessary that they both arrive (or potentially arrive) within the timing window. The experimental design of optical Bell's inequality violations, as discussed in section 2.5.4, must compensate for this factor.

7.2. Distributed clock synchronization

Synchronization of clocks over a distance is an important task in many systems. The GPS system depends on nanosecond-level accuracy between the receiver and the satellites, high-frequency trading systems depend on accuracy in the very low microseconds and experimental physics creates very high-precision clocks and also uses them to measure other physical phenomena. Current international clock standards have a precision of 10^{-15} [TAK 05], and experiments continue to push the boundaries toward 10^{-18}[HIN 13]; the field is evolving rapidly. With these capabilities, the strength of gravitational fields and physical constants can be tested [YE 08, HÄN 06, CHI 00], and position can be examined, as well [GIO 01]. Fiber networks can today send a frequency reference with an instability of 10^{-14} over modest distances, so repeater networks will have a very high standard to beat [FOR 07].

Coincidentally, Chuang and the group led by Jozsa each proposed quantum clock synchronization algorithms in 2000. Both algorithms use the phase evolution of Larmor precession or its equivalent, rather than attempting to compensate for it. In this section, we look at Chuang's algorithms, followed by the JADW algorithm in the next section, then end with some analysis and follow-on work.

7.2.1. *Chuang's algorithms*

Chuang proposed two algorithms. The first is a handshake protocol, and the second is a distributed quantum algorithm using *quantum phase estimation* in a hybrid digital computation/analog sensor approach [CHU 00].

Ticking qubit handshake. Assume Alice and Bob each have a clock that ticks at the same rate, but the phase difference between them is unknown. The core routine in Chuang's algorithm is the six-step *ticking qubit handshake*, $\text{TQH}(\omega, |\psi\rangle)$, where ω is the oscillation frequency of the qubit $|\psi\rangle$. Alice and Bob cooperate, with Alice first preparing the qubit, then sending it to Bob, who performs some local operations then returns the qubit to Alice, who performs a final step then measures the qubit. Each message includes a classical timestamp and, if not already known, ω. The local operations done depend on the value of the participant's clock. We will not detail the local operations necessary, but this round-trip process allows Alice to transform her original qubit $|\psi\rangle$ into $e^{-2i\omega Z\Delta}|\psi\rangle$, where Z is the usual Pauli operator and Δ is the phase difference between Alice's and Bob's clocks.

To make use of this routine, Alice begins with $|\psi\rangle = |+\rangle$ and executes $\text{TQH}(\omega, |\psi\rangle)$, getting the return value

$$|\psi'\rangle = \frac{1}{\sqrt{2}}(e^{-2i\omega\Delta}|0\rangle + e^{2i\omega\Delta}|1\rangle). \qquad [7.5]$$

After applying a Hadamard to this qubit, she has

$$|\psi''\rangle = \frac{e^{-2i\omega\Delta} + e^{2i\omega\Delta}}{2}|0\rangle + \frac{e^{-2i\omega\Delta} - e^{2i\omega\Delta}}{2}|1\rangle. \qquad [7.6]$$

Measuring this state in the computational basis gives the result 0 with probability $\cos^2(2\omega\Delta)$. If Alice and Bob repeat this operation a large number of times, they can determine Δ, even in the presence of an unknown and variable (within limits) propagation delay between them.

However, this process is simply equivalent to classical interferometry, and offers no gain in execution time relative to a classical approach. To determine Δ to n bits, Alice and Bob would have to repeat this routine 2^{2n} times, giving execution time and communication cost exponential in n. We can do much better by using quantum phase estimation.

Quantum phase estimation (QPE) is a quantum Fourier transform (QFT)-based subroutine lying at the heart of quantum algorithms for tasks as diverse as factoring of large numbers and calculation of energy levels in quantum chemistry. For an extended description, see Chapter 5 of Nielsen & Chuang [NIE 00].

Chuang's insight that gives an exponential reduction of the resources required for clock synchronization is the use of the QFT and phase estimation, allowing n bits of Δ to be determined by repeating TQH m times, $m \approx 2n$. Rather than using single qubits and simply collecting classical statistics, Alice begins with $m + 1$ qubits. She uses one to shuttle back and forth to Bob, and keeps the others as a single m-qubit register. Before beginning, a Hadamard gate is applied to each qubit in the m register, giving it the total value $2^{-m/2}\sum_{j=0}^{2^m-1}|j\rangle$.

Before the jth TQH round, a CNOT is performed between the shuttle qubit and the jth qubit in the m register, with the register qubit being the control and the shuttle qubit the target. The oscillation frequency for the shuttle qubit in the jth round (numbering rounds starting from 0) must be adjusted to be $\omega_j = 2^j\omega_0$, where ω_0 is the frequency in the initial round. The operation executed becomes $\text{TQH}(2^j\omega_0, |\psi_j\rangle)$.

In this extended process, the m register picks up a phase related to the value in the register. Because m began as a superposition of all possible m-bit numbers, we have a superposition of all phases, as well. The final step is to run an inverse QFT on m and measure m. The measurement should give us the number $\Delta\omega$, and because ω is known, we can easily calculate Δ.

The precision that is practically achievable is related to the highest multiple of ω. In optical frequency ranges, this will give precision of a fraction of a femtosecond, less than 10^{-12} s.

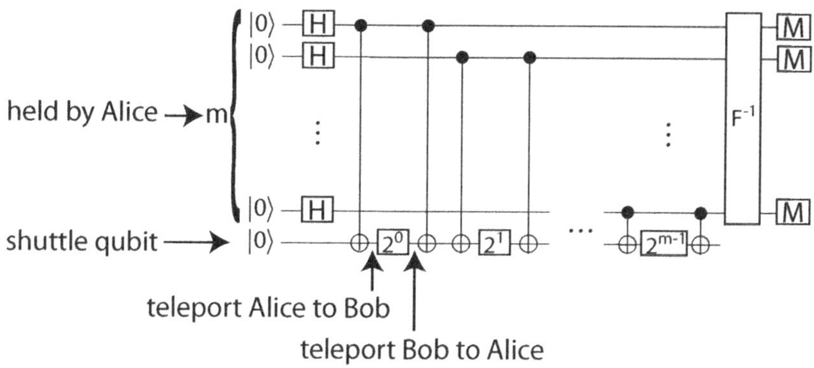

Figure 7.1. *Circuit for Chuang's clock synchronization. M boxes are measurement, 2^i boxes are the corresponding TQH operator*

Network resources: as formulated, the algorithm assumes direct transmission of the shuttle qubit, rather than teleportation, and assumes that the qubit will undergo precession at exactly the same frequency throughout its round trip. Chuang acknowledges that the algorithm would have to be modified to account for these challenges, especially taking into consideration that teleportation allows a qubit to move from one physical representation to another. Assuming these changes can be easily accommodated, the algorithm will require $2m$ Bell pairs to accomplish the m rounds of the algorithm.

The biggest challenge to implementing this protocol is the need to calibrate the system, and as with the Jozsa protocol, bootstrapping a reference. Moreover, this application requires the ability to vary the oscillation rate of a qubit across a very broad range with very high precision.

7.2.2. *Jozsa et al.'s clock synchronization*

In the same issue of *Physical Review Letters* as Chuang's algorithm, a separate paper appeared by Jozsa, Abrams, Dowling and Williams, proposing *quantum clock synchronization* (QCS) [JOZ 00]. Rather than numerically determining the phase offset between two already running clocks, QCS is a mechanism for directly, physically synchronizing the precession of Alice's and Bob's qubits.

This paper begins with a uniquely quantum approach by recognizing that the stability of the singlet in equation [7.4] is destroyed when the entanglement is broken. If Alice and Bob share the Bell pair, and Alice measures her qubit, then Bob's qubit will begin to precess, and under the right circumstances, this can be used to determine the time at which Alice's measurement was performed. Naturally, it is not quite that simple, and as in teleportation the need for supporting classical communication prevents the technique from being used to violate special relativity.

The authors refer to a shared singlet Bell pair as a *preclock* state, primed and ready for use but not yet triggered for independent oscillation. The singlet state can be rewritten in the $\{|+\rangle, |-\rangle\}$ basis,

$$|\Psi^-\rangle = \frac{|0\rangle|1\rangle - |1\rangle|0\rangle}{\sqrt{2}} = \frac{|+\rangle|-\rangle - |-\rangle|+\rangle}{\sqrt{2}} \qquad [7.7]$$

from which it is easy to see that Alice has a 50% probability of finding $|+\rangle$ and a 50% chance of finding $|-\rangle$, when she measures her qubit in the $\{|+\rangle,|-\rangle\}$ basis. For each $|+\rangle$ Alice finds, Bob will find a $|-\rangle$, and vice versa.

Alice and Bob start by sharing a large number of Bell pairs in $|\Psi^-\rangle$. Alice begins the protocol by measuring all of her qubits in the $\{|+\rangle,|-\rangle\}$ basis, breaking the entanglement. Once the entanglement is broken, each qubit begins to evolve independently, as in equation [7.2].

Alice sorts her qubits into two groups, those she found in $|+\rangle$ and those she found in $|-\rangle$, and sends the list to Bob. Bob would like to sort his, as well, but cannot independently figure out which qubits are which, and so must wait until Alice can inform him. Alice and Bob will both use the same population, e.g. $|-\rangle$, and know that Alice's group is precessing in the same fashion as Bob's; because they began in precessing at the same time, their phase will be identical. We now have a population of synchronized clocks at both sites.

To use the clocks, of course, we need to extract classical information from them. Alice and Bob measure their qubits (again in the $\{|+\rangle,|-\rangle\}$ basis), a few at a time, and watch the evolution of the population as it oscillates from $|+\rangle$ to $|-\rangle$ and back. The probability of finding the plus and minus states, respectively, are

$$P(+) = \frac{1}{2}(1 + \cos(\Omega t)), \quad [7.8]$$

$$P(-) = \frac{1}{2}(1 - \cos(\Omega t)) \quad [7.9]$$

where Ω is defined in equation [7.3].

From measuring enough qubits, we can determine the phase, and hence the zero point of the evolution, when the entanglement was broken. Because we are tracking these statistically, we must measure enough to gain confidence in our value of the phase. This is true even in the case of perfectly pure states; with fidelity $F < 1$, we require even more.

In fact, we are measuring only modulo a factor of $2\pi k$ for some integer k; if we do not inherently know the offset to within one precession period, our result is very ambiguous. This can be corrected by repeating the protocol using two or more separate populations that precess at different frequencies, measuring them all at the same time to create the same starting point, and looking for the beat pattern in the populations as they evolve.

The authors also point out that the physical gyroscopic nature of our qubits implies that we require solid physical reference points, as we discussed in section 7.1.1. Measuring relative to these anchors will be complex in some solid-state technologies. The authors also point out that spin-1 devices will behave differently from spin-1/2 devices; in this book, we have concentrated primarily on the latter. As an additional limitation, they note the need for the same relativistic frame, or a means of compensating for the difference.

Network resources: the authors do not give any explicit analysis of the number of Bell pairs required to determine the zero point of time with a chosen precision. For example, with a Larmor frequency of 10 GHz, finding the time to 10^{-14} s would require the ability to detect a phase difference of $\approx 2\pi \times 10^{-4}$ radians, which will require sampling the population at above 10^5 rates, with enough samples at each time step to be statistically reliable, including the imperfect fidelity. This sampling will have to continue for a number of complete cycles. If we use 100 Bell pairs at each sample point and continue for 10 complete cycles, we can estimate very crudely that a single execution of QCS will require at least some 10^8 Bell pairs. Moreover, as similar techniques are actually used to calibrate the Bell pairs themselves, the statistical analysis will be complex, and likely will raise the sample requirements.

The basic form of the experiment assumes that all Bell pairs are measured at the same time, to start the synchronous oscillation. Experimentally buffering that many Bell pairs is beyond proposed experimental capabilities for the foreseeable future, and more importantly, creating that many Bell pairs using a repeater network will take a very long time, during which the Bell pairs are subject to environmental forces altering their precession and coherence.

7.2.3. *Further work*

The three algorithms just presented served as a stimulus for a range of follow-on work by others as well as by the original authors. They addressed difficulties in the original algorithms, proposed new algorithms, and examined the need for entanglement.

The difficulties in the general problem of clock synchronization are subtle, but we must get away from an underlying requirement for an already shared reference frame for time, direction, frequency standards or local conditions such as magnetic field strength [BUR 01, JOZ 01]. Compensating for different reference frames as a result of the slowing of time caused by general relativistic effects is both important and hard, as we noted in section 7.1.1. Most of the work in this section assumes that Alice and Bob are in the same relativistic reference frame; hence, the clocks they are attempting to synchronize can be assumed to ultimately tick at the same rate. However, the group of Nobel laureate Dave Wineland has even discovered that it is necessary to compensate for the difference in oscillation frequencies induced by the general relativistic effects of Earth's gravity with an altitude difference of as little as 30 cm.

It is worth noting that the Jozsa scheme requires the use of distributed entanglement; whether that entanglement arises through a repeater network or by physically moving qubits is irrelevant in the theory, but matters in practice. The other schemes require qubits to be relocated, but entanglement is not required at the application level. De Burgh and Bartlett have shown that the ultimate limit of the precision is the same with or without using entanglement at the application level [DE 05].

The protocols that physically transport a qubit from place to place, including versions of these clock protocols and Rudolph and Grover's algorithm for determining a reference frame [RUD 03], require that this be done slowly enough that any relativistic slowing of the qubit's oscillation will be too small to affect the results. Of course, in wide-area usage, rather than direct transport, qubits are likely to be moved via teleportation over a repeater network. This leaves us with the recursive problem of establishing a strong enough purification regime within the network that our movement of qubits can be trusted.

7.3. Very long baseline optical interferometry

Creating interference of optical signals is one of the first experiments that undergraduates with an interest in optics perform on a laboratory bench, and forms the basis of holography. Interference serves as the mathematical underpinning of quantum algorithms. It is also a critical tool in astronomy.

The resolution of a telescope is limited by the ratio of the instrument's aperture to the wavelength being studied, known as the *Rayleigh diffraction limit*. The inbound light *diffracts*, or bends, at the edges of telescope's objective (the largest light-gathering lens or mirror), resulting in a fuzzy appearance for effectively point sources such as stars.

The resolution of radio telescopes, which study long wavelengths, is generally diffraction limited, even with the large dish antennas commonly seen. To improve the resolution, the signals from a group of individual antennas spread over a larger area are combined, in a technique known as *long baseline interferometry* (LBI). Figure 7.2 shows three dish antennas that are part of an LBI array. This technique can be used to study light even when the original signals are not coherent. With LBI, the resolution is roughly

$$R \approx 1.2\frac{\lambda}{B}, \qquad [7.10]$$

where R is in radians and B is the distance between the antennas or telescopes in the array. For example, a baseline $B = 1$ km for $\lambda = 1$ cm would give $R = 1.2 \times 10^{-5}$ radians, or approximately 2.5 arcseconds. Leading-edge radio telescopes such as the Atacama Large Millimeter/submillimeter Array (ALMA), coming online as this book is being written, will have resolutions as low as 6 milliarcseconds in some frequencies, achieved using baselines of tens of kilometers.

One approach to LBI is, of course, to route the incoming waves to a central location and directly interfere them. This requires low-loss paths from the antennas to the correlator (the measurement device), and paths whose lengths are stable to a fraction of a wavelength. We must be able to compensate for the phase difference induced by this distance, and by the difference in the lengths of the paths from the astronomical source to the antennas. However, as the waves come in, we do not necessarily need to know *which* peak or trough impinges on the separate antennas; the phase difference can be $\phi + 2\pi k$ for any unknown integer k, provided we can stably measure and compensate for ϕ.

Figure 7.2. *Photo of three antennas, part of a radio telescope array for long-baseline interferometry at the National Astronomical Observatory of Japan, Nobeyama Radio Observatory*

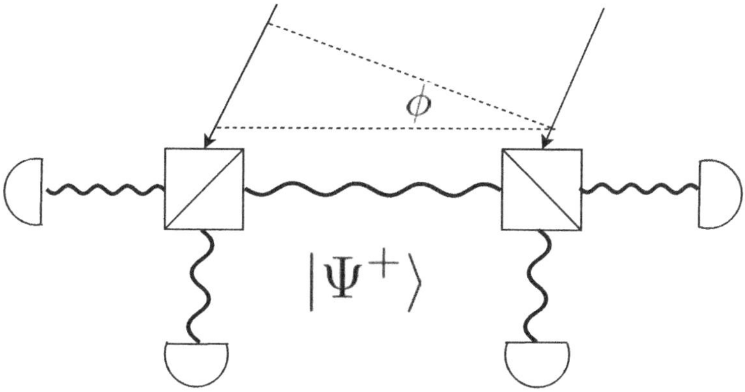

Figure 7.3. *Entanglement can improve resolution for astronomical optical interferometry*

A second approach is to record or digitize the signals at each location, then calculate the interference digitally, which requires phase-sensitive measurement at each location as well as accurate recording of the physical location, to determine the difference in expected propagation time. This digital interferometry is conducted using high-precision GPS clocks and hydrogen masers (a form of atomic clock and frequency reference) for timing signals. A complete digitization of the incoming signal would require taking digital samples at more than twice the signal's own frequency, known as the *Nyquist rate*, to prevent aliasing of the signal. However, the sampling rate can be reduced using *heterodyne* detection, in which a reference signal is subtracted from the astronomical signal. Heterodyne detection and the related homodyne detection are used in some of the physical entanglement mechanisms discussed in Chapter 8. This allows the interferometry to be done with a sampling rate of 1–10 GHz, even for signals with much higher frequencies. Long baselines requires corrections for the effects of general relativity – clocks at higher altitudes run slightly faster than clocks at lower altitudes. To date, the technique can be applied for frequencies as high as 90 GHz, a radio wavelength of 3.3 mm, with plans for substantially higher frequencies. Under some circumstances, this kind of interferometry can be applied between pairs of telescopes on separate continents, or even in space, leading to the name very long baseline interferometry (VLBI).

The frequency of optical light is $\sim\ 4 \times 10^{14}$ Hz, and direct digital sampling accurate enough for interferometry would require a sample rate more than twice that, well beyond what is technically feasible in the near future. Optical LBI over intermediate distances is possible today using heterodyne detection and analog amplification of the signals, but with some limitations.

In 2012, Gottesman, Jennewein and Croke proposed quantum long baseline optical interferometry [GOT 12]. They propose that effective use of Bell pairs between two telescopes L and R can produce quantum interference effects in detectors set behind beamsplitters used to erase which-path information telling us whether a detected photon arrived from outer space or was part of the Bell pair. This allows us to collect light in a phase-sensitive fashion that builds a signal equivalent to having a single telescope with an aperture equal to the baseline. In effect, the Bell pairs serve as the reference signal for the heterodyne detection done in the classical interferometry.

If successful, this scheme would extend interferometry into short wavelengths and low signal strengths (small numbers of photons) in ways that have not been possible before. However, this scheme has some significant drawbacks.

First, each original incoming photon must arrive at both antennas in a coherent superposition spanning the distance between them, which means it must pass through interstellar space without having been measured for which-path information. The input density matrix is of the form

$$\rho = p_A \frac{1}{2} \begin{bmatrix} 0 & 0 & 0 & 0 \\ 0 & 1 & \mathcal{V}^* & 0 \\ 0 & \mathcal{V} & 1 & 0 \\ 0 & 0 & 0 & 0 \end{bmatrix} + (1 - p_A) \begin{bmatrix} 1 & 0 & 0 & 0 \\ 0 & 0 & 0 & 0 \\ 0 & 0 & 0 & 0 \\ 0 & 0 & 0 & 0 \end{bmatrix}, \qquad [7.11]$$

written in the basis $|0\rangle_L |0\rangle_R$, $|0\rangle_L |1\rangle_R$, $|1\rangle_L |0\rangle_R$, $|0\rangle_L |0\rangle_R$, where p_A is the probability of getting any photon at all in the detector. (Recall that

$$\rho = \frac{1}{2} \begin{bmatrix} 0 & 0 & 0 & 0 \\ 0 & 1 & 0 & 0 \\ 0 & 0 & 1 & 0 \\ 0 & 0 & 0 & 0 \end{bmatrix} \qquad [7.12]$$

would represent the purely classical mixture of $|0\rangle_L |1\rangle_R$ and $|1\rangle_L |0\rangle_R$, that is, a 50/50 classical probability of the photon arriving at the right or left telescope, while

$$\rho = \frac{1}{2} \begin{bmatrix} 0 & 0 & 0 & 0 \\ 0 & 1 & 1 & 0 \\ 0 & 1 & 1 & 0 \\ 0 & 0 & 0 & 0 \end{bmatrix} \qquad [7.13]$$

is the d.m. corresponding to the pure state $|+\rangle = (|1\rangle_L + |1\rangle_R)/\sqrt{2}$, the superposition of a single photon being in the left path and the right path.) The visibility $\mathcal{V}$ is determined by these environmental factors, and our ability to compensate for path length differences: unlike classical interferometry, the factor of $2\pi k$ matters here, and we must be able to interfere the halves of our Bell pair with the *correct* incoming photon at each end. This indistinguishability depends on the detector timing windows, as well as correct software tracking of the Bell states and adjustment for the differences in travel times. The details of this, being very specific to the hardware technologies involved, are beyond the scope of this book. A quantum derivation of this visibility was shown by Berthold-Georg Englert in 1996 [ENG 96].

As noted in section 7.1, qubits in the environment serve as miniature gyroscopes. We need to retain this characteristic through the quantum repeater network, such that interference at each end will happen relative to the correct physical orientation of input signals.

Network resources: the authors estimated that approximately 10^{11} entangled photon pairs per second are needed in order to operate on stars with apparent magnitude 7.5, for a wavelength of 800 nm. This figure will vary significantly depending on many factors in the system, including telescope aperture and wavelength, but it is clear that this will be a very demanding application in terms of required Bell pair production rate. However, it only needs to operate over a few tens of kilometers in order to provide new astronomical observing capabilities.

7.4. Conclusion

In this chapter, we have covered two applications, clock synchronization and quantum astronomical interferometry, that illustrate some of the ways in which distributed quantum states and effects can be used as sensors in real-world cybernetic environments. Superposition and entanglement can improve the sensitivity, or reduce the convergence time, of scientific instruments, but only when the relevant phase and oscillations can be tracked with enough precision to give an accurate reference signal.

The protocols described here are described as two-party operations. As this book was being completed, the Lukin group submitted a paper on a large network of quantum clocks, using a multi-party GHZ state, discussing precision improvements as the size of the GHZ state and number of participants grows [KÓM 13]. Their proposal even discusses the robustness of the application to misbehaving nodes.

Coupled with the sensitive detection of entanglement or measurement provided by QKD networks, we can see some of the range for such applications. The fields of quantum imaging and activitation are actually far broader. Quantum effects have been proposed for improving the sensitivity of imaging [LLO 08], and two-photon absorption (a most decidedly quantum technique) has been proposed for fine-feature

VLSI photolithography [BOT 00]. In one fascinating thought experiment, Elitzur and Vaidman proposed that a quantum interferometer and presence or absence of interference could be used to test whether a photon-sensitive bomb trigger is "live" or a "dud". As of the writing of this book, the annual SPIE conference in the United States features a symposium on quantum imaging.

PART 3

Lines of Repeaters

Chapter 8

Physical Entanglement and Link-Layer Protocols

The science of creating entanglement between distant solid-state or atomic qubits using light would warrant a full physics textbook of its own – except that the state of the art is advancing so rapidly that any such book must inevitably be out of date. Here, we will introduce a few of the many proposed and experimental approaches, while minimizing the need to understand complex physics. For a much more thorough, physics-oriented explanation of the topics in this chapter, see the book by Christopher Gerry and Peter Knight [GER 05]. The collection edited by Bouwmeester, Ekert and Zeilinger in 2000 remains an excellent introduction [BOU 00], and Nielsen and Chuang of course is a comprehensive and rigorous text with good material about the physical phenomena of quantum information [NIE 00].

The emphasis here will be on a qualitative understanding of the physical process, including what factors limit the probability of success and the fidelity represented in the output density matrix. Most importantly, we will study the common characteristics of the operations that affect link protocol design and ultimately network design. With this basic grounding in the concepts, readers should be able to ingest new developments in the field and assess their relative value in building complete networks.

8.1. Creating entanglement using light

8.1.1. *Quantum states of light*

Up to this point, we have discussed quantum information in the abstract, in terms of the basis states $\{|0\rangle, |1\rangle\}$. Those are our logical values, which we like to compute upon, as introduced in section 2.3.1. These logical values must be matched to the states of physical phenomena in order to completely realize them.

In section 2.5.4, we introduced horizontal and vertical linear polarization as one form of qubit, $\{|H\rangle, |V\rangle\}$. Polarization also comes in circular or helical form so that we can encode a qubit in the right-circular and left-circular polarizations instead, which we can write as $\{|R\rangle, |L\rangle\}$. Conversion from linear to circular polarization is straightforward, and circular polarization is more robust against misaligned optical elements or the disruptions of fiber.

Besides polarization, photons have many characteristics that can be used to represent a data value. Light can be run through different paths in the apparatus such that we might define $|1\rangle_L |0\rangle_R$ and $|0\rangle_L |1\rangle_R$, where the two expressions mean a single photon in the left and right arms, respectively. This is sometimes called a *dual-rail* qubit. The subscript (sometimes omitted) identifies the path. This notation allows for more than one photon in each path, e.g., $|0\rangle |2\rangle$, potentially allowing us to operate on photon number as a multi-level (ternary or higher) representation, and is useful for describing physical arrangements. However, the mapping to binary logical states when we wish to discuss algorithms defined in the gate model sometimes becomes tedious. This which-path approach works well in a lab, but not over long distances.

A more useful form for long-distance transmission is *time bin*, in which one state is advanced or retarded in time relative to the other $\{|t\rangle, |t + \Delta t\rangle\}$. Combined with classical time synchronization signals in the channel, this approach is robust against many types of fluctuation in the channel. Polarization can easily be turned into a which-path encoding using a polarizing beam splitter, and which-path can be converted to time bin by making one path longer than the other and then recombining into a single channel for transmission. Time bin can, in theory, be used to represent more than a single qubit by increasing the number of bins.

Some of the mechanisms discussed below generate two different wavelengths of light depending on the state of the qubit used to generate the light; therefore, they are frequency-defined qubits $\{|\omega_0\rangle, |\omega_0 + \Delta\omega\rangle\}$.

States that exhibit "quantumness" are not limited to single photons. Those focusing on some quantum effects but not individual photons often refer to *non-classical* states of light. The state of light can be described in terms mathematically similar to "position" and "momentum", and Heisenberg's uncertainty principle then applies. Although the uncertainties in the two terms are equal in a classical state, in a quantum state, we can trade off increased uncertainty in one term for decreased uncertainty in the other, creating what are known as *squeezed states* [WAL 83]. These squeezed states can then be used as analog, rather than digital, quantum variables, known as *continuous variable* operation.

Stronger coherent light pulses have been proposed as variables and are useful because stronger light can control atoms more quickly or reliably than single photons. The state of the light itself can be modified to carry quantum information by an interaction with a matter qubit that alters the phase of the light in a fashion dependent on the state of the matter qubit, as in the *qubus* system using weak nonlinearities proposed by Munro, Nemoto, Spiller and their collaborators [MUN 05, SPI 06].

One additional possibility to receive attention recently is *orbital angular momentum* (OAM), in which the wavefront of the light spirals as it propagates [ALL 92, MOL 07, YAO 11]. This differs from polarization in that the peak intensity of the wavefront is not at the center of the propagating beam, but is offset and revolves around the center. It actually has more in common with the path modes discussed above, and it is possible to physically convert between OAM and the path. OAM can be used to encode several qubits in a single photon [RAY 10].

8.1.2. *Emission*

Quantum states of light (a single photon, a small group of photons or sometimes a moderately strong pulse of light) can be created by transferring a qubit held in a stationary device to our light pulse or more directly using physical mechanisms that reliably emit one photon or a pair of photons we can interpret as qubits.

One common approach to creating pairs of entangled photons is known as *parametric down conversion* (PDC), which sends a high-energy photon into a special type of crystal that absorbs the single photon and emits an entangled pair of lower-energy photons as shown in Figure 8.1. The form of PDC common in quantum information experiments was developed by Paul Kwiat, Klaus Mattle and Harald Weinfurter, working at the time in Anton Zeilinger's group, and Alexander Sergienko and Yanhua Shih at UMBC [KWI 95]. It produces entangled photon pairs of the form

$$|\psi\rangle = \frac{|HV\rangle + e^{i\alpha}|VH\rangle}{\sqrt{2}} \qquad [8.1]$$

where α is a relative phase that can easily be corrected, introduced by the crystal. Because the two photons are spatially separated, it is easy to generate any of the four Bell pairs using a rotation on one of the photons. The PDC functionality can be incorporated into one end of the link or can be situated halfway between the two endpoints in a node without memory of its own. PDC is particularly common in all-optical experiments.

Some types of stationary qubits (explained below), such as individual ions in an *ion trap*, *quantum dots* or *nitrogen vacancy centers* in diamond, can directly emit photons whose states are dependent upon the state of the qubit. For example, a $|1\rangle$ state may be engineered to emit a photon while $|0\rangle$ does not or $|0\rangle$ and $|1\rangle$ may emit photons of different wavelengths. The resulting photon is entangled with the state of the qubit as in Figure 8.3. This approach has a variety of experimental challenges, including controlling the wavelength of the photon and the exact timing of its emission and convincing the photon to enter the communication channel (e.g., optical fiber) properly.

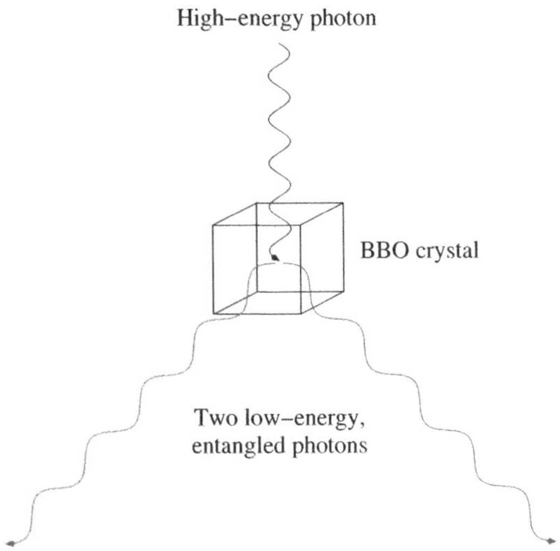

Figure 8.1. *PDC is one common method of creating pairs of entangled photons. PDC creates only generic Bell pairs; it cannot be used to copy or transmit arbitrary quantum states. An entangled pair such as this can be used to initiate entanglement between stationary memories in distant nodes*

8.1.3. *Transport*

Naturally, once the light is created, we need to transmit that light to the next station. We have two options for our optical channels: free space transmission managed using lenses and mirrors and waveguides such as optical fibers.

No amplifiers or classical signal repeaters are allowed in the channel. Amplifiers "copy" the state in a fashion forbidden by the no-cloning theorem (section 2.6). Thus, a key factor in channel design is the tradeoff between longer distances between stations, which is economically and logistically desirable, and the higher loss in a longer channel, which dramatically reduces performance. For strong signals, engineers often discuss the amount of signal that is lost in terms of decibels (dB), defined as $L = \log_{10} S_{\text{out}}/S_{\text{in}}$, giving the loss L in dB for input and output signal strengths S. -10 dB corresponds to loss of 90% of the signal, and -20 dB corresponds to loss of 99%. When considering individual photons, these correspond to 90% probability and 99% probability of loss of the photon, respectively. Physicists sometimes refer instead to the *attenuation length* l_0, the distance over which the signal strength (or reception probability) falls by a factor of $1/e$ or roughly -4.3 dB.

Over benchtop distances, air is adequately transparent for all wavelengths of interest here, provided the air is kept still. Losses are primarily due to unwanted reflections from lenses and detector surfaces, misaligned optical elements etc. Over kilometer-or-longer distances, turbulence in the atmosphere, refractive effects from eddies of different temperature, and scattering due to particles in the air result in photon loss and changes to the phase. The natural dispersion of the light also presents a problem, demanding the use of carefully collimated telescopes at both the sending and receiving ends, with a moderately large aperture at the receiver.

Free-space experiments have been conducted over a variety of types of links and distances [TUN 10]. A free-space link formed part of the original DARPA QKD network [ELL 05a]. The Zeilinger group has demonstrated quantum communication over a distance of 144 km between two peaks in the Canary Islands [URS 07, FED 09].

Once light leaves the atmosphere, all of the effects except dispersion disappear; transmission to a satellite or the International Space Station is feasible [ASP 03, VIL 08, PEN 05]. Naturally, pointing accuracy is a problem and tracking of the distance accurately enough to monitor phase is not directly possible, necessitating continuous use of calibration pulses. However, experimental tests have begun on space-based systems already, and indications are that it will work. Satellite-based operations have the unique characteristic that their physical security is guaranteed in a way that is impossible on Earth's surface, although of course any control systems remain potentially vulnerable to hacking through classical communication channels.

In today's Internet, the majority of long-distance links use optical fiber, a type of waveguide. A waveguide consists of a core that carries the light and a cladding with a substantially different *index of refraction*. This difference results in the light being reflected internally, forcing the light to propagate along the waveguide, provided that the waveguide is not bent to too sharp an angle. For microwave signals, the earliest type of waveguide was a simple aluminum tube where the core was air. For optical and infrared photons, both the core and cladding are types of glass or plastic. Waveguides can be fabricated directly in silicon (which is almost transparent to infrared light) or in conductors such as aluminum (for microwave) using standard VLSI techniques.

Light moves substantially slower through optical fiber than air or vacuum, typically at about $c_{\text{fiber}} = 0.7c$. (A common misperception among newcomers to optics is that this is due to reflections within the waveguide, but it is simply that light moves more slowly through glass than air; in fact, this is the very definition of the index of refraction). One-way latency through fiber is about $5\mu s$ per kilometer.

Fiber comes in two main flavors, single-mode and multi-mode. Multi-mode fibers have a larger core and hence are easier to robustly couple to a light source, but single-mode fibers restrict the behavior of the light more tightly and hence are better for high-precision purposes such as quantum experiments. Classical signals typically rely primarily on the signal strength and do not require perfect coherence of the signal. Some newer methods, such as OAM, do require at least some coherence. Therefore, phase-preserving and polarization-preserving fibers are gradually being deployed.

The loss in fiber is wavelength-dependent. Telecommunications fiber is typically optimized for transmission of infrared light around $1.5\mu m$. Loss is typically around 0.2 dB/km; high-quality fibers with loss of 0.17dB/km are available, and in laboratories, loss is as low as 0.12 dB/km. At 0.17 dB/km, the attenuation length l_0 is about 25 km. This value is commonly used in quantum repeater simulations. Conventional wisdom holds that this loss will limit effective distances between repeater stations to a few tens of kilometers, although NTT recently accomplished the remarkable feat of distributing time-bin entangled photons through 300 km of fiber [INA 13].

A problem arises when we consider creation of the photons: most of the attractive technologies generate light with shorter wavelengths than $1.5\mu m$. We must have some means of converting short-wavelength photons into longer-wavelength photons without loss of the quantum state. Ikuta *et al.*, in the Nobuyuki Imoto group at Osaka University, have recently demonstrated conversion of input $F = 0.97$ visible photons into $F = 0.93$ infrared photons [IKU 11, IKU 13]. De Greve *et al.*, in the Stanford Yamamoto group, proved that they could successfully entangle the state of a quantum dot with a photon, including the conversion to telecom wavelength [DE 12]. In an exciting twist on the problem, Fekete *et al.* demonstrated PDC-created entangled pairs where one photon is suitable for a interaction with a solid-state memory, and the other is suitable for telecom [FEK 13].

The final, large factor in the efficiency of our channel is the coupling between the quantum memory and the channel itself. Some memory types emit photons in random directions, while others emit in a defined direction. Capturing all of the photons and guiding them toward the fiber is a severe challenge in either case. This collection process results in the loss of several dB; in some experimental setups, 10 dB or more.

8.1.4. *Detection*

Emitting quantum states is half the problem; the other half is absorbing or detecting them at the receiving end. When the light pulse is received, some mechanisms attempt to physically absorb the light into a static memory. Others reflect the light off of a static structure, then measure the state of the light. One approach is to use two incoming photons and *interfere* them, then measure the resulting state, as shown in Figure 8.2. As with PDC, this functionality can be incorporated into one end of the link or can be situated halfway between the two endpoints in a node without memory of its own.

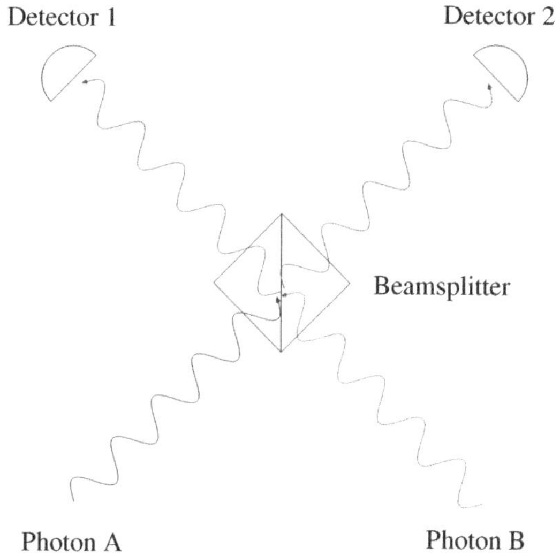

Figure 8.2. *If two photons arrive at the beamsplitter at the same time, they can be made to* interfere *such that the detectors are uncertain which photon came from the left and which from the right. If each photon was entangled with a stationary memory when it was emitted, the result can be entanglement of those two memories*

Two common types of physical detectors are *avalanche photodiodes* (APDs) and *superconducting single photon detectors* (SSPDs), serving the same role as the charge-coupled device (CCD) or CMOS detector in a digital camera. Both operate by having the absorption of a single photon generate a cascade of electrons (current). Both work better at low temperatures, but SSPDs in particular generally require cryogenic operation. Factors affecting their value for our purposes include not just the wavelengths to which they are sensitive, but also their *cycle time*, how long it takes them to prepare for a subsequent detection event after the cascade has been released. The *detection efficiency* is the percentage of arriving photons that are correctly detected. Detectors are also ocasionally triggered by the arrival of spurious photons or other random events. Therefore, the *dark count* of photons detected even when no signal light is present directly degrades the fidelity of the output state in normal operation.

Discrimination of signal states is often accomplished by comparing to a reference signal, using a *homodyne* or *heterodyne* detection setup. For some types of technologies (e.g. qubus), fidelity and success probability have a direct tradeoff. Discarding a higher fraction of ambiguous states gives a higher output fidelity, but lower success rate.

8.2. Memory and transceiver qubits

The devices being explored as memory and logic for quantum computation come in a bewildering variety, and even a moderately complete treatment of the topic is well beyond the scope of this book. Beyond the classic references mentioned at the start of this chapter, recent work on a few of the types being aggressively studied is surveyed by Ladd *et al.* [LAD 10].

In section 8.1.1, we saw examples of the *state variable* possibilities for light. Memory comes in a variety of types of structures, but we typically deal with one of only a few types of physical phenomena as the state variable. States can be the up or down *spin* of an electron, $\{|\uparrow\rangle\,|\downarrow\rangle\}$, aligned or anti-aligned with a local reference magnetic field, as discussed in section 7.1.1. Some phenomena, such as the spin of some atomic nuclei, are more naturally ternary, requiring more care in mapping to and from the binary states we more commonly use. Besides spin, we can also use the *ground state* (lowest energy state) and *first excited state* (created by absorbing a single quantum of energy, e.g. a photon) of an electron orbiting an atom or a quantum dot, for which we can write $\{|e\rangle\,,|g\rangle\}$.

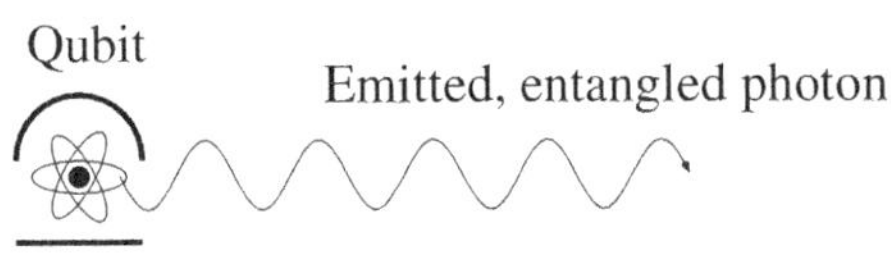

Figure 8.3. *Some types of qubits can be coerced into emitting a photon entangled with the state of the qubit*

These state variables can be created and manipulated using a variety of structures. *Ion traps* hold individual, ionized atoms in a vacuum, suspending them in place using radio-frequency magnetic fields and cooling them using lasers. The state variable is generally $\{|e\rangle\,,|g\rangle\}$, manipulated and read out using lasers. A *quantum dot* is a small block of semiconductor material with an electrical potential set up to trap an electron, much as it orbits an atomic nucleus. The state variable can be spin, energy level or position in a pair of quantum dots. A *nitrogen vacancy center* is a particular type of defect in a diamond that can trap a single electron around it, allowing us to use the spin of the electron as a qubit. NV centers in diamond are especially attractive due to the possibility for room-temperature operation, whereas the other solid-state technologies require millikelvin temperatures. Room-temperature operation does pose significant problems even for NV diamond, however. Depending on the details of the structure and the state variable, both quantum dots and NV centers can be controlled using electrical or magnetic fields or optically. All of the above can emit (or, with more difficulty, absorb) photons of optical or near-optical wavelengths that can then be converted to telecom wavelengths using the techniques referred to above.

One additional type of structure being studied for use in repeaters is *superconducting Josephson junction flux* qubits, in which the states are a quantum of magnetic flux, generated by current flowing clockwise or counter-clockwise in a small loop of superconducting wire. A flux qubit would emit a microwave photon, rather than optical, but flux qubits are extremely flexible and are being experimentally coupled to other types of qubits, making them useful for the logic in a larger device with a different type of qubit as transceiver.

Another fundamental tool in the toolkit is the use of *cavity QED*, in which light (e.g. from a laser) interacts with the state of a qubit of one of the above types held in a small reflector cavity, which amplifies the interaction. If the system is designed so that the change to the state of the light depends on the state of the qubit, then two stationary, distant memories can become entangled when the light is measured appropriately, as in Figure 8.4.

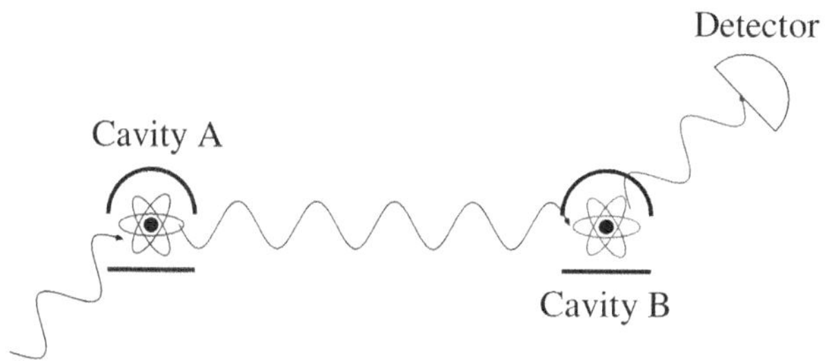

Figure 8.4. *Interaction of light with a qubit in a cavity (either an atom, as shown or a quantum dot) can entangle the light with the qubit. Interacting the light first with the qubit in cavity A, then sending the light down the channel and doing the same with cavity B before measuring the light pulse can create entanglement of the two memories*

8.2.1. *Gate noise*

The operations we perform on qubits are inherently noisy; we cannot execute them perfectly. The simplest and most general noise model is *white noise*, in which information "leaks" out of our principal component and into every other component.

With a gate noise parameter x, the individual elements $\rho_{j,k}$ in our d.m. ρ for a Bell pair evolve as

$$\rho'_{j,k} = (1-x)\rho_{j,k}, \text{ for } j \neq k, \qquad [8.2]$$

$$\rho'_{j,j} = (1-x)\rho_{j,j} + x/4. \qquad [8.3]$$

This results in gradual decay of the state toward a completely mixed state with each diagonal element $1/4$ and all off-diagonal elements 0. This white noise operator is the same when the d.m. is written in either the computational or Bell basis. (Recall that in the computational basis, all superposition, and hence entanglement, is lost when the off-diagonal elements become zero; in the Bell basis, the basis vectors themselves are entangled states. However, the diagonal $1/4$ state is the same mixed, unentangled state in both bases.)

8.2.2. *Single-qubit decoherence*

Physical memories are inevitably imperfect; their contents decay over time as they interact with the local environment. This decay, called *decoherence*, poses one of the most formidable challenges to the development of quantum repeaters.

A qubit, if left alone for a very long time, will eventually settle into some classical distribution of the basis states $|0\rangle$ and $|1\rangle$. This process is known as *thermalization*. If the energy levels of the two basis states are *degenerate* (the same energy), the final distribution will be 50/50,

$$\rho_\infty = (|0\rangle\langle 0| + |1\rangle\langle 1|)/2. \qquad [8.4]$$

Otherwise, the balance varies with the difference in the energy levels of the two states and the temperature. Without going into detail, we can assume that a population of idle qubits would have a *polarization* of p, where $0 \leq p \leq 1$ is the fraction of qubits that settle into the $|0\rangle$ state. This gives us

$$\rho_\infty = p|0\rangle\langle 0| + (1-p)|1\rangle\langle 1| = \begin{bmatrix} p & 0 \\ 0 & 1-p \end{bmatrix} \qquad [8.5]$$

as the density matrix for each qubit. By convention, it is common for the lower-energy state to be the $|0\rangle$ state. At absolute zero, the energy relaxation rate T_1 is the time it takes a known $|1\rangle$ state to decay to a $|0\rangle$ state with probability $1/e$. Conversely, at infinite temperature, it would be the time for a known $|0\rangle$ state to decay to a $|1\rangle$ state with probability $1/e$.

In most technologies, the amplitude of a state is affected by different physical processes than the phase; so, we also must consider the phase relaxation rate known as T_2. T_2 is the time it takes a $|+\rangle$ state to flip to a $|-\rangle$ state with probability $1/e$. With both the T_1 and T_2 processes, the decay is exponential; because the probability of a qubit changing in the coming instant is independent of whether or not it changed in the last instant, these are called *memoryless* decay processes.[1]

Here, we first look at decoherence in the computational basis for a single qubit, then study the more complex case of a Bell pair, written in the Bell basis.

There are a variety of mathematical approaches to the problems of decoherence. We will consider the model known as *generalized amplitude damping* (GAD) to describe what happens to the relative weights of the $|0\rangle$ and $|1\rangle$ states over time and *phase damping* to describe what happens to the phase of the state. (See Nielsen and Chuang, section 8.3, for a more complete discussion of this process.)

For general amplitude damping, using $\gamma = 1 - e^{-t/T_1}$, where t is the elapsed time, the total operation we apply to our density matrix is

$$\mathcal{E}_{\mathrm{GAD}}(\rho) = \sum_{j=0}^{3} E_j \rho E_j^\dagger \qquad [8.6]$$

where

$$E_0 = \sqrt{p}\begin{bmatrix} 1 & 0 \\ 0 & \sqrt{1-\gamma} \end{bmatrix} \qquad [8.7]$$

$$E_1 = \sqrt{p}\begin{bmatrix} 0 & \sqrt{\gamma} \\ 0 & 0 \end{bmatrix} \qquad [8.8]$$

$$E_2 = \sqrt{1-p}\begin{bmatrix} \sqrt{1-\gamma} & 0 \\ 0 & 1 \end{bmatrix} \qquad [8.9]$$

$$E_3 = \sqrt{1-p}\begin{bmatrix} 0 & 0 \\ \sqrt{\gamma} & 0 \end{bmatrix}. \qquad [8.10]$$

We call $\mathcal{E}_{\mathrm{GAD}}$ a *superoperator*, and this manner of writing it as a sum is called the *operator sum representation* or *Kraus representation*. This representation is not unique, but will be sufficient for our purposes here.

It is easy to verify that the steady state

$$\mathcal{E}_{\mathrm{GAD}}(\rho_\infty) = \rho_\infty \qquad [8.11]$$

as we would expect.

1 This has nothing to do with the "memoryless" repeater architecture we will see in Chapter 12.

The above is amplitude damping, which changes the values of the qubit according to the rate T_1. It also mutes the phase of the qubit to a certain degree, but the primary contribution of phase damping is a separate process governed by the rate T_2. The superoperator and operators for phase damping can be written as

$$\mathcal{E}_{\rm PD}(\rho) = E_0 \rho E_0^\dagger + E_1 \rho E_1^\dagger \qquad [8.12]$$

where

$$E_0 = \sqrt{\frac{\alpha}{2}} \begin{bmatrix} 1 & 0 \\ 0 & 1 \end{bmatrix} \qquad [8.13]$$

$$E_1 = \sqrt{\frac{1-\alpha}{2}} \begin{bmatrix} 1 & 0 \\ 0 & -1 \end{bmatrix}. \qquad [8.14]$$

In these equations, $\alpha = (1+\sqrt{1-\lambda})/2$, where λ is the probability of a phase flip error occuring in our time interval t such that $\sqrt{1-\lambda} = e^{-t/2T_2}$.

This pair of operators results in the decay of the off-diagonal elements, which is perhaps best seen with an example. If we begin with $|\psi\rangle = |+\rangle$,

$$\rho = |\psi\rangle\langle\psi| = \frac{1}{2} \begin{bmatrix} 1 & 1 \\ 1 & 1 \end{bmatrix} \qquad [8.15]$$

and applying our phase flip superoperator gives

$$\mathcal{E}_{\rm PD}(\rho) = E_0 \rho E_0^\dagger + E_1 \rho E_1^\dagger \qquad [8.16]$$

$$= \frac{1}{2} \begin{bmatrix} \alpha & \alpha \\ \alpha & \alpha \end{bmatrix} + \frac{1}{2} \begin{bmatrix} 1-\alpha & \alpha-1 \\ \alpha-1 & 1-\alpha \end{bmatrix} = \begin{bmatrix} \frac{1}{2} & \alpha - \frac{1}{2} \\ \alpha - \frac{1}{2} & \frac{1}{2} \end{bmatrix}. \qquad [8.17]$$

8.2.3. *Two-qubit decoherence*

The decoherence of a two-qubit state where the qubits are physically separated is, in principle, the product of the independent decoherence processes. We must take into account that the two qubits may be encoded in different physical carriers, meaning we need to worry about Alice and Bob having different T_1, T_2 and p, which we will write with a superscript A or B.

Let the individual operators E_j^A be the operators on Alice's qubit and E_k^B be the operators on Bob's qubit. For two isolated qubits with completely independent relaxation processes, then,

$$\mathcal{E}(\rho) = \sum_j \sum_k (E_j^A \otimes E_k^B) \rho (E_j^{A\dagger} \otimes E_k^{B\dagger}). \qquad [8.18]$$

For the generalized amplitude damping, our new Kraus operators are just the 16 combinations of the operators at Alice and Bob, and for phase damping, they are the four combinations of operators for individual phase damping.

Much of our work in this book will deal with Bell pairs with the density matrix written in the Bell basis. These operators are written in the computational basis. We will not work out the Bell basis representation of these operators here, but doing so would make a good exercise for the reader.

With this ability to manage the decoherence of the qubits in a Bell pair somewhat independently, we are prepared to discuss distributed management of density matrices. We will tackle this important problem in section 8.5, but first we will study the structure of a link more closely.

8.3. Link structure

Parts from this toolbox of techniques can be assembled in a variety of ways. Depending on this choice of physical mechanism, the state of the light pulse can be (a) a generic state, useful for later work but with no relationship to any other data in our application, (b) a state entangled with the stationary qubit or (c) the recipient of the stationary qubit's state in its entirety. How the light state is terminated to create entanglement also affects link organization.

We can call links using PDC in the middle and memories at the ends $M \leftarrow P \rightarrow M$ links, with the arrow representing the direction of propagation of the photons. Those using interference in the middle of a link, with the photons starting life entangled with memories, can be called $M \rightarrow I \leftarrow M$ links, as shown in Figure 8.5. Those that interact light with a cavity, then transmit or measure or that incorporate the PDC functionality into a sending node or the interference functionality into a receiving node, we can call simply $M \rightarrow M$ links. This distinction in physical capability impacts our network architecture as we will see later.

In the link design, arrangements are often assumed to be $M \rightarrow M$, memory to memory, although the may be implemented underneath as $M \rightarrow I \leftarrow M$, as in Figure 8.6. Each memory at the transmitter side is entangled with a light pulse in flight. Roughly, the number of memories assigned at the transmitter side determines how many light pulses may be outstanding at a time, although a recent proposal cleverly compresses that number. At the receiver, as few as one qubit may be enough, though the receiving repeater as a whole typically requires more. Newer designs may eliminate the need for the receiver to have memory at all by measuring the light pulse (or receiving qubit) directly after reception.

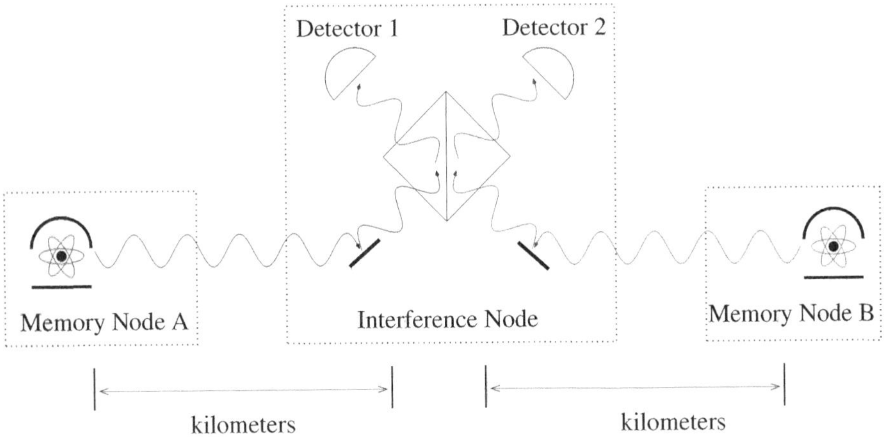

Figure 8.5. *Link structure for $M \rightarrow I \leftarrow M$*

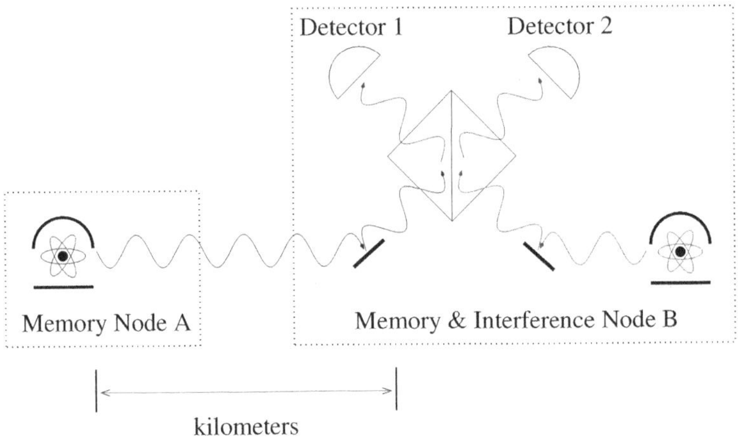

Figure 8.6. *Link structure for $M \rightarrow I \leftarrow M$ behaving as an $M \rightarrow M$ link*

Whether the links are $M \rightarrow M$, $M \rightarrow I \leftarrow M$ or $M \leftarrow P \rightarrow M$ affects timing, but otherwise should have little impact on the session architecture. An open question is whether I and P nodes should be named as first-class network nodes or left hidden in the link.

The availability of physical resources impacts scheduling and purification choices and hence the session architecture. The earliest purify-and-swap repeaters required a few tens of qubits in each node, growing logarithmically with the number of hops to be spanned. While this number sounds modest, experimental capabilities even today have not yet reached such levels. An adapted version of this approach reduced the hardware requirements to the dead minimum two qubits per memory-enabled node by carefully scheduling the purification and swapping [CHI 05].

Characteristics of the link also affect the fundamental goal of the network. The low success probability and the imperfect fidelity at the link level drove much of the original purify-and-swap proposal, in which the network is organized to create end-to-end Bell pairs (failure of which is acceptable) rather than to directly transmit data states used by end-node applications. The availability of very high fidelity links would enable hop-by-hop teleportation of valuable data states; coupled with a high reception probability, even link-level direct transmission of states would become possible. High fidelity may be achieved by applying error correction in the memory buffer, but doing so on the link requires clever adaptations to tolerate high loss rates. A flexible network architecture will accommodate future evolution along these lines.

8.4. State machines and protocol interactions

Figure 8.7 shows the message sequence for creating base-level entangled pairs. The wavy lines in the figure (labeled PE, for Physical Entanglement) indicate the optical pulses that interact directly with the qubits, while the straight lines are classical communication. At the sender, an optical pulse is entangled with each separate physical qubit, then multiplexed into the long-distance fiber. The pulses are very short compared to the propagation delay of tens to hundreds of microseconds (τ_l in the figure). Therefore, we can treat the pulses as effectively being instantaneous. Upon arriving at the receiver, the pulses are demultiplexed, and an attempt is made to entangle each one with a free qubit. Certain properties of the pulse are then measured [CHI 05, CIR 97, DUA 04, ENK 97]. The measurement results tell us if the entangling operation succeeded. If so, we have created a *Bell pair*, entangling a qubit at the sender with a qubit at the receiver. The receiver prepares ACK/NAK "keep" flags for each qubit and sends them back to the sender, letting the sender know which operations succeeded. This measurement and flag preparation is τ_o in the figure and the return message is labeled AEC (Acknowledged Entanglement Control).

For a simple physical layer, we need only three states to describe the behavior as shown in Figure 8.8:

– *Uninitialized.* Buffer memory qubits in an unknown state or otherwise not prepared for entangling operations are held in the Uninitialized state.

– *Unentangled.* After initialization, a buffer qubit moves to the Unentangled state, where it is held only briefly. If the qubit's fidelity falls below a threshold while waiting, it becomes unusable and will be returned to the Uninitialized state. In general, qubits will be in either Uninitialized or Unentangled only briefly.

– *Interim Entangled.* This state is reached after entanglement is attempted. The qubit will remain in this state until it receives an answer from the remote station or a local timer times out. (Technically, this state will normally represent the period in which our stationary qubit is entangled to a flying qubit propagating toward the far end of the link.)

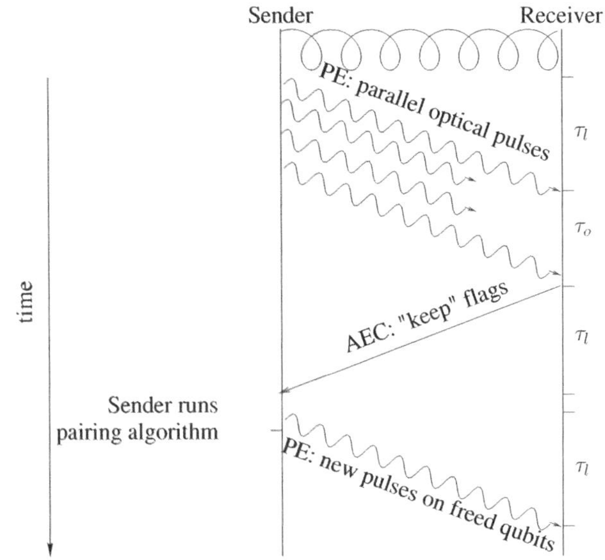

Figure 8.7. *Messaging sequence for the lowest level of Bell pair creation, using an $M \rightarrow M$ channel. For physical layers using single photons or weak pulses, the probability of correctly receiving the pulse is low. For stronger signals, such as the qubus approach, the probability is relatively high*

The timer to move back to Uninitialized is based on the lifetime of the quantum memory used and in practice will be set to limit the impact on the fidelity to around 1%. If an Entanglement Failure message is received or no answer is received before the timer expires, the qubit will be moved to Uninitialized to start over again. If an Entanglement Success message is received, the qubit will be moved to the next higher protocol layer.

8.5. Managing density matrices in distributed software

As this is the first chapter dealing in concrete fashion with network elements, we shall describe some of the key aspects of managing density matrices in a distributed fashion. The principles and data structures described here and in the next section apply broadly across all levels of the system. This section will be referred to extensively in later chapters.

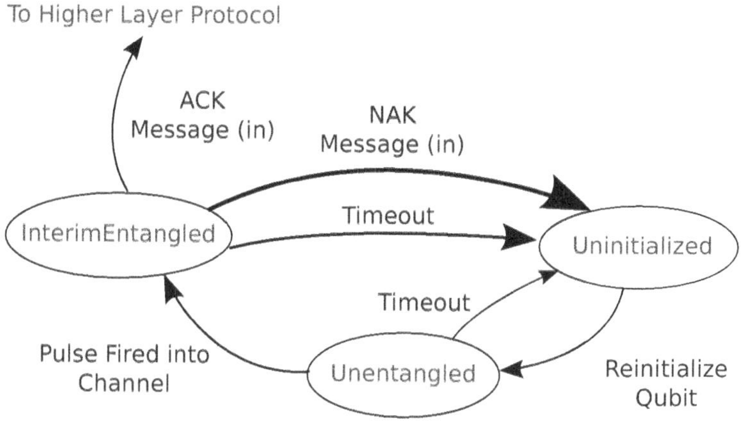

Figure 8.8. *Finite state machine for the transmitter in the simplest acknowledged entanglement protocol*

A density matrix (d.m.) is an abstraction describing our best understanding of the quantum state. It is, in essence, a statistical construct, derived empirically from experimental data and known interactions and imperfections in the devices comprising our system.

The d.m. is represented as a software data structure on the classical side *of each node*. Each qubit memory at each end of the link must have a d.m. associated with it, incorporating the node's knowledge about the state. Until fully error-corrected, logical qubits become ubiquitous at repeater nodes, key parameters are the qubit's (or Bell pair's) *age* (the time since the state was created) and the *type of memory* in which it is held, both of which affect its fidelity.

The fidelity of the initial Bell pair created depends on the condition of the channel. The characteristics of the physical channel will change over time. If the channel is optical fiber, the rate of change is minutes to hours, as for example the fiber stretches due to changes in the temperature. Thus, the d.m. for a Bell pair produced by the channel will be different at noon than it was at 6:00 am. Free-space channels may change at the millisecond level due to atmospheric conditions, a far more challenging problem. The latency of a satellite link even varies as the satellite moves. Much of this is the domain of the physical layer, at a level we will not concern ourselves with in this book. Here, we will assume that a separate channel management protocol tracks the condition of the channel itself such that both nodes have a coordinated notion of the fidelity of a quantum state resulting from using the channel.

Moreover, a node is free to move a qubit from one memory location to another in order to manage its buffer space, optimize the memory lifetime or because certain required operations (e.g. measurement) are only possible on certain memory locations. A common operation in this class might be encoding a physical qubit into an error-corrected logical one for longer-lifetime storage, after entangling with a quantum pulse in the channel. In theory, at least, a node should not have to inform any partners of such actions, which ideally would be purely internal matters. Classical Internet routers, after all, are free to store packets in any memory. However, in a quantum system, every action affects the fidelity of the Bell pair (or larger entangled state), and it is important for all partners to have an understanding of the actual state that is both as accurate as possible and closely coordinated with the other partners.

The obvious approach is for a node to inform its partner(s) of every action that it takes in the form of either operators to apply to the density matrix or a complete copy of the density matrix. However, this approach is exceedingly verbose, and more importantly fails to take into account the distributed nature of the system. Transmission of a d.m. or operator takes time, during which the local memory will degrade further, and the node at the far end may be applying operators of its own and even making independent decisions about the next higher-level action to be taken (which we will see in detail in sections 9.4 and 10.4).

At the link level, we can constrain this problem. A given node or NIC (network interface card) may have two types of memory, one for the physical transceiver, a second for longer-term memory. A link-level management protocol can monitor the link for low-frequency changes, as noted above, and the set of operations to be applied at each end can be intentionally circumscribed. Together, Alice and Bob can determine the d.m. that arises from an entangling operation and the link latency. Alice then can promise Bob that she will maintain her qubit in a memory with a certain dephasing operator that mimics the time-based degradation of memory and likewise Bob can promise Alice.

With the addition of high-precision, synchronized clocks, each end can calculate the state of the Bell pair with fair accuracy and be assured that the other end will have exactly the same idea of the state at the same point in time. Alice's only uncertainty or constraint is that she will not know for a full round-trip time after her initial transmission whether or not the actual entangling operation succeeded. Bob has only a limited understanding of when an entangling pulse will arrive, but he does know before Alice when an entangling operation has succeeded. Let us look at this process in a little more detail.

8.5.1. *Link-Level tracking of memory*

Consider a link in which Alice entangles a solid-state memory with a photon she fires down the channel toward Bob, an $M \rightarrow M$ channel. For this example, we will assume the photon states are the linear polarization bases $|H\rangle$ and $|V\rangle$, although the conditions outlined here apply to any type of link architecture or qubit basis.

Let us define the *zero-time pair* ρ_0 as the Bell pair that arises from the entanglement operations at both ends, where the storage time in memory is zero. Alice's memory qubit is first entangled with the photon. She immediately measures her memory, while the photon propagates toward Bob. When the photon arrives, Bob entangles it with his memory, then immediately measures the memory. The results of this sequence are *post-selected*, keeping only the successful ones. Repeating this operation (with minor variations) gives us enough statistical data to write down ρ_0 accurately. This complete process nicely subsumes errors in the photon state during propagation and those induced by the entanglement and measurement processes.

In normal operation, often Alice keeps her qubit without measuring it. Typically, she must hold the qubit for a round-trip time while waiting for confirmation of entanglement success. During this time, she may transfer the qubit from one memory to another. Bob must be informed of the memory decoherence operation and any voluntary changes such as memory transfer. Overall, to guarantee that Alice and Bob have the same idea about the Bell pair state at all important decision making points, we must ensure that the following criteria are met:

1) Alice and Bob eventually must agree on the original creation time of the Bell pair;

2) Alice and Bob must understand the memory dephasing operator or sequence of operators and any other actions that affect the fidelity at each end; and

3) the operators that Alice applies must commute with the operators that Bob applies.

The first criterion is met by using GPS clocks at each end and Alice including timestamps in the appropriate messages or Bob being able to determine directly when a Bell pair was created. The third criterion is met naturally by having Alice and Bob focus only on their own qubit, except when informed of actions by the partner. Alice and Bob can each apply operators only to the qubits they hold, and the mathematical operators therefore inherently commute. The second criterion requires that Alice and Bob must explicitly inform each other about actions that affect the fidelity and that Alice and Bob's classical messaging channel provides reliable, in-order message delivery, at least at an error level well below that of the quantum system itself. For the link layer, we will often demand bounded-delay arrival of messages as well.

A typical link-level sequence of events and evolution of the d.m. is shown in Table 8.1, where τ_l is the one-way latency from Alice to Bob and the reverse path is assumed to be the same. This corresponds to the successful entanglement case in Figure 8.7. Because the photon's decoherence during propagation is incorporated into ρ_0, Bob's T_1 and T_2 are calculated as infinity until the entanglement is transferred to Bob's qubit. This sequence ends with the Bell pair still held in memory. The next step will vary depending on higher layer protocols.

Time	Event	Operator/state
0	Alice entangles & transmits	(memory entangled with photon)
τ_l	Bob receives & entangles	$\rho_1 = \mathcal{E}_{\text{GBD}}(\rho_0, \tau_l, T_1^A, T_2^A, \infty, \infty)$
$2\tau_l$	Alice receives Bob's ACK	$\rho_2 = \mathcal{E}_{\text{GBD}}(\rho_1, \tau_l, T_1^A, T_2^A, T_1^B, T_2^B)$
$2\tau_l + \Delta$	Keep for time Δ	$\rho_3 = \mathcal{E}_{\text{GBD}}(\rho_2, \Delta, T_1^A, T_2^A, T_1^B, T_2^B)$

Table 8.1. *Events controlling the evolution of the density matrix of a link-level Bell pair*

8.5.2. *Synchronizing higher layers*

The natural layered software architecture will promote Bell pairs as quickly as possible from the physical layer to the next higher protocol layer, where they will be stored in a list for further operations. However, if implemented sloppily, this software boundary can result in mis-operation at the higher level. Therefore, it is important to define the semantics of this pair promotion. Consider the example illustrated in Figure 8.9. The sender and receiver already have one $F = 0.85$ shared Bell pair, and more $F = 0.75$ base-level Bell pairs are being made. Assuming reliable classical messaging, both ends see events in the same order: first the pre-existing pair, then one new pair and finally the third pair.

However, if the receiver and sender are free to execute the higher-level algorithm at a time of their choosing, the two ends may execute the algorithm on different lists of available Bell pairs. Here, the receiver runs the algorithm on the list $F = 0.85, F = 0.75, F = 0.75$, but the sender runs it on the list $F = 0.85$, $F = 0.75$.

A common next-layer protocol will be purification, which we will discuss in more detail in the next chapter. The basic form of purification uses two Bell pairs and operations at both ends. Alice and Bob *must* choose the same two Bell pairs for the operation to succeed. The purification protocol, to operate properly, must see the same list of Bell pairs at each end. The physical layer, of course, knows nothing of the details of this process, but appropriate choices here can ensure correct operation there.

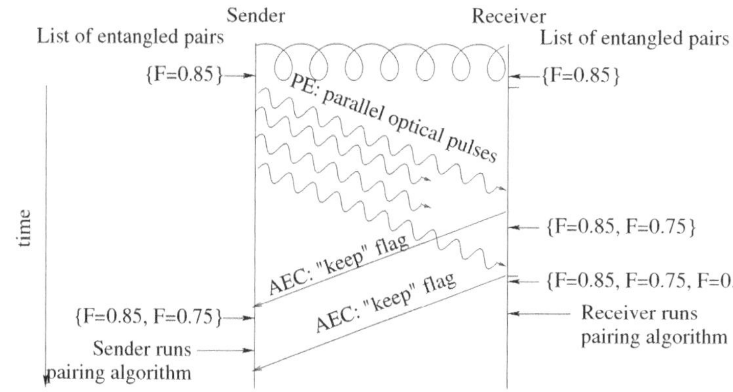

Figure 8.9. *The pairing mismatch problem. The higher-level protocol run at the sender will see a different list of Bell pairs than the receiver. Note that the actions of several protocol layers are represented here: Physical Entanglement (PE) and Acknowledged Entanglement Control (AEC) messages form our link, while the pairing action occurs at the higher software layer of Purification Control, discussed in more detail in the next chapter*

In order to guarantee that Bob's higher-layer protocols make decisions that will not surprise Alice and that Alice can replicate, we must fulfill four conditions: the three listed above and

4) Alice and Bob must run higher-level algorithms *at the same point in the sequence.*

Fulfilling these conditions will guarantee that Alice and Bob will construct not just the same d.m. for a given Bell pair at all meaningful points in time, but also the same *set* of Bell pairs at appropriate points in time.

The fourth criterion is the difficult one: arguably, it is solely the responsibility of that higher-layer protocol, and the physical layer can remain blissfully ignorant of such issues. However, a simple protocol change here may be useful: a flag that Bob can set in his Entanglement Control messages that indicates when a batch of Bell pairs has been delivered to a higher-level protocol. Bob can choose when to set the flag – on every Bell pair, after a fixed time interval or even ad hoc, delivering all ready-to-go Bell pairs when the software routine is executed. If Alice honors the same sequence, both ends will deliver the same groupings of Bell pairs to the higher-level protocol. We will see this technique in operation again in later chapters.

8.6. Examples

Our aim in this section is not to comprehensively cover quantum repeater-related experimental work; rather, it is to give the reader the ability to read the literature and place it in context by discussing some important historical signposts. Some simulated links and three valuable theoretical models are also included.

The ultimate goal at this level is, of course, light-mediated entanglement of two stationary memories separated by a distance, allowing us to perform purification, entanglement swapping and logic in order to support the applications discussed in Chapters 5 through 7. An important mid-term milestone is demonstration of the entanglement of the photonic pulse and the stationary memory. If entanglement is to be created using interferometry ($M \rightarrow I \leftarrow M$), interference effects may be demonstrated before enough statistical data is collected, or a high-enough fidelity is achieved, to demonstrate CHSH inequality violation.

Here, for the sake of completeness, we will also mention a few seminal experiments with direct entanglement of matter qubits. Many photon–photon entanglement experiments have been conducted, but we will not detail these here. Attempting to keep up with experimental results is difficult even via the Web, and impossible in a print book, but here we will try to highlight some of the historically important results as well as recent experiments, as of the writing of this book in Fall 2013.

In 2012, the Nobel Prize for Physics was awarded to Dave Wineland (NIST, United States) and Serge Haroche (Collège de France and École Normale Supérieure, Paris), both of whom have worked on the quantum interaction of light and matter and on experimental demonstration of entanglement. Besides the optical/atomic work, Wineland's group first demonstrated deterministic generation of entanglement, done between two ions in a single trap using shared vibrational modes of the system [TUR 98]. Haroche's group has worked with atoms held in resonating cavities, which strengthens their interaction with electromagnetic fields, and hence light, as discussed above [RAI 01].

Several groups working with a variety of materials have demonstrated entanglement of a photon with a stationary qubit. The group of Nicolas Gisin in Geneva has worked extensively with coupling atomic ensembles to photons, especially time bin photonic qubits [SAN 11]. Ion traps are excellent experimental systems for coaxing the emission of light, and Chris Monroe's group demonstrated entanglement of an ion with a single emitted photon [BLI 04].

Quantum dots are a very promising solid state technology, and quantum dots are expected to emit infrared photons of useful wavelengths. Experimental progress here has been rapid in recent years, with quantum dot to single photon entanglement demonstrated by the group of Yoshihisa (Yoshi) Yamamoto [DE 12], the Imamoglu group [GAO 12] and the Gammon and Steel group [SCH 13]. The Yamamoto group result includes conversion to telecom wavelengths. Interestingly, the Imamoglu group result uses two different wavelengths of photon as the $|0\rangle$ and $|1\rangle$ states; this makes state-specific filtering easy, but may make maintaining an interferometer difficult.

The Gammon/Steel group result generates photons entangled with the quantum dot with $F = 0.59 \pm 0.04$ at a rate of 3 kHz.

NV centers in diamond have been studied extensively by the Harvard group and others, and entanglement between the NV center and an optical photon was demonstrated in 2011 [TOG 11, SIP 12].

As of the end of 2013, only three technologies have convincingly demonstrated stationary-to-stationary qubit coupling via light: ion traps, neutral atoms in cavities and NV centers in diamond.

In 2007, Monroe again displayed the strengths of ion traps with entanglement of two ions in separate traps one meter apart, achieved by interfering indistinguishable photons from the two traps, in what we would call an $M \rightarrow I \leftarrow M$ link [MOE 07]. An approximation of the density matrix for this experiment written in the Bell basis is shown as the ion trap line in Table 8.2. (The full d.m. would include off-diagonal elements and is perhaps better written in the computational basis, but we will see in Chapter 9 that it is more useful for our purposes here to write it in the Bell basis and that off-diagonal elements have only a modest impact on the behavior of purification in that basis.) All of the entries in the table should be considered rough, as they depend on tuning parameters (listed in the table) for the experiment and are of course subject to improvement over time.

In 2012, the Rempe group showed entanglement of two atoms held in separate cavities, with fidelity to $|\Psi^-\rangle$ of $F = 0.85$ [RIT 12]. An approximation of the density matrix for this experiment written in the Bell basis is shown as the cavity QED line in Table 8.2. In this experiment, increased post-selection can improve fidelity.

The line in the table with NV center data is from a 2013 result by the Delft group [PFA 13, BER 13, NÖL 13]. The Bernien paper gives a good summary of the experimental difficulties with NV diamond. Only 3% of photons are emitted in the zero-phonon line (no effect of vibrations of the crystal), which is necessary for the photons to be indistinguishable when sent through the interferometer. Microfabricated solid immersion lenses are needed to collect a high percentage of the photons. Dynamic decoupling and reinitialization of the NV center are necessary to suppress various undesired behaviors. The experiment is repeated at a rate of 20 kHz, and the ultimate success rate of detecting entanglement is only 10^{-7}, giving only one Bell pair every ten minutes. The main rate-limiting factor is the initialization process, which is 9μs. Collecting the photons into a fiber is difficult, and the success probability will fall exponentially with the length of fiber. This experiment was conducted over three meters. The output fidelity of this experiment is $F = 0.58$ to $F = 0.73$, depending on which Bell state is prepared and some details of the post-processing process. The table entry is estimated based on graphs in the supplementary information for the Bernien paper.

Mechanism	Link org.	E/S/A	Example output d.m.	Tuning parameters
Ion trap (spin), single photon (polarization) [MOE 07]	$M \to I \leftarrow M$	E	$0.11 \lvert\Phi^+\rangle\langle\Phi^+\rvert + 0.15 \lvert\Psi^+\rangle\langle\Psi^+\rvert + 0.11 \lvert\Phi^-\rangle\langle\Phi^-\rvert + 0.63 \lvert\Psi^-\rangle\langle\Psi^-\rvert$	detector window width
Neutral atom in resonator (spin), single photon (polarization)	$M \to I \leftarrow M$	E	$0.055 \lvert\Phi^+\rangle\langle\Phi^+\rvert + 0.04 \lvert\Psi^+\rangle\langle\Psi^+\rvert + 0.055 \lvert\Phi^-\rangle\langle\Phi^-\rvert + 0.85 \lvert\Psi^-\rangle\langle\Psi^-\rvert$	
NV center in diamond, single photon [NÖL 13]	$M \to I \leftarrow M$	E	$0.08 \lvert\Phi^+\rangle\langle\Phi^+\rvert + 0.10 \lvert\Psi^+\rangle\langle\Psi^+\rvert + 0.08 \lvert\Phi^-\rangle\langle\Phi^-\rvert + 0.73 \lvert\Psi^-\rangle\langle\Psi^-\rvert$	detector window width
Weak nonlinearity [LAD 06]	$M \to M$	S	$0.633 \lvert\Phi^+\rangle\langle\Phi^+\rvert + 0.244 \lvert\Psi^+\rangle\langle\Psi^+\rvert + 0.061 \lvert\Phi^-\rangle\langle\Phi^-\rvert + 0.061 \lvert\Psi^-\rangle\langle\Psi^-\rvert$	laser strength, detector window width
Displacement (low photon number) [MUN 08]	$M \to M$	S	$(1-F) \lvert\Phi^+\rangle\langle\Phi^+\rvert + F \lvert\Phi^-\rangle\langle\Phi^-\rvert$	F tunable via average photon number, detector window width
White noise (Werner)		A	$F \lvert\Phi^+\rangle\langle\Phi^+\rvert + (1-F)\rho_{\text{white}}$	fidelity F
Phase flip noise		A	$F \lvert\Phi^+\rangle\langle\Phi^+\rvert + (1-F) \lvert\Phi^-\rangle\langle\Phi^-\rvert$	fidelity F
Bit flip noise		A	$F \lvert\Phi^+\rangle\langle\Phi^+\rvert + (1-F) \lvert\Psi^+\rangle\langle\Psi^+\rvert$	fidelity F

Table 8.2. *Examples of physical layers for generating entanglement. "E" indicates experimentally demonstrated, "S" is simulated and "A" is analytic model. The rows below the double line are abstract models of imperfect systems*

Table 8.2 includes two simplified analytic models that are commonly used in analysis of repeaters, the Werner (white noise) model and the phase noise model. The Werner model is

$$F\,|\Phi^+\rangle\langle\Phi^+| + (1-F)\rho_{\text{white}} \qquad [8.19]$$

where ρ_{white} is $\frac{1}{4}I$, an equal mixture of all four Bell states. Care must be taken in the algebra; the Werner model is sometimes written as

$$F\,|\Phi^+\rangle\langle\Phi^+| + \frac{1-F}{3}(|\Psi^+\rangle\langle\Psi^+| + |\Phi^-\rangle\langle\Phi^-| + |\Psi^-\rangle\langle\Psi^-|). \qquad [8.20]$$

We will generally use this latter definition.

This white noise model is the most difficult form of error for purification schemes to correct. Therefore, it can be used in a worst-case analysis. At the opposite end, some models assume a single type of well-characterized error. Note that the more realistic models in the table include a primary noise term and smaller amounts of the other terms.

8.7. Conclusion

Even as this book is being written, the experimental state of the art in generation of entanglement, as with teleportation, continues to advance [SLO 13]. New types of quantum memories, new representations of qubits in optical states, and new means of forging the entanglement continue to be proposed and tested. In this chapter, we have only been able to give a shallow overview spanning the range of work, but the examples should provide the reader with some ability to recognize the strengths and weaknesses of new technologies as they come online, as well as some vocabulary and concepts for conversations with experimentalists.

Rather than the physics itself, we have focused on the engineering of quantum links, including the classical communication protocols necessary to support them. The output of this layer is a series of Bell pairs entangled over a single hop, which we call *base pairs*. Each Bell pair is described using a density matrix and will be used by some higher-level software.

One critical factor in the success of quantum networks will be adaptability and continuous, real-time monitoring of the network will play a key role. The quality of the actual quantum state created will vary over time; hence, the density matrix reported out from the AEC software layer to the next layer will vary. The interface between AEC and the next highest layer therefore must support not just transfer of control of the Bell pair, but a full copy of the d.m. matched to this specific Bell pair. Typically, the next software layer will be purification, which we can see in the next chapter.

Chapter 9

Purification

The density matrix represents our *knowledge* about the state: is it truly in the state that we are trying to create? *Purification* is the process of improving our knowledge about the state, by testing propositions about it. This improvement is reflected as an increase in the fidelity in our density matrix.

A complete purification protocol can be described in terms of:

1) the number and type of *input states*;

2) the test procedure for certain *propositions*, consisting of

   a) the *quantum operations* performed for the basic operation,

   b) the tests for success and failure,

   c) the *actions* taken on success and failure, and

3) the *scheduling algorithm* used to select states for participation in purification.

Most commonly, the required input states are two imperfect Bell pairs, with the goal being to produce one output Bell pair of higher fidelity. Unless stated otherwise, it can be assumed that we are talking about such Bell state purification.

9.1. Measurement revisited

So far, we have discussed measurement primarily in qualitative terms, looking at the state vectors and reasoning from them. From this point forward, we are going to need a more mathematical treatment of measurement and will be focusing on the density matrix representation of the system state, rather than the state vector, because imperfections in the state are of paramount importance.

The simplest measurement we make in quantum computation is a *projective measurement* in the Z, or computational, basis. Each possible measurement outcome has a measurement *operator* associated with it. As the two possibilities for a single qubit are zero and one, our basic density matrix is the 2×2 matrix

$$\rho = \begin{bmatrix} a_{0,0} & a_{0,1} \\ a_{1,0} & a_{1,1} \end{bmatrix} \quad [9.1]$$

and the corresponding projection operators are

$$\mathcal{P}_0 = |0\rangle\langle 0| = \begin{bmatrix} 1 & 0 \\ 0 & 0 \end{bmatrix} \quad [9.2]$$

and

$$\mathcal{P}_1 = |1\rangle\langle 1| = \begin{bmatrix} 0 & 0 \\ 0 & 1 \end{bmatrix}. \quad [9.3]$$

When applying a measurement, we need to know two things: the probability of a given outcome and the resulting state. The probability of finding $|0\rangle$ (or more precisely, finding the eigenvalue corresponding to that state) is

$$\Pr(0) = \mathrm{Tr}(\mathcal{P}_0\rho), \quad [9.4]$$

and the resulting state is

$$\rho' = \frac{\mathcal{P}_0\rho\mathcal{P}_0}{\mathrm{Tr}(\mathcal{P}_0\rho)}, \quad [9.5]$$

that is, applying the corresponding operator $\mathcal{P}_0$ to the density matrix, then dividing by the probability in order to renormalize. The operator for $|1\rangle$ is similar. We noted in section 2.3.6 that the diagonal elements of the density matrix are the probabilities of the corresponding state, and this is easily seen in equations [9.1]–[9.5]. In fact, this calculation is so straightforward it can be done in your head, without working out the details. However, the formalism becomes more necessary when we are dealing with *imperfect* measurement operators.

If the measurement is perfect, then repeated measurements using the same operator will give the same result, $\rho'' = \rho'$. With an imperfect measurement operator, the reality is somewhat different and the results of consecutive measurements on the same state may differ. In practice, this is represented in simulation software by inserting a small amount of noise before and after the measurement operation.

9.2. Basic purification

In the most basic purification operation, illustrated in Figure 9.1, we begin with two Bell pairs, $|\Phi^+\rangle_1$ and $|\Phi^+\rangle_2$. Alice and Bob each hold one half of each pair. We will use pair 2 to test the following proposition: *Pair 1 is in the state* $|\Phi^+\rangle$.

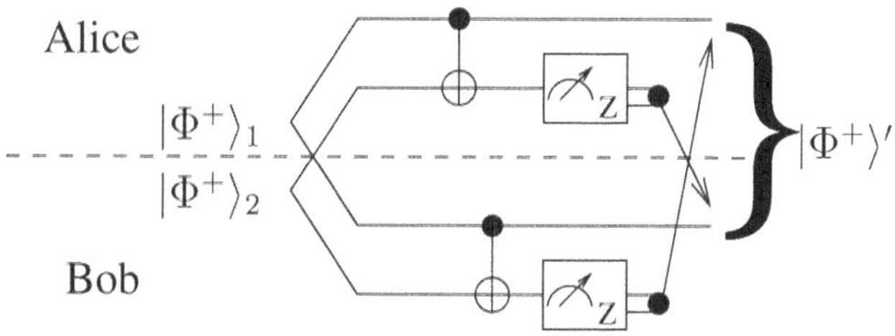

Figure 9.1. *Circuit for basic purification. The arrows indicate classical messages exchanged between Alice and Bob, with follow-up actions as described in Tables 9.1 and 9.2*

As shown in the figure, Alice and Bob each perform a CNOT gate using their half of pair 1 as the control and their half of pair 2 as the target. Recall that CNOT is equivalent to XOR, so that after executing the gate, the target qubit will hold the parity of the control and target. Alice has now calculated the parity of the two qubits she holds, and Bob calculates the parity of the two qubits he holds. The parity will be zero if the original parity was even, and the parity will be one if the original parity was odd.

Next, Alice and Bob each measure their member of Bell pair 2 and exchange the measurement results. If the measurement results agree, then the parity was the same at both Alice and Bob, as it should be, and the Bell pair is kept. If they disagree, pair 1 is discarded. Pair 2, having been measured and therefore no longer entangled, is always discarded. We call pair 2 the *sacrificial pair.*

In more detail, when the Bell pairs are both perfect ($F = 1.0$) and the gates and measurements are also perfect, the two CNOT gates cancel. Dispensing with normalization,

$$
\begin{aligned}
|\Phi^+\rangle_1 |\Phi^+\rangle_2 &= (|00\rangle + |11\rangle)(|00\rangle + |11\rangle) && [9.6]\\
&= |00\rangle|00\rangle + |00\rangle|11\rangle + |11\rangle|00\rangle + |11\rangle|11\rangle && [9.7]\\
&\xrightarrow{\text{CNOT}_A} |00\rangle|00\rangle + |00\rangle|11\rangle + |11\rangle|10\rangle + |11\rangle|01\rangle && [9.8]\\
&\xrightarrow{\text{CNOT}_B} |00\rangle|00\rangle + |00\rangle|11\rangle + |11\rangle|00\rangle + |11\rangle|11\rangle && [9.9]\\
&= |\Phi^+\rangle_1 |\Phi^+\rangle_2 && [9.10]
\end{aligned}
$$

and we still have two unentangled $|\Phi^+\rangle$ pairs. When the second pair is measured, Alice and Bob each have a 50% chance of finding 0 and 50% chance of finding 1, but when they exchange their measurement results they will always find the same value.

9.2.1. *Bit flip errors*

The more interesting case occurs when the Bell pairs are imperfect. Let us consider the case where a Bell pair may have suffered a bit flip error, $\rho = F\,|\Phi^+\rangle\langle\Phi^+| + (1 - F)\,|\Psi^+\rangle\langle\Psi^+|$. If we could test pair 1 directly, with probability F, we would find that is indeed a $|\Phi^+\rangle$ pair. With probability $1 - F$, we would find that it is a $|\Psi^+\rangle$ pair, which is still useful information because flipping one bit will take us back to our desired $|\Phi^+\rangle$ state.

Because we cannot test the first pair directly, we use the second pair as a test tool. If both pairs are in fact in $|\Phi^+\rangle$, the circuit operates as mentioned previously. This happens with probability F^2. Examining the circuit again, it is easy to see that if *either* of the Bell pairs has a bit flip error, Alice and Bob will find *different* values when they measure their qubits. Because we cannot tell if the error was in pair 1 or pair 2, we have no choice but to discard pair 1, even though it might be good. If *both* Bell pairs have an error, Alice and Bob will find the *same* value. With probability $(1 - F)^2$, the error in pair 1 goes undetected due to the error in pair 2.

This is the essence of purification. The operation "succeeds" (including the false positive case engendered by two errors) with probability $F^2 + (1 - F)^2$, failing with probability $2F(1 - F)$. When it succeeds, the resulting fidelity is

$$F' = \frac{F^2}{F^2 + (1 - F)^2} \qquad [9.11]$$

and our final state is

$$\rho' = F'\,|\Phi^+\rangle\langle\Phi^+| + (1 - F')\,|\Psi^+\rangle\langle\Psi^+|\,. \qquad [9.12]$$

A moment's thought will confirm that $F' > F$ when $F > 0.5$. This behavior is summarized in Table 9.1. The input and output fidelity are plotted in Figure 9.2. For input fidelity greater than approximately 0.8, the improvement in fidelity is dramatic.

Pair 1	Pair 2	Prob.	Agree?	Action	Result	Comment
$\|\Phi^+\rangle$	$\|\Phi^+\rangle$	F^2	Y	Keep	$\|\Phi^+\rangle$	True positive
$\|\Phi^+\rangle$	$\|\Psi^+\rangle$	$F(1-F)$	N	Discard	—	False negative
$\|\Psi^+\rangle$	$\|\Phi^+\rangle$	$F(1-F)$	N	Discard	—	True negative
$\|\Psi^+\rangle$	$\|\Psi^+\rangle$	$(1-F)^2$	Y	Keep	$\|\Psi^+\rangle$	False positive

Table 9.1. *Possible combinations of our two Bell pairs in purification, when the only errors are bit flip errors. Pair 1 is control and Pair 2 is the target of the* CNOT *at each end. Pair 2 is measured, and Pair 1 is kept*

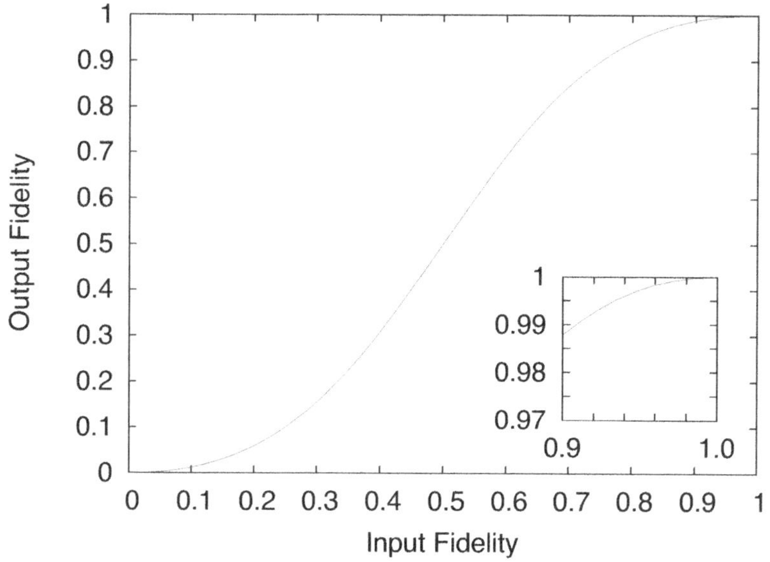

Figure 9.2. *Output fidelity as a function of input fidelity for basic purification of two identical Bell pairs with bit flip errors only, and perfect purification operations*

9.2.2. *Generalizing: incorporating phase flip errors and different Bell pairs*

The protocol as presented is not designed to detect when either pair has a phase error. Worse, in fact, a phase error on pair 2 will propagate onto pair 1,

$$\left|\Phi^+\right\rangle_1 \left|\Phi^-\right\rangle_2 = |00\rangle|00\rangle - |00\rangle|11\rangle + |11\rangle|00\rangle - |11\rangle|11\rangle \quad [9.13]$$

$$\xrightarrow{\text{CNOT}_A} |00\rangle|00\rangle - |00\rangle|11\rangle + |11\rangle|10\rangle - |11\rangle|01\rangle \quad [9.14]$$

$$\xrightarrow{\text{CNOT}_B} |00\rangle|00\rangle - |00\rangle|11\rangle - |11\rangle|00\rangle + |11\rangle|11\rangle \quad [9.15]$$

$$= \left|\Phi^-\right\rangle_1 \left|\Phi^-\right\rangle_2, \quad [9.16]$$

just as a phase error on a single qubit propagates from the target to the control of a single CNOT gate.

As discussed in section 8.2.3, we can write our quantum states using the Bell pairs as a basis set, rather than the computational basis. Rather than the simpler analysis above using two identical Bell pairs, let us now consider two different Bell pairs,

$$\rho_i = A_i \left|\Phi^+\right\rangle\left\langle\Phi^+\right| + B_i \left|\Psi^+\right\rangle\left\langle\Psi^+\right| + C_i \left|\Psi^-\right\rangle\left\langle\Psi^-\right| + D_i \left|\Phi^-\right\rangle\left\langle\Phi^-\right| \quad [9.17]$$

for $i \in \{1, 2\}$. If only one of the error terms B, C or D is non-zero, we will call the state a *binary state*; the 'bit flip only' case just discussed is one such binary state. If all three B, C and D are equal, it is a Werner state, as in equation [8.20].

With four non-zero terms in this equation, there are 16 possible combinations for ρ_1 and ρ_2, enumerated in Table 9.2. The probability of successful purification, including the false positives, is

$$\text{P(success)} = (A_1 + D_1)(A_2 + D_2) + (B_1 + C_1)(B_2 + C_2). \quad [9.18]$$

The new values for our post-purification d.m. ρ' are

$$A' = \frac{A_1A_2 + D_1D_2}{\text{P(success)}} \quad [9.19]$$

$$B' = \frac{B_1B_2 + C_1C_2}{\text{P(success)}} \quad [9.20]$$

$$C' = \frac{B_1C_2 + C_1B_2}{\text{P(success)}} \quad [9.21]$$

$$D' = \frac{A_1D_2 + D_1A_2}{\text{P(success)}}. \quad [9.22]$$

It is easy to see that bit flips are strongly suppressed. If the two pairs are similar, our fidelity A_1 is high, and bit flips are more common than phase flips, then $B' \approx B_1^2$. The probability of a bit flip remaining undetected in the output state declines with the square of the initial bit flip probability.

However, the probability of undetected phase flips in the output state can grow, $D' \approx 2D_1$. If the input Bell pairs are Werner states as in equation [8.20],

$$\rho = F\left|\Phi^+\right\rangle\left\langle\Phi^+\right| + \frac{1-F}{3}\left(\left|\Psi^+\right\rangle\left\langle\Psi^+\right| + \left|\Phi^-\right\rangle\left\langle\Phi^-\right| + \left|\Psi^-\right\rangle\left\langle\Psi^-\right|\right), \quad [9.23]$$

the success probability is

$$\text{P(success)} = \left(F + \frac{1-F}{3}\right)^2 + \left(\frac{2(1-F)}{3}\right)^2 \quad [9.24]$$

and the resulting d.m. is

$$A' = F' = \left(F^2 + \left(\frac{1-F}{3}\right)^2\right) \times \frac{1}{\text{P(success)}} \quad [9.25]$$

$$B' = C' = 2 \times \left(\frac{1-F}{3}\right)^2 \times \frac{1}{\text{P(success)}} \quad [9.26]$$

$$D' = 2F \times \left(\frac{1-F}{3}\right) \times \frac{1}{\text{P(success)}}. \quad [9.27]$$

Pair 1	Pair 2	Prob.	Agree?	Action	Result	Comment
$\|\Phi^+\rangle$	$\|\Phi^+\rangle$	A_1A_2	Y	Keep	$\|\Phi^+\rangle$	True positive
$\|\Phi^+\rangle$	$\|\Psi^+\rangle$	A_1B_2	N	Discard	—	False negative
$\|\Psi^+\rangle$	$\|\Phi^+\rangle$	B_1A_2	N	Discard	—	True negative
$\|\Psi^+\rangle$	$\|\Psi^+\rangle$	B_1B_2	Y	Keep	$\|\Psi^+\rangle$	False positive
$\|\Phi^+\rangle$	$\|\Phi^-\rangle$	A_1D_2	Y	Keep	$\|\Phi^-\rangle$	True positive, phase error propagates
$\|\Phi^+\rangle$	$\|\Psi^-\rangle$	A_1C_2	N	Discard	—	False negative
$\|\Psi^+\rangle$	$\|\Phi^-\rangle$	B_1D_2	N	Discard	—	True negative
$\|\Psi^+\rangle$	$\|\Psi^-\rangle$	B_1C_2	Y	Keep	$\|\Psi^-\rangle$	False positive, phase error propagates
$\|\Phi^-\rangle$	$\|\Phi^+\rangle$	D_1A_2	Y	Keep	$\|\Phi^-\rangle$	True positive, phase error remains
$\|\Phi^-\rangle$	$\|\Psi^+\rangle$	D_1B_2	N	Discard	—	False negative
$\|\Psi^-\rangle$	$\|\Phi^+\rangle$	C_1A_2	N	Discard	—	True negative
$\|\Psi^-\rangle$	$\|\Psi^+\rangle$	C_1B_2	Y	Keep	$\|\Psi^-\rangle$	False positive, phase error remains
$\|\Phi^-\rangle$	$\|\Phi^-\rangle$	D_1D_2	Y	Keep	$\|\Phi^+\rangle$	True positive, phase errors cancel
$\|\Phi^-\rangle$	$\|\Psi^-\rangle$	D_1C_2	N	Discard	—	False negative
$\|\Psi^-\rangle$	$\|\Phi^-\rangle$	C_1D_2	N	Discard	—	True negative
$\|\Psi^-\rangle$	$\|\Psi^-\rangle$	C_1C_2	Y	Keep	$\|\Psi^+\rangle$	False positive, phase errors cancel

Table 9.2. *Possible combinations of our two Bell pairs in purification, taking into consideration both bit flip and phase flip errors. Pair 1 is the control and Pair 2 is the target of the* CNOT *at each end. Pair 2 is sacrificed (measured), and Pair 1 is kept*

The output of a single round of purification of a Werner state is shown in Figures 9.3(a) and 9.3(b). The very minimal improvement in fidelity in this case compared with the large quadratic improvement with bit flip noise only demonstrates the difficulty of dealing with white noise. However, the output d.m. concentrates the noise almost entirely into one error state (note the different vertical scale for the B' and C' graph vs. the D' graph). If that error state is the bit flip error state $|\Psi^+\rangle$, a second round of purification will behave more like Figure 9.2. We turn to multi-round purification next.

9.2.3. *Multiple rounds and error redistribution*

Purification's ability to drive the fidelity of a Bell pair toward the ideal $F = 1.0$ is founded on iteration. With a Werner state as input, the fidelity of the output state has improved by only a few percent. To continue improving the state, we can take two purified Bell pairs, and run the same operation on them. However, we quickly realize that there is a problem: the remaining error has been concentrated in $|\Phi^-\rangle$, and a second round of purification exacerbates this problem, making the fidelity *worse* instead of better. Bennett *et al.* solved this problem by introducing *twirling*, a non-unitary depolarization operation in which the three error terms are rebalanced, allowing a second round of purification to make forward progress [BEN 96a, BEN 96b, BEN 96c]. We will not discuss this protocol in detail, except to note that it requires the main component to be in $|\Psi^-\rangle$ rather than the $|\Phi^+\rangle$ on which we have focused, but this is easily achieved through local operations.

More importantly, this protocol (sometimes called the IBM protocol; we can take the name BBPSSW from the authors' initials) fails to take advantage of the fact that one round of purification may not have improved the fidelity by much, but has pushed the state toward a binary state, which we have already seen is easier to purify. Deutsch *et al.* introduced a more general variant (sometimes called the Oxford protocol; we use the name DEJMPS), from which our previously given description is derived [DEU 96].

The key to efficient iterative purification is recognition that the single-qubit operations performed at each end can reorder the diagonal terms in our d.m. Performing a Hadamard gate on each qubit, $H_{\text{Alice}}H_{\text{Bob}}$, will swap $|\Psi^+\rangle$ and $|\Phi^-\rangle$ *only*,

$$\{A, B, C, D\} \rightarrow \{A', B', C', D'\} = \{A, D, C, B\}. \quad [9.28]$$

Likewise, a $\pi/2$ rotation about the Z axis on each qubit, $S_{\text{Alice}}S_{\text{Bob}}$, will swap $|\Phi^+\rangle$ and $|\Phi^-\rangle$ *only*. We can use these two operations to reorder the elements of the resulting d.m., so that the principal component remains $|\Phi^+\rangle$ and the largest error component is once again $|\Psi^+\rangle$, and we are ready to execute a second round of purification.

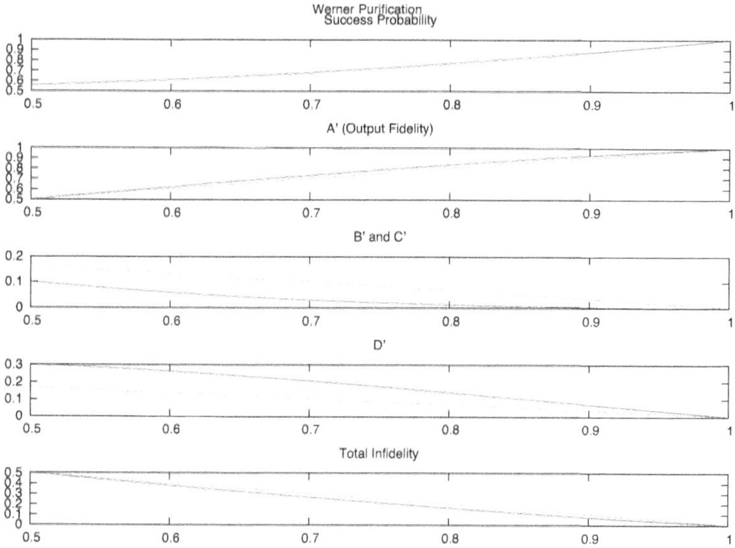

a) Purification of a Werner state.

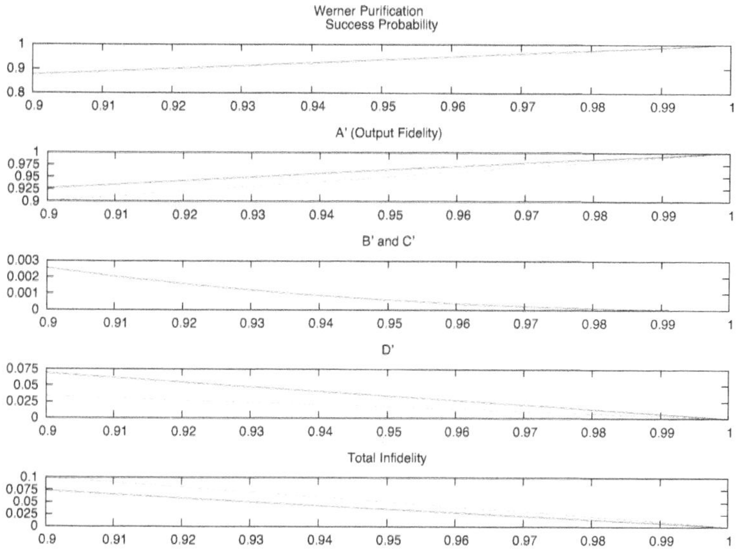

b) Detail for $F > 0.90$.

Figure 9.3. *Output as a function of input fidelity for basic purification of two identical Bell pairs with white noise, known as a Werner state. Purification operations are assumed perfect. The dashed line is the pre-purification value*

Details of this procedure were worked out by Dehaene *et al.* [DEH 03]. The effective use of this approach is made possible by careful characterization of the physical channel, memory and purification processes. The set of operations is fixed for a particular setup, rather than varying on a Bell-pair-by-Bell-pair basis.

With Werner states as inputs, in practice, the first round of purification gains little in fidelity (Figure 9.3(a)) but leaves us well positioned to do another round. That second round then gives us a great boost in fidelity (Figure 9.2).

9.2.4. *Resource consumption in multiple rounds*

Perfect purification is not possible; even under ideal conditions, we can only asymptotically approach $F = 1.0$. In practice, applications will demand Bell pairs of some particular fidelity suitable for their requirements. We will run multiple rounds of purification to reach this fidelity, which we will call F_{final}. (We will discuss multiple rounds more completely in the next section.) If M is the number of base Bell pairs of fidelity F_b, we need to reach F_{final}, M will be exponential in the number of rounds of purification, but the number of rounds is small and logarithmic in the infidelity.

Because Werner noise is quickly converted to binary noise, we only need to consider binary states. Beginning with states of fidelity F_b,

$$F' = \frac{F_b^2}{F_b^2 + (1 - F_b)^2} \qquad [9.29]$$

after one round. If we then purify again, using two pairs of fidelity F', naturally we obtain

$$F'' = \frac{(F')^2}{(F')^2 + (1 - F')^2}. \qquad [9.30]$$

This second round of purification has now required the creation of four base Bell pairs. For r rounds of purification, obviously $M = 2^r$, raising the concern that resource growth will be exponential and hence unsustainable. In practice, only a few rounds of purification will be required, so that the actual number of Bell pairs for a given case is more important than the asymptotic growth for an arbitrary number of rounds. Nevertheless, let us confirm that exponential resource consumption will not occur, by looking at the infidelity. For one round,

$$1 - F' = \frac{(1 - F)^2}{F^2 + (1 - F)^2}. \qquad [9.31]$$

As $F \to 1$, over multiple rounds $1 - F' \to (1 - F_b)^{2r}$. The infidelity is suppressed more quickly than the resource consumption grows.

In section 10.3.1, we will see a similar argument for performing purification over multiple hops.

9.3. Scheduling purification

Purification is not itself our end goal; the point of building a network is to deliver entangled pairs to an application that can use them. We must choose a *delivery threshold* above which purification is terminated, and the Bell pair handed off to some other subsystem for further processing or use. That threshold will depend on a number of operational issues, including the application needs, as discussed in Chapters 5–7, as well as long-distance repeater operation, which is discussed in the remainder of this book. Our goal, then, is to make the best possible use of the available resources to deliver Bell pairs of an externally defined fidelity as quickly as possible.

Both the BBPSSW and DEJMPS protocols assume that purification is performed using two identical Bell pairs. Such protocols are often called *recurrence* protocols; we also refer to them as *symmetric*. The right-hand element of Figure 9.4(a) shows such a process, drawn as the history tree of a final pair.

Recurrence protocols require a number of base Bell pairs (the pairs generated by the physical entanglement mechanism) that grows exponentially in the number of purification rounds. Efficient implementation requires large amounts of memory, so that all of the purifications in the first round (with the largest number of Bell pairs) can be performed simultaneously. However, as can be seen in the timeline drawn in the figure, it is not always the case that enough base Bell pairs are available at the right time, and the purification process can be forced to wait. This is especially true when the probability of successful purification is low.

An alternative procedure was introduced by Wolfgang Dür, Hans Briegel, Ignacio Cirac and Peter Zoller in the same papers in which they introduced the core concepts of repeaters [DÜR 99, BRI 98]. Using the minimum two-qubit memories at each repeater node, we can perform *entanglement pumping*, in which one pair is kept over several rounds and repeatedly purified using a base pair, as shown in Figure 9.4(b). This procedure's simplicity and minimal resource requirements have led to some groups adopting it as their preferred method, notably Mikhail (Misha) Lukin's group at Harvard [CHI 05, CHI 06].

Entanglement pumping has a serious drawback: when the difference in fidelity between the two pairs is large, the probability of success is low and the increase in fidelity on success is small. Recurrence protocols asymptotically approach $F = 1.0$, but pumping may have a "fixed point" of well less than 1.0. This can be solved using *nested pumping*, in which a Bell pair is pumped until its fidelity saturates. Then a second stage of pumping can be performed, replacing the base Bell pairs with these first-stage-saturated pairs. If necessary, a third stage can be performed using second-stage-saturated pairs, and so on.

The generalization of this issue can be called the *purification scheduling problem*: given a limited set of physical memories, and the stochastic arrival of base Bell pairs on some or all of those memories, which purification operations will result in producing the largest number of high-fidelity Bell pairs in a fixed period of time?

Recurrence and pumping can be viewed as two extremes in a spectrum of scheduling policies. We can also consider using Bell pairs as rapidly as they are created, in a *greedy* policy, as shown in Figure 9.4(c). This policy was used implicitly in simulations by Ladd, van Loock, Nemoto, Munro and Yamamoto [LAD 06], and later identified and named by Van Meter, Ladd, Munro and Nemoto [VAN 09]. Each time the purification algorithm is run, it matches pairs of Bell pairs and assigns them for purification. If the list of currently available Bell pairs is maintained in order of fidelity, we can take the first two pairs from the head of the list, then the second two pairs, etc.

The processing of this ordered list of pairs turns out to have a large impact on the overall system performance. Each additional round of purification represents a substantial investment in resources, both in terms of the number of base Bell pairs consumed and the time a pair has been kept in memory, which itself is a precious resource. Thus, it pays to be conservative about committing to the use of the highest-fidelity Bell pairs. We discovered that processing the list bottom-up, from the lowest-fidelity Bell pairs, gives a total throughput of 2–4 times that of processing the list top-down. This arises largely from the stochastic nature of many of these processes, giving us at random either an even or odd number of Bell pairs available each time the pairing algorithm is run. When the number of pairs is odd, in bottom-up the highest-fidelity pair is left to await a more deserving partner, whereas in the top-down case a low-fidelity pair is left idle.

The greedy algorithm is almost completely unstructured, in contrast to the highly regulated recurrence and pumping algorithms. At an intermediate level of regulation, Van Meter, Ladd, Munro and Nemoto introduced *banded purification*, illustrated in Figure 9.4(d). In banded purification, the fidelity space is divided into a series of bands, and Bell pairs within the same band are matched for purification.

The optimal number of bands and the placement of their boundaries is highly dependent on the base Bell pair fidelity and target fidelity. Fortunately, our experience suggests that best choice of bands is independent of the distance or number of hops, leaving us with a large but manageable combinatoric space to search for optimal settings. Careful setting can improve throughput as much as a factor of 100. An example of banding behavior is explored in more detail in Chapter 10.

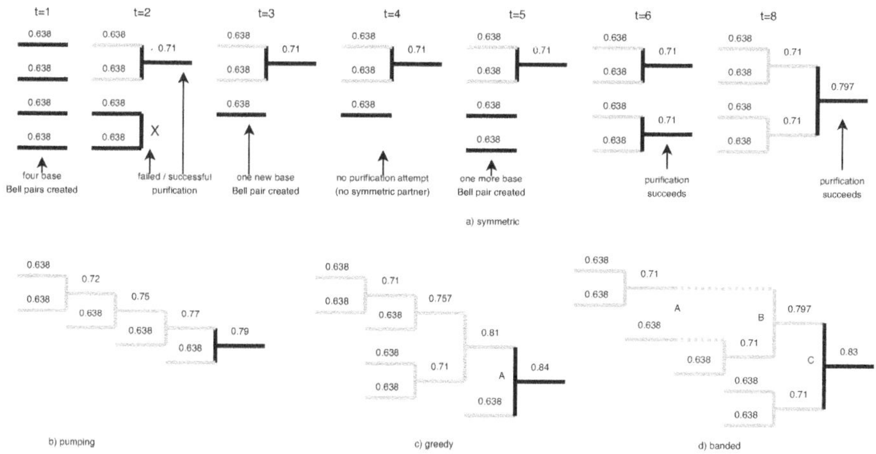

Figure 9.4. *Different purification scheduling algorithms. Gray bars represent the history of the pair. Black horizontal bars represent currently entangled Bell pairs. Numbers show the fidelity of the Bell pair. a) Logical evolution of a Bell pair grown using a recurrence protocol. b) History tree of a Bell pair grown using the entanglement pumping algorithm. c) History tree of a Bell pair grown using the greedy algorithm. d) An example history of the evolution of a Bell pair using the banded purification algorithm. If the boundary between two bands is placed at e.g. 0.66, at point A, the pairs 0.71 and 0.638 will not be allowed to purify. Dashed lines represent time that Bell pairs are forced to wait for a suitable partner to be created*

9.4. State machines and protocol interactions

In this section, we describe a general framework for the classical network protocols required to control the operation of purification in a distributed fashion. The state of each qubit within a repeater can be tracked using a finite state machine (FSM). The FSM for the basic Purification Control protocol is illustrated in Figure 9.6. The states are:

– *Entangled*: when control of a qubit is first handed to the purification control layer, it is placed in the entangled state. This state indicates that the qubit is known to be entangled to another qubit in a distant station. It can be entered from a lower layer such as AEC (just after entanglement is produced, see Figure 8.8), or as a result of entanglement swapping (discussed in Chapter 10). As purification is often iterated, it may also return to this state after completion of a prior purification operation. It may also arrive in the PC layer as a result of the expiration of a timer while waiting in some higher-layer protocol, when the fidelity of the state is understood to have fallen below a pre-established threshold.

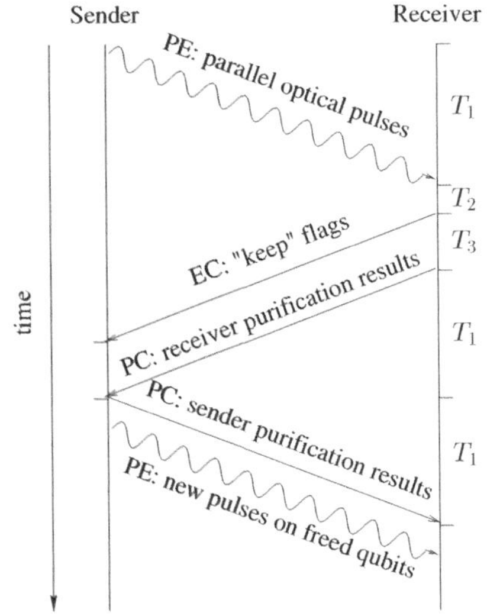

Figure 9.5. *Messaging sequence for the lowest level of Bell pair creation coupled with purification. As in Figure 8.7, physical pulses are sent in a heavily pipelined fashion, but are shown with only one line here for clarity. Purification operations at each end are asynchronous and depend on the node's ability to accurately discern the next valid operation. Composing this asynchrony across multiple hops of varying latencies is an important engineering optimization problem*

Ideally, a qubit's residence in this state will be brief. The preferred exit from this state is for control to be transferred to a higher protocol layer right away, if the fidelity is high enough, or to be assigned for purification as soon as the purification scheduling algorithm allows.

If the Bell pair generation rate is low, the residence time may be long. If the qubit remains in this state for a long time, the fidelity will drop due to decoherence. If a timer expires indicating that the fidelity has fallen below the minimum desired threshold, the entangled state will be discarded and the qubit buffer memory returned to the pool of uninitialized qubits for reuse.

– *PurifyAttempt*: the purification scheduling algorithm selects one or more entangled states to purify, and one or more entangled states to sacrifice. Those that are to be purified move into this state. Residence in this state should be only the amount of time it takes to execute the physical operations for purification. Once a message with the parity has been sent, the state is immediately moved to *MyHalfPurified*. Because both ends of the state may be starting purification asynchronously, a message from the far end may arrive while we are in the process of performing purification. In this case, we transition to the temporary state *HerHalfPurified* until we complete the purification operations.

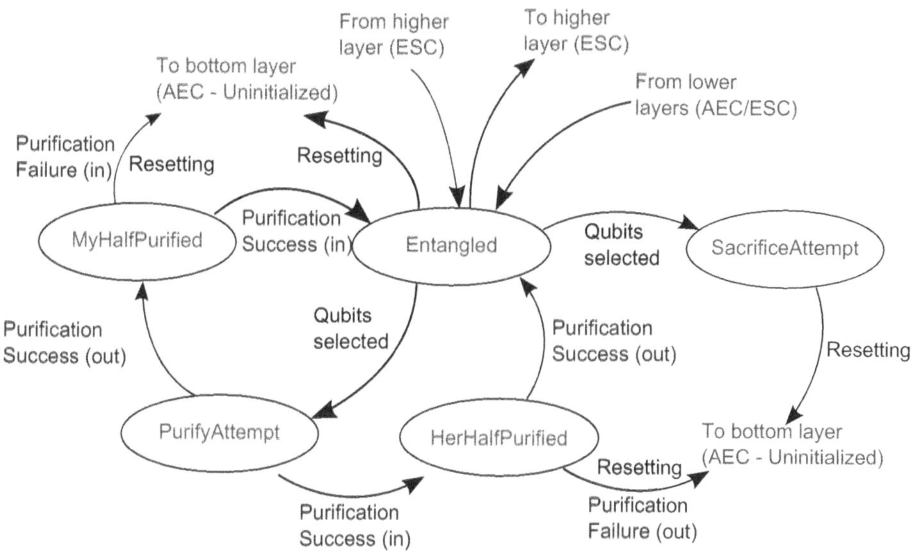

Figure 9.6. *Finite state machine for purification control, with transitions to typical higher- and lower-level protocols marked. Messages sent and received are indicated with (out) and (in). Other transitions are triggered by local events, such as the expiration of timers*

– *SacrificeAttempt*: the second entangled state chosen for purification is assigned to be sacrificed to improve the fidelity of the first. As with *PurifyAttempt*, this state is occupied only while the purification operations are being conducted. Regardless of the result of the purification, this qubit will always be returned to the control of the lowest protocol layer, generally placed in the *Uninitialized* state. No message to the partner node is required.

– *MyHalfPurified*: upon completion of the purification operations, the qubit is moved into this state to await the arrival of a message from the partner. When the message is received, the parities are compared. On success, the state will be moved to *Entangled*, or to the lowest layer on failure. Residence in this state, in theory, could be as long as one classical round trip time, plus the purification operation time at the far node, if the partner is unable to independently determine which states to purify.

– *HerHalfPurify*: this state indicates that the station received a message notifying it that the remote station has started the purification process. If the operation on the local station succeeds and the local parity matches that received from the partner, the qubit will be moved to *Entangled*, after sending a message to the remote station. If the comparison fails, the state will be sent to the lowest layer, after sending a PurificationFailure message to the remote station. Note that the content of the PurificationFailure and PurificationSuccess messages are actually the same, containing only the locally measured parity, except in the case where the local purification operation itself fails.

A major problem for the design of our control protocols is "managed obsolescence" of the Bell pairs. Each Bell pair is held jointly by two parties, or by one party with the other half of the Bell pair in transit as a photon, as discussed in section 8.4. The discussion above repeatedly mentioned fidelity-related timers. These timers must be externally set, and they may vary depending on the intended consumer of the Bell pair or other entangled state. The value chosen may vary over time, as network conditions change, as discussed in Chapter 13. Worst of all, it may change for a given Bell pair at either node, independently, as each node retains the right to autonomously migrate its member of the Bell pair from memory location to memory location.

As a second example, consider a basic example of how we can go wrong in something as simple as bottom-up, unbanded scheduling: variations in timers and workloads at different nodes can mean that the pairing algorithm can run at different points in the sequence, as shown in Figure 8.9. Assume pair #1 is $F = 0.85$, and has been waiting for a while, then pair #2 and pair #3 arrive in short order, both base-level pairs of $F = 0.75$. Both Alice and Bob agree on the order of arrival, so we have not violated the second condition in section 8.5. However, it is possible that Alice runs the pairing algorithm *between* the arrivals of pairs #2 and #3, but Bob runs the algorithm *after* the arrival of pair #3. This will result in Alice purifying using pairs #1 and #2, and Bob purifying using pairs #2 and #3: a disaster.

This situation is avoided by requiring the two nodes to execute the purification pairing algorithm on the same lists of Bell pairs. Entanglement pumping naturally does this, but other scheduling protocols will require more careful coordination. It is worth noting that experimental entanglement generation rates remain low enough that this conflict is unlikely to occur in practice until rates have improved by several orders of magnitude.

9.5. More complex purification protocols

So far, we have only discussed purification of two Bell pairs, in which bidirectional communication is used. However, very early in the history of quantum error correction and purification, it was recognized that the protocols could be generalized to $N \rightarrow M$ operations for $M < N$, starting with N multi-party states and using $n = N - M$ of them as sacrificial test states, ending with M states of higher fidelity [BEN 96c, DÜR 07]. Moreover, managed properly, purification can be done using one-way communication, albeit generally with a penalty in the resulting fidelity.

Intuitively, one-way entanglement purification protocols (called 1-EPP) can be understood as error correcting codes applied unconditionally to the state, rather than error detection codes in which the remaining state is discarded upon failure, which forces the two-way communication.

As a simple example, 1-EPP is even possible with the protocol of Figure 9.1. Bob does not send Alice a message, so the information flow is one way, reducing operational latency in some cases. In the results in Tables 9.1 and 9.2, rather than discarding on disagreement, we always keep the remaining Bell pair. When the results disagree, we know that one Bell pair had a bit flip. Provided the fidelity of pair 2 was higher than that of pair 1, we can assume that the error was in pair 1, and Bob can correct the error by performing a bit flip on his qubit from pair 1, without bothering to inform Alice. However, in general, the new fidelity of pair 1 will not exceed that of the original pair 2.

Instead of using just two Bell pairs, if we use a larger number, we can perform parity checks on large groups of Bell pairs, in exactly the same way as a full classical error correction scheme checks the parity of groups of bits and uses the syndrome information to correct individual errors. If we set aside n Bell pairs for syndromes and establish n groups of Bell pairs among the remaining $N - n$ Bell pairs, we can detect and correct errors, giving us an $N \to M$ protocol that typically does improve the fidelity. This approach, known as *hashing*, has an asymptotic yield of 1. As the fidelity $F \to 1$, we need fewer check bits, our block size N can grow and $M/N \to 1$.

A particularly attractive multi-state purification procedure is *double selection* purification [FUJ 09], illustrated in Figure 9.7. In double selection, two Bell pairs are sacrificed to purify one Bell pair. The normal CNOT is performed to check the parity of the two Bell pairs, then the second CNOT checks for phase errors between the second and third Bell pairs. This approach has a lower success probability than single selection under most circumstances, and converges more slowly than single purification, but is more robust against local gate errors, works with a lower initial fidelity and saturates at a higher final fidelity. Testing for two types of errors in a single round intuitively would seem to reduce the number of round-trip latency waits, but the lower success probability outweighs this benefit, so throughput is actually quite a bit lower under most circumstances. However, the strengths of double selection mean that it is likely to have a place in networks.

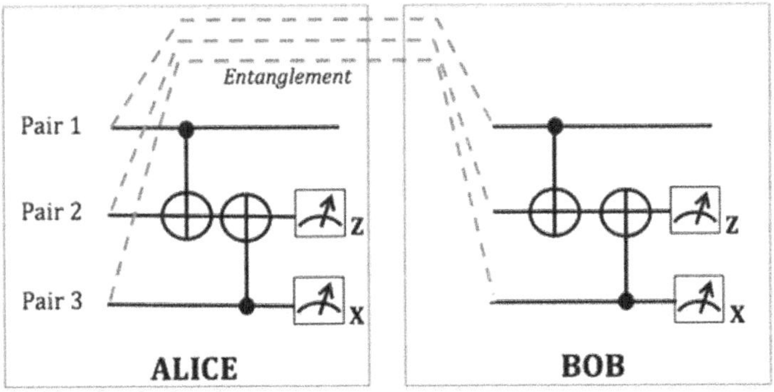

Figure 9.7. *The double selection circuit*

9.6. Experimental demonstrations

Purification has been experimentally demonstrated in various forms. Here, we will discuss three experiments, using photons [PAN 03] and atoms in an ion trap [REI 06].

Above, we described purification in terms of computational gates. Mapping these procedures to physical implementations sometimes involves more than simply assigning meaning to states. Jian-Wei Pan and others in Anton Zeilinger's group at the University of Vienna developed a variant of the CNOT-based circuit in Figure 9.1, more appropriate for a purely optical implementation using polarization qubits, which they demonstrated in 2003 [PAN 03]. Rather than using a CNOT, which is difficult to implement directly in linear optics, an optical setup at Alice sends both of her photons through an appropriately configured beamsplitter, and Bob does the same, giving a total apparatus with four input ports and four output ports.

They considered Bell pairs in binary mixed states,

$$\rho = F\left|\Phi^+\right\rangle\left\langle\Phi^+\right| + (1-F)\left|\Psi^-\right\rangle\left\langle\Psi^-\right|. \qquad [9.32]$$

When both Bell pairs are in $|\Phi^+\rangle$ or both pairs are in $|\Psi^-\rangle$, 50% of the time one photon will come out of each of the four ports. When one pair is in $|\Phi^+\rangle$ and the other is in $|\Psi^-\rangle$, the four-way coincidence never occurs; those errors are filtered out and discarded. To complete the process, one of the pairs is measured and the results compared in order to determine if a phase flip operation is needed, leaving us with one Bell pair in the state

$$\rho' = F'\left|\Phi^+\right\rangle\left\langle\Phi^+\right| + (1-F')\left|\Psi^+\right\rangle\left\langle\Psi^+\right| \qquad [9.33]$$

where $F' = F^2/(F^2 + (1-F)^2)$.

In two experiments, they succeeded in raising the fidelity, $F = 0.75 \rightarrow F' = 0.92$ and $F = 0.80 \rightarrow F' = 0.94$. However, the overall success rate was low; their SPDC system generated Bell pairs at a rate of 17,000 per second, but after all considerations (loss and inefficient detectors, as well as the limitations of their procedure) one purification succeeded every 2–3 s. Because the confirmation of detection of exactly one photon is difficult, their experimental setup also required destructively measuring all four photons in the end, rather than retaining the second pair for further operations.

These experiments used the polarization states of two photons as the Bell pair, $(|H\rangle |H\rangle + |V\rangle |V\rangle)/\sqrt{2}$. The Gisin group has proposed a protocol where a Bell pair consists of only a *single* photon, where the basis states are the presence of the photon at Alice or at Bob [SAN 08]. We can write the basis states as $|1\rangle_A$ and $|1\rangle_B$, resulting in a Bell pair of $(|1\rangle_A |0\rangle_B + |0\rangle_A |1\rangle_B)/\sqrt{2}$. In this notation, $|0\rangle_A |0\rangle_B$ would indicate the loss of the photon, and hence of our qubit. These two choices of state will result in somewhat different physical setups for experiments, which allows us to make engineering tradeoffs, but both purify one optical Bell pair using another optical Bell pair, without storing the state into some sort of stationary memory.

In 2006, Dave Wineland's group at NIST demonstrated purification of two Bell pairs, where each Bell pair consists of two ions [REI 06]. The Bell pairs are not created using photons; rather, the ions are physically placed together, entangled, then separated by a small distance. The entire experiment takes place within one apparatus, rather than over a distance. In contrast to optical experiments, their experiment succeeded more than 35% of the time, but the initial fidelity was lower and the gain in fidelity as a result of purification was small. As one of the first experiments to purify matter qubits, rather than photonic, it helped to pave the way for future solid-state implementations.

9.7. Conclusion

Purification gives us an interim tool for improving the fidelity of quantum states, before we can effectively execute true error correction. Indeed, as we have just seen, purification has already been demonstrated in the laboratory although true error correction has not yet been accomplished.

Purification is particularly useful in distributed environments because it meshes nicely with the Bell pairs that network links typically create below and with the applications' need to have high-fidelity Bell pairs above.

In this chapter, we have discussed purification from the practical point of view of the procedures themselves, with an eye toward actual implementations in software with the accompanying classical communication protocols. It is important to recognize the impact of the scheduling algorithm for choosing Bell pairs to assign for purification. As noted at the end of the previous chapter, this software layer will receive a full d.m. from the lower layer software (e.g. AEC). Likewise, it will ultimately report out a complete d.m. to higher layers, such as entanglement swapping. In Chapter 10, we take up entanglement swapping, and see our first complete chain of quantum repeaters.

Chapter 10

Purification and Entanglement Swapping-Based Repeaters

Finally, the first full designs for repeaters can be presented, or rather, the first design for a line of repeaters, with focus on the communication session architecture. This chapter begins by asking why the most obvious technique fails: why can't we simply send our qubit via hop-by-hop teleportation, just as we would send an IP packet over the Internet? We then cover the original proposal of Dür and Briegel, with a few more modern extensions. The protocol state machine approach will be emphasized, and the impact of purification scheduling on performance will be discussed. The following two chapters will cover the other primary architectures for managing a communication session.

10.1. Hardware architectures

Before delving more concretely into the repeater operations, we should pause to consider the hardware architecture. In Chapter 8, we saw abstractly how a single qubit at each end of a link can be entangled, using a variety of mechanisms. Repeater hardware is more than a link with two qubits, one at each end. In practice, the internal architecture will affect the operation of purification (Chapter 9) or quantum error correction (section 11.1) and entanglement swapping (section 10.2.2). Moreover, it will affect how we can connect repeaters to create interesting networks. However, here we will limit ourselves to an abstract, generic model.

The basic link structure may be a configuration such as $M \rightarrow I \leftarrow M$, as in Figure 8.5, or a real or *de facto* $M \rightarrow M$ link as in Figures 8.4 and 8.6. Because our interest is in the Bell pairs in memories, we can treat most of these as $M \rightarrow M$ links except for the most detailed timing simulations, so from this point in the book, we will primarily consider links to be $M \to M$. The reader should be able to easily extend to other timing models as needed.

Generically, the hardware components of interest here consist of transmitters, buffer memories, detectors, controllable qubit couplers (point-to-point or shared bus, used for the internal logic necessary for purification, error correction and internal buffer management), optical multiplexers and of course the channels. In practice, we can reduce the set of components that we need to model to just memories and channels and the interactions among them. For the base Bell pairs, the effects of all of the components in the optical path will be incorporated into the d.m. reported out from the AEC layer, as discussed in Chapter 8.

We can roughly divide nodes into one of three categories, based on their role within a network: end nodes (one network connection), repeaters (two connections, i.e. only appears in a line), and routers or switches (three or more connections, i.e. can create complex topologies). We will generically refer to all of these as "repeaters", unless it is necessary to distinguish their role in the network.

To perform purification, at minimum, a node must be able to couple two qubits in its memory in order to perform a CNOT gate, then measure one of the resulting qubits. Entanglement swapping requires similar capabilities, and the organization of the system affects our operations.

In classical systems, the part of the system coupling the computer's main memory to the network channel is called an NIC. We can adopt the same terminology here, though the details of how qubits are coupled to the channel and to each other will vary. We can call the set of transceiver qubits coupled to a channel our quantum NIC. In the two middle repeaters in Figure 10.1, the left-hand column of qubits represents one NIC and the right-hand column represents a second NIC.

All nodes except the end nodes must be able to couple a qubit from one NIC to a qubit from another NIC. In Figure 10.1, this would be one qubit from the left-hand column with one qubit in the right-hand column inside a repeater node. Some researchers have assumed that initial hardware models will restrict the pairings, such that only those qubits facing each other directly in this figure would be able to couple, then gone on to investigate relaxing that constraint [COL 07, ABR 14]. However, other researchers, including Ladd *et al.*, began with the assumption that any qubit within a node can be coupled with any other [LAD 06]. In this book, we have worked with this assumption.

The need for any-to-any coupling becomes more readily apparent when more than two NICs are considered. For some technologies, the coupling can be accomplished using internal waveguides. For others, it will require use of the SWAP gate (equation [2.66]), moving qubits around from memory location to memory location within the node.

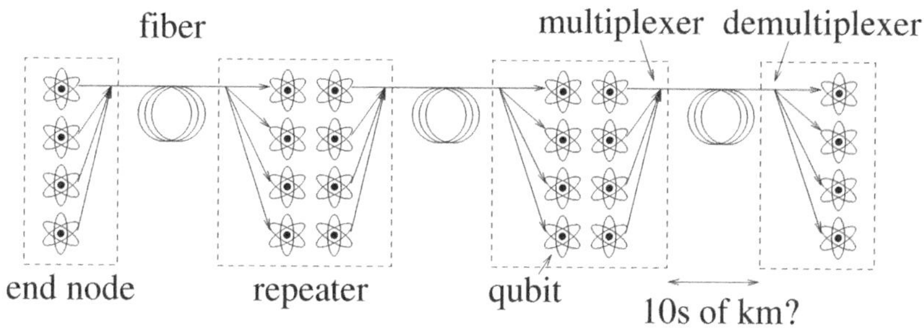

Figure 10.1. *Generic view of the hardware of a line of repeaters. Qubit memories are represented by the atom symbol, regardless of physical device type*

These factors reflect the reality of engineering repeater nodes, but are rightly the domain of engineers on a particular project; hence, they are beyond the scope of this book. A complete physical-level simulation would include all of these operations, paying attention to the graph of couplings possible within the node and tracking both operation timing and impact on fidelity. For most purposes, this level of detail is not necessary, and a simple mathematical model of purification or entanglement swapping that incorporates gate and measurement errors will be adequate.

The simulation results presented in this chapter all assume optical fiber connections with 0.17 dB/km loss and a signal propagation speed of $c_{\text{fiber}} = 0.7c$, corresponding to telecommunications fiber. Real-world fiber installations will generally have worse loss characteristics than this.

10.2. Getting from here to there

10.2.1. *Hop-by-hop teleportation*

Before going into the more complex entanglement swapping, let us examine the most obvious approach to sharing quantum data across a network: hop-by-hop teleportation of the data, as in a classical network.

Consider teleporting a qubit first from node A to node B, then on to node C. Node A begins with the data qubit to be transported, and one end of a Bell pair begins with node B. Node B holds one end of each of two Bell pairs, with one coupled to a qubit at node A, the other to a qubit at node C, which we will call $|\Psi^-\rangle^{(AB)}$ and $|\Psi^-\rangle^{(BC)}$, respectively. The teleportation operation is as shown in Figure 4.1.

The Bell state measurement can be represented as being composed of a CNOT and measurement of A's two qubits, the Bell pair qubit (the target of the CNOT) in the Z basis and the data qubit (the control qubit of the CNOT) in the X basis, as shown in Figure 2.4. The results of the measurements are sent to the partner node for Pauli frame corrections to complete the teleportation.

To cover longer distances, we can repeat the teleportation. Figure 10.2 shows a chain of teleportations conducted in this fashion. The teleportation operations can be performed *in any order*, across any link even before the Bell pairs in neighboring links have been created.

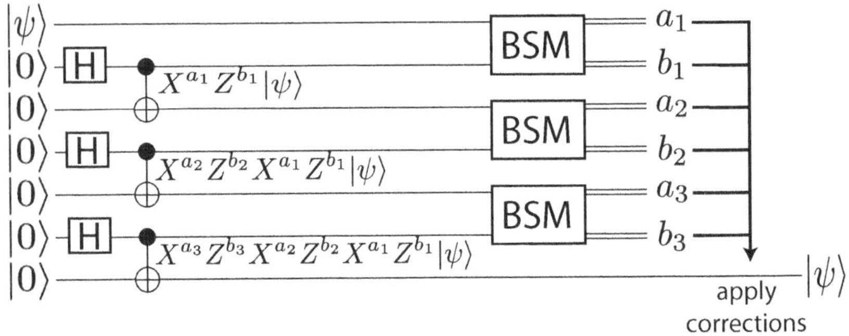

Figure 10.2. *Teleportation operations can be chained to move a qubit over multiple hops. The Pauli frame corrections can be accumulated and applied end to end*

The Pauli frame corrections do not have to be applied at each stage, before the next teleportation is performed. Instead, the corrections can be accumulated along the chain and applied at the receiver. Two Z corrections or two X cancel so that the final correction applied at the receiver is the parity of all of the Z or X corrections. The end-to-end teleportation is ultimately finalized when the receiver collects all of those corrections, which can happen no faster than the one-way latency from the sender to the receiver, t_{E1}.

The problem arises when taking into consideration that our systems are imperfect. For the moment, we will assume that the Bell pairs are realistic with $F < 1.0$, but that the gates and measurements for teleportation are perfect. To isolate the issues of the network from those of the application, we will assume that the qubit to be teleported is of fidelity $F = 1.0$. We assume that our Bell states are diagonal when the density matrix is written in the Bell basis, e.g.,

$$\rho_{AB} = a_{AB}\left|\Phi^+\right\rangle\left\langle\Phi^+\right| + b_{AB}\left|\Psi^+\right\rangle\left\langle\Psi^+\right| + c_{AB}\left|\Phi^-\right\rangle\left\langle\Phi^-\right| + d_{AB}\left|\Psi^-\right\rangle\left\langle\Psi^-\right|. \quad [10.1]$$

With the data qubit's original state being the pure state $\rho_D = \langle\psi|\psi\rangle$, let us write the error states as

$$\begin{aligned} |\psi^*\rangle &= Z|\psi\rangle && [10.2] \\ |\bar{\psi}\rangle &= X|\psi\rangle && [10.3] \\ |\bar{\psi}^*\rangle &= XZ|\psi\rangle. && [10.4] \end{aligned}$$

After teleportation, the output state at node B will be

$$\rho'_D = a_{AB}|\psi\rangle\langle\psi| + b_{AB}|\bar{\psi}\rangle\langle\bar{\psi}| + c_{AB}|\psi^*\rangle\langle\psi^*| + d_{AB}|\bar{\psi}^*\rangle\langle\bar{\psi}^*|, \quad [10.5]$$

reducing our original $F = 1.0$ to $F' = a_{AB}$.

Repeating the operation and teleporting our qubit from node B to node C, some of the error terms accumulate and some cancel, giving us

$$\begin{aligned} \rho''_D = \; & (a_{AB}a_{BC} + b_{AB}b_{BC} + c_{AB}c_{BC} + d_{AB}d_{BC})|\psi\rangle\langle\psi| \quad [10.6] \\ & + (a_{AB}b_{BC} + b_{AB}a_{BC} + c_{AB}d_{BC} + d_{AB}c_{BC})|\bar{\psi}\rangle\langle\bar{\psi}| \\ & + (a_{AB}c_{BC} + b_{AB}d_{BC} + c_{AB}a_{BC} + d_{AB}b_{BC})|\psi^*\rangle\langle\psi^*| \\ & + (a_{AB}d_{BC} + b_{AB}c_{BC} + c_{AB}b_{BC} + d_{AB}a_{BC})|\bar{\psi}^*\rangle\langle\bar{\psi}^*|. \end{aligned}$$

If both Bell pairs are of high fidelity, the b, c and d terms will be small, and the output fidelity over two hops is $F' \approx a_{AB}a_{BC}$. If we have a chain of repeater hops, all with the same fidelity $F \approx 1.0$, the output fidelity over a moderate number of hops n will be $F' \approx F^n$. As n continues to grow, the fidelity declines to a useless pure noise state with $F = 0.5$ even if the original infidelity $1 - F$ was small.

This observation that repeated teleportation degrades quantum information leads to the pessimistic conclusion that long-distance networks are untenable. The solution is two-fold: the discovery of how to extend entanglement to reach nodes that have never directly interacted, and a method for combining that with the purification discussed in Chapter 9.

10.2.2. *Basic entanglement swapping*

To create entanglement between nodes that have never directly interacted, we take advantage of teleportation and operate on generic states, rather than our valuable data qubits. The use of teleportation in repeaters, known as *entanglement swapping*, lengthens distributed Bell pairs by teleporting the state of one member of a Bell pair over progressively longer distances, until the pair stretches from end to end. Teleportation consumes Bell pairs; the repeaters are responsible for replenishing their supply of shorter-distance pairs in order to make the end-to-end Bell pairs.

The term "entanglement swapping" was introduced by Żukowski, Zeilinger and Ekert [ŻUK 93], originally in the context of photonic Bell pairs created via PDC and coupled using beamsplitters. The concept of entanglement swapping via Bell measurement was introduced by Żukowski, Zeilinger, Horne and Ekert, in 1993 [ŻUK 93]. Here, we will discuss the concept in the more general terms.

The principle is shown in Figure 10.3. Node B holds one end of each of two Bell pairs, one coupled to a qubit at node A, the other to a qubit at node C, which we will call $|\Psi^-\rangle^{(AB)}$ and $|\Psi^-\rangle^{(BC)}$, respectively. B decides to lengthen the pair on the left using the pair on the right, in an operation exactly analogous to one stage of Figure 10.2. B performs a Bell state measurement (BSM), as described in section 2.5.2, on qubits $|\Psi^-\rangle_B^{(AB)}$ and $|\Psi^-\rangle_B^{(BC)}$. The results of this operation must be communicated to node C, allowing C to recreate the state of $|\Psi^-\rangle_B^{(AB)}$, resulting in a new Bell pair $|\Psi^-\rangle^{(AC)}$.

In theory, A never needs to be told that the operation has occurred. Although C must apply corrective operations to complete the reconstruction, A is entirely passive, merely storing its half of the Bell pair in a buffer memory. However, A is very likely waiting on the completion of the swapping operation in order to perform some other action; at the very least, an application at node A is waiting to use the end-to-end Bell pair. Optimizing the performance of the network while providing robust, correct operation is a key problem in protocol design: when can we *guarantee* that a node knows enough to make a decision about that next action that will be *consistent* with decisions made by other nodes? We will return to this question of the minimum waiting time and maximum amount of action that nodes can take, both in section 10.3.2 and in Chapter 12.

Let us look in detail at the mathematics of a swapping operation. First, we assume that the AB state is diagonal when the density matrix is written in the Bell basis,

$$\rho_{AB} = a_{AB}\left|\Phi^+\right\rangle\left\langle\Phi^+\right| + b_{AB}\left|\Psi^+\right\rangle\left\langle\Psi^+\right| + c_{AB}\left|\Phi^-\right\rangle\left\langle\Phi^-\right| + d_{AB}\left|\Psi^-\right\rangle\left\langle\Psi^-\right|. \quad [10.7]$$

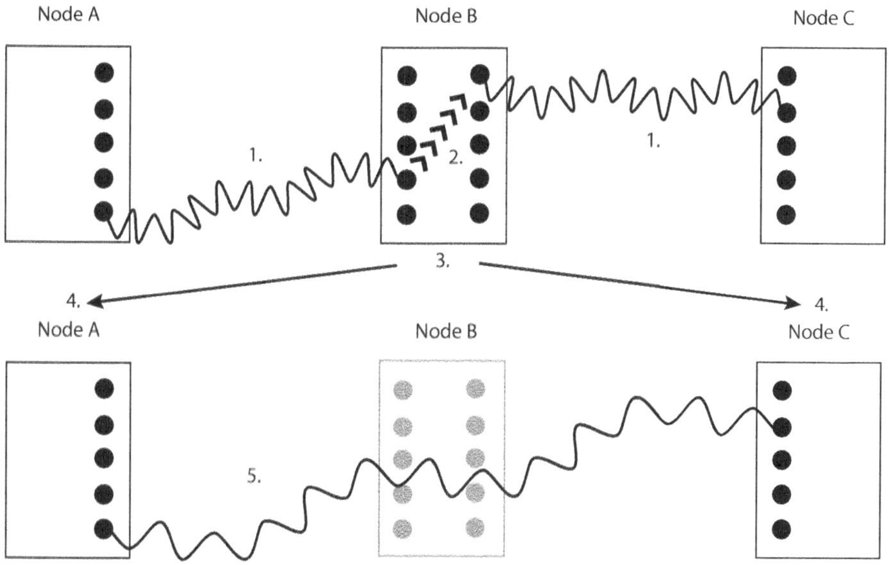

1. Nodes begin with two entangled pairs, AB and BC.
2. Node B selects pairs to teleport, performs local operations, measures one qubit of each pair.
3. B communicates measurement results and new entanglement status to A and C.
4. Receive partner's measurement result and new entanglement status, including node/qubit addresses.
5. Result is single lower-fidelity, longer-distance Bell pair.

Figure 10.3. *Teleportation can lengthen one Bell pair using another*

The Bell state measurement can be represented as being composed of a CNOT and measurement of B's two qubits, the target qubit in the Z basis and the control qubit in the X basis, as in Figure 2.4. Assuming these operations are perfect and that the AB pair is the CNOT target and the BC pair is the control, our resulting density matrix is

$$\rho_{AC} = a_{AC} \left|\Phi^+\right\rangle\left\langle\Phi^+\right| + b_{AC} \left|\Psi^+\right\rangle\left\langle\Psi^+\right| + c_{AC} \left|\Phi^-\right\rangle\left\langle\Phi^-\right| + d_{AC} \left|\Psi^-\right\rangle\left\langle\Psi^-\right| \quad [10.8]$$

where the d.m. diagonal elements are

$$\begin{aligned} a_{AC} &= a_{AB}a_{BC} + b_{AB}b_{BC} + c_{AB}c_{BC} + d_{AB}d_{BC} \\ b_{AC} &= a_{AB}b_{BC} + b_{AB}a_{BC} + c_{AB}d_{BC} + d_{AB}c_{BC} \\ c_{AC} &= a_{AB}c_{BC} + b_{AB}d_{BC} + c_{AB}a_{BC} + d_{AB}b_{BC} \\ d_{AC} &= a_{AB}d_{BC} + b_{AB}c_{BC} + c_{AB}b_{BC} + d_{AB}a_{BC} \end{aligned} \quad [10.9]$$

after node C receives the measurement results and applies the corrections shown in Table 4.1.

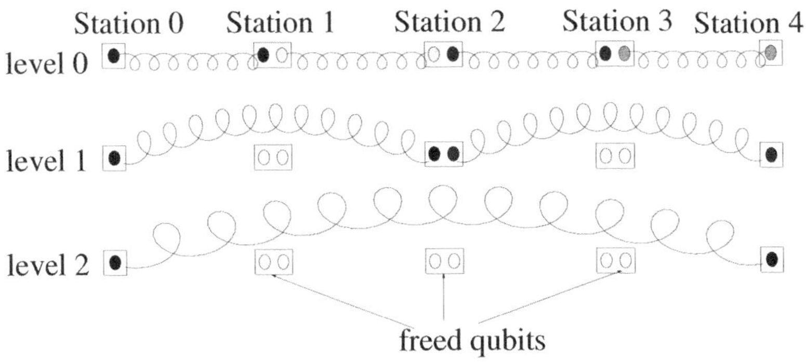

Figure 10.4. *Entanglement swapping. Spiral lines represent distributed Bell pairs, and straight lines are classical communication*

The simplest mathematical approach to modeling imperfections of the operations within a repeater node is to apply a *noise operator* to each qubit before each operation, including the CNOT and measurement operations in both purification and swapping, as discussed in section 8.2.1. The output fidelity is

$$F_{AC} = a_{AC} \approx F_{AB}F_{BC}. \quad [10.10]$$

10.2.3. *Multi-hop swapping*

Multi-hop swapping is functionally equivalent to hop-by-hop teleportation over the same number of links, except that the floor for fidelity for a two-qubit state is $F = 0.25$ rather than $F = 0.5$, which is the floor for teleportation and decoherence of a pure single qubit. Over n hops each with fidelity F, the output fidelity is

$$F' \approx F^n. \quad [10.11]$$

If $F < 1.0$, we are unable to sustain a high fidelity over more than a few hops when using direct teleportation. Note that this is independent of the order in which we do the teleportations; an n-hop hop-by-hop teleportation chain results in exactly the same fidelity as using entanglement swapping to build successively longer Bell pairs.

The value of entanglement swapping becomes clear when we recognize that purification can be done over multi-hop Bell pairs, independent of the distance and requiring only classical communication. Armed with this insight, we begin searching for an efficient manner of combining purification with teleportation or swapping. Dür, Briegel, Cirac and Zoller hit upon the idea of repeating the purification after entanglement swapping, publishing the seminal paper in quantum repeaters in 1999 [DÜR 99]. Purification then consumes one two-hop Bell pair to purify another two-hop Bell pair. After purification of this generic Bell pair, teleportation of a data qubit can be done over the two-hop distance with higher fidelity than teleporting it twice over single-hop Bell pairs. This gives us a capability we do not have when thinking only about the data qubit to be teleported.

Single-stage entanglement swapping was successfully demonstrated in 2001 by Zeilinger's group [JEN 01]. Multi-stage entanglement swapping of photonic Bell pairs was successfully demonstrated by Pan's group in 2008 [GOE 08].

10.3. Nested purification session architecture

We have just seen how to splice two shorter Bell pairs into one longer one. This operation is independent of the length of the Bell pairs, measured in either hops or kilometers. This allows us to apply the concept recursively, as shown in Figure 10.4. Rather than one hop at a time, as in Figure 10.2, we can organize a set of swaps into a logarithmic number of rounds. We can begin with four one-hop Bell pairs, perform entanglement swapping to be left with two two-hop Bell pairs, and again to reach the goal of a single four-hop Bell pair.

Repeating this process, we can purify, then swap, then purify again, in a nested arrangement, as shown in Figure 10.5. Each layer in this architecture corresponds to one of the roles described in section 1.3.2. The topmost, application layer is the consumer of our end-to-end Bell pairs, presumably one of the applications presented in Part 2 of this book. At the bottom is one of the physical layers discussed in Chapter 8, topped by the Acknowledged Entanglement Control protocol. In between AEC and the application layer, we see repeated instances of Purification Control (PC) (section 9.4) and Entanglement Swapping Control (section 10.4).

10.3.1. *Proof of polynomial resource growth*

As in Figure 10.4, each "level" of swapping doubles the distance so that at level n, we have Bell pairs spanning 2^n hops. If one round of purification is done at each level, and purification always succeeds, generating one purified level n Bell pair requires two lower-fidelity level n Bell pairs (one purified, one sacrificed). The low-fidelity level n Bell pairs are created by swapping two high-fidelity level $n-1$ Bell pairs, each of which was created using two low-fidelity level $n-1$ pairs.

Even with always-successful purification and swapping, simple hop-by-hop teleportation would cost us 2^n base-level Bell pairs, of course, but would not complete with high fidelity. One round of symmetric purification at any level doubles the number of Bell pairs consumed, so that one level n Bell pair spanning 2^n hops, purified once at each level, costs us 2×4^n base-level, one-hop Bell pairs.

We do not necessarily have to perform purification after every swapping operation. More generally, if we do entanglement swapping to create Bell pairs over L hops before purification, generalizing from the $L = 2$ above, and purification at each level consumes M Bell pairs spanning L hops, again generalizing from the $M = 2$ above, then building a single n-level Bell pair covers L^n hops and consumes

$$R = (LM)^n \tag{10.12}$$

base-level Bell pairs. Here, n grows logarithmically in the total distance covered; L is a engineering parameter we can choose; and M will be determined by the purification mechanism we choose, its probability of success, and the number of rounds of purification required.

The operational approach to using this scheme is to first purify base-level Bell pairs to a "working fidelity", perform swapping, then repeat purification until the working fidelity is restored. As we saw in section 9.2.4, M will be exponential in the number of rounds of purification, but the number of rounds is small and logarithmic in the infidelity.

This relationship demonstrates that the growth of resources for a chain of repeaters is polynomial in distance, which is a critical element of the proof of the viability of long-distance entanglement creation.

Taking into account various systems-related and reality-related constraints, analyzing the full behavior of a chain of purify-and-swap repeaters is complex, and does not lend itself easily to analytic expression; simulation is often the best way to see characteristics of operation. We will discuss reality in more detail after analyzing some of the logic problems that can occur, and the software state machines used to provide robust operation.

10.3.2. *Problems to avoid*

Careless implementation of the classical network protocols in Figure 10.5 will result in incorrect and/or inefficient operation. Each node requires enough information to make decisions that will be consistent with those made by other nodes.

Figure 10.6 shows one possible case that protocol designs must be careful to avoid, which we call the "leapfrog" problem. In this four-hop configuration, the goal is to create a Bell pair from node A to node E, necessitating swapping operations at B, C and D. If the CD link is spanned by two separate Bell pairs, then node C and node D must each make a decision about which of the two Bell pairs to use in swapping. If C chooses one pair and D chooses the other, then rather than joining to create a single Bell pair spanning BE, the entanglement will leapfrog to create a BD pair and a CE pair, leaving us with no clear path to building the desired AE pair.

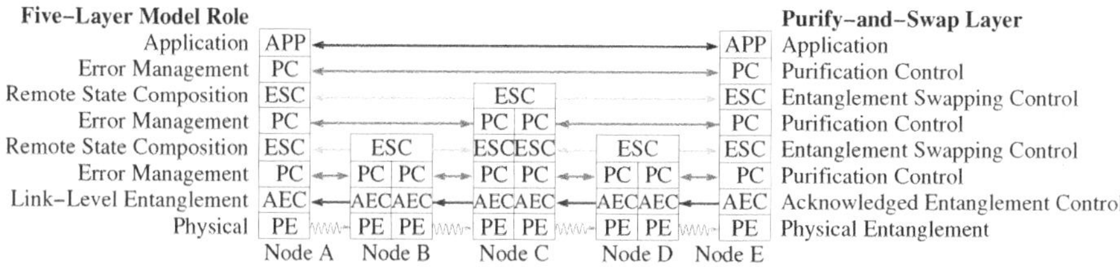

Figure 10.5. *Protocol layers and their interaction in purify-and-swap repeaters, in a five-node, four-hop chain. The labels on the left indicate the model layer represented, and the labels in the boxes and on the right indicate the protocol name for purify-and-swap repeaters. Double-headed arrows indicate bidirectional classical communication is required. The only quantum portion of the stack is the physical layer, shown with all links propagating left to right*

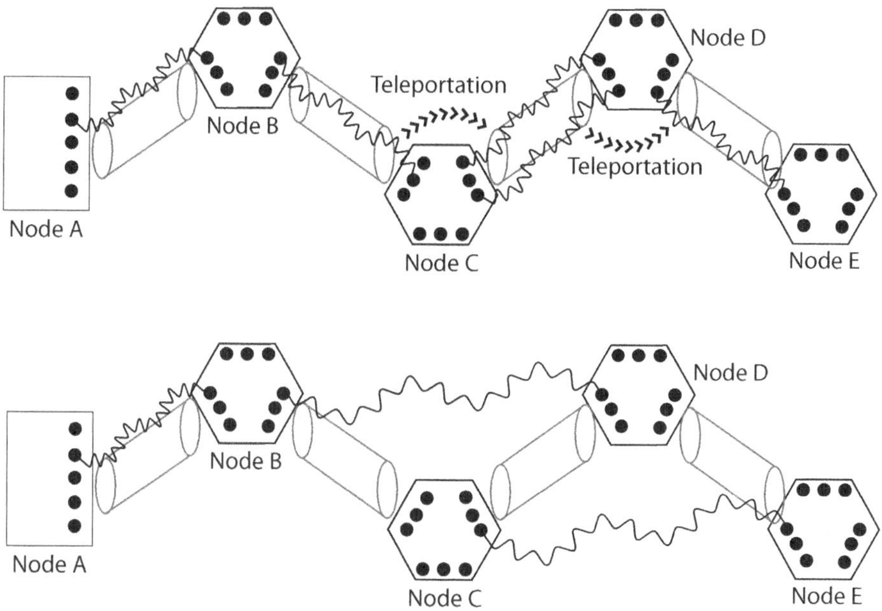

Figure 10.6. *Conflicting teleportation choices by nodes C and D in the top figure may result in Bell pairs "leapfrogging" each other, as in the bottom figure, leaving no easy path to connect A to E*

This problem may be solved by guaranteeing that both C and D use a deterministic selection algorithm, based on the same information. The same set of rules proposed in sections 8.5.1 and 8.5.2 can prevent misbehavior, especially the admonishment that each node use the same list of Bell pairs.

The second major problem, akin to one discussed in section 9.4, is "managed obsolescence" of Bell pairs. Both nodes holding a Bell pair must again make a coordinated decision on the fate of the Bell pair. For entanglement swapping, typically one node passively holds its qubit, whereas the other actively searches for a swapping partner, and is therefore effectively in control of the Bell pair. However, in later chapters we will see that this interaction pattern changes, requiring more care.

The banded and symmetric purification algorithms (section 9.3) are potentially subject to *deadlock*. Consider a link where each repeater has seven physical qubits, and purification is configured with seven bands (or seven rounds of purification for the symmetric case). After an extended period of operation, we might reach a condition in which each purification band contains one Bell pair. Each pair would have no possible purification partner, and no free qubits would be available to create new pairs to add to the bottom band. For the banded algorithm, we can restrict the number of bands used to a number that will not deadlock for a given hardware configuration. For the symmetric algorithm, we can likewise limit the number of rounds of purification configured, although this configuration allows less flexibility.

For chains of repeaters, each swapping level is independent, so the minimum number of qubits per station must actually be the number of bands times the number of levels, plus one, for the receive half and send half of the repeater. For example, over $2^6 = 64$ hops (six swapping levels) with seven purification bands, we must have a minimum of $6 \times 7 + 1 = 43$ qubits in each node to guarantee deadlock-free operation. Because this number increases as paths lengthen, management of resources in large networks is a difficult problem.

10.4. State machines and protocol interactions

Figure 10.7 shows the simple structure of the protocol state machine for a qubit at a mid-path node, where entanglement swapping will occur. A little thought reveals several problems that must be addressed.

In the top center of the figure is the transition, "Assigned for swapping". The algorithm that makes this assignment will ultimately have an enormous impact on the effectiveness of the network. What information is necessary for this node to make a decision that will advance the process of retiring an end-to-end, application-level request? As illustrated in Figure 10.6, poor decision-making here can allow inconsistent actions to be taken at the distant nodes, retarding rather than advancing retirement of requests.

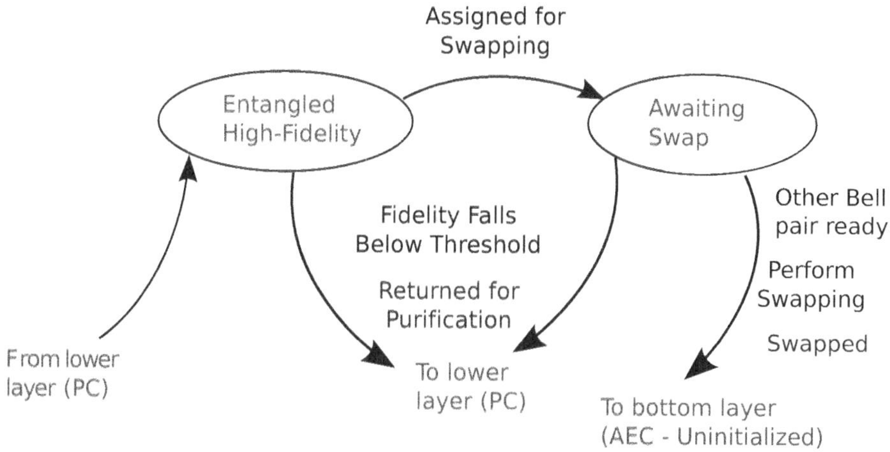

Figure 10.7. *Finite state machine for the Entanglement Swapping Control (ESC) protocol for a qubit at a mid-path node*

The managed obsolescence of Bell pairs is highlighted by the two transitions in our figure triggered by a Bell pair falling below a designating fidelity threshold. This is marked as resulting in the transmission of a RETURNEDFORPURIFICATION message. Here once again, coordination with the holder of the far end of the Bell pair is essential.

In normal operation, purification and swapping (PC and ESC) are repeated at each level until the top, end-to-end level is reached, as shown in Figures 10.5 and 10.8. At that final distance, purification may be repeated one more time to create the final end-to-end pair of the fidelity required by the application. Of course, purification can be omitted or repeated at any level, depending on the fidelity of the Bell pairs. In Figure 10.8, purification at level 0 is shown happening twice on the left. The actual timing of messages may vary somewhat; PC(0) can only be initiated after the status of qubits has been established by EC, as in Figure 9.5. Because the stations run a deterministic algorithm to select which pairs to purify, PC does not need to negotiate which operations to perform, only inform its partner of the outcomes.

The notation in the example in section 10.2.2, with ρ_{AB} and ρ_{BC} disappearing and being replaced with ρ_{AC}, hints at one of the challenges in designing protocols for repeaters: tracking the qubit states (or Bell pairs) as they move around the network. A qubit held at node A may first be entangled with B, then later with C and on across the network. *Naming* the state then becomes tricky, especially when we realize that there are many quantum states throughout the network that must be uniquely identified in order to communicate decisions about their fates.

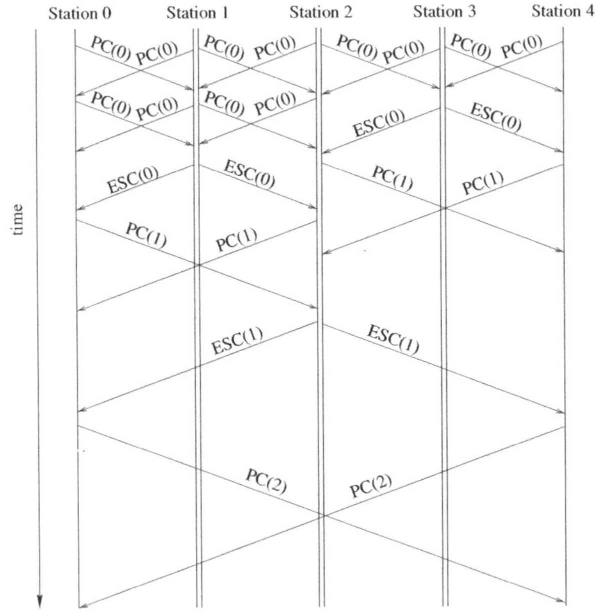

Figure 10.8. *Example message sequence for PC and ESC. Numbers in parentheses are the level or distance*

Ultimately, as shown in Figure 8.9, many issues can be resolved by ensuring that operations are performed on the same list of states, which, in turn, implies that each entry in the list is some form of identifier. We can study this issue in detail in section 15.2.2, when we establish a more complete network architecture.

Composition of links to establish a path through the network, including allocation of resources, is an issue in real-world networks that does not appear in analytical work or simulations of simple chains of repeaters. Likewise, dealing with a non-power-of-two number of hops in purify-and-swap architectures requires care. We will see these and other network-related issues in Part 4.

10.5. Putting it all together

Combining nested purification and swapping, and using the purification scheduling techniques we discussed in section 9.3, we can analyze the behavior of a full purify-and-swap, Dür-Briegel-Cirac-Zoller-style chain of repeaters. Here, we present one example architecture, including the full messaging schemes of the protocols in Figure 10.5. We cannot describe the structure of existing repeaters, since there are none, so we will describe simulations of chains of repeaters that track physical effects and message exchanges in detail.

10.5.1. *Simulating lines of repeaters*

We will see simulations of a particular form of qubus scheme, using a weak nonlinear interaction with a quantum dot in a cavity to create Bell pairs, as in Figure 8.4 [MUN 05, SPI 06, LAD 06, LOO 06]. These simulation results were published by Van Meter, Ladd, Munro and Nemoto in 2009 [VAN 09].

Using the qubus scheme, the probability of successfully creating a Bell pair is high, but even when the operation succeeds the fidelity of the created Bell pair is low (these two parameters represent an engineering trade off, which is not discussed here). For the parameter settings we have chosen, Bell pairs are created with fidelities of $F = 0.77$ or $F = 0.638$ for 10 km and 20 km distances, respectively, and the creation succeeds on 38–40% of the attempts [LAD 06].

For 20 km links at c_{fiber}, the one-way latency for signals is just under 100 μs, so the "clock rate" for these simulations is about 10 kHz. The pulses are very short compared to the propagation latency, so their duration is ignored.

As discussed in section 9.3, in real systems with multiple qubits per node, purification scheduling will have a large impact on performance. Symmetric purification would take our starting fidelity of 0.638 to a target fidelity of 0.98 after five rounds. If purification always succeeded, 32 (2^5) base-level Bell pairs would be required: $32 \times 0.638 \rightarrow 16 \times 0.71 \rightarrow 8 \times 0.797 \rightarrow 4 \times 0.867 \rightarrow 2 \times 0.952 \rightarrow 1 \times 0.988$. Unfortunately, purification is a state-dependent, probabilistic operation. When using our starting state, the first step (0.638 + 0.638) will succeed only 57% of the time, whereas the last step will succeed 92% of the time. In total, symmetric purification actually consumes, on average, more than 450 base-level Bell pairs to make one Bell pair of 0.98 fidelity.

The principal drawbacks to the symmetric algorithm are the inflexible use of available resources, both time and space (as shown by e.g. the wait at $t = 4$ in Figure 9.4), and the fact that the truly symmetric history tree is effectively impossible to achieve. Memory degradation over time causes two pairs that arrived at different times to have different fidelities, so forcing exact matches only is impractical.

Recall that the purification operations can fail, but their probability of success increases as the fidelity of the pairs involved increases. Any attempt to predict the exact best sequence of purification operations from a given state, therefore, must take into account which resources are currently busy, the fidelities of all available Bell pairs, the probability of success of possible purification choices and the probability that currently unentangled qubits will be successfully entangled in the near future using the physical entanglement mechanism.

The metric we use to evaluate quantum networks is the throughput, measured in Bell pairs per second of a certain fidelity over a given distance. We have chosen a target fidelity of $F = 0.98$, and simulate for distances up to 20,000 km. Unless otherwise specified, the simulations presented here are for 64 links of 20 km each, with 100 qubits per station (50 for receive and 50 for send, except at the end points where all 100 can be used for one direction). With these settings, in the first time step, each station will attempt to entangle 50 qubits, successfully creating about 20 base-level Bell pairs on each link. In successive time steps, the number of attempts on each link is capped by the number of available, unentangled qubits at each station.

Our code is capable of simulating imperfect local gates, but to isolate the individual factors presented here, the simulations in this chapter assume perfect local quantum operations and memory. Our simulations have shown that gate errors of 0.1% result in about a factor of two reduction in the performance of the system, with performance degrading rapidly and a final fidelity of 0.98 being unattainable with gate errors of 0.3%.

As a rough approximation, the gate error rate can be considered to be the *combination* of both local gate errors and memory errors. With one-way latency in fiber of approximately 6 ms at 1,280 km, memory must be able to retain its state for times on the order of seconds to meet the above constraint. Hartmann *et al.* have examined the role of memory errors in quantum repeaters [HAR 07], finding that memory that can successfully retain a quantum state for about 1 s can support ultimate repeater distances of 5–20,000 km, albeit it at a large cost in resources and with a cap on the achievable fidelity. If memory times are shorter, then local quantum error correction should be added, which will add substantial additional complexity to the system design.

For each banded data point in the graphs presented here, extensive runs over large parameter spaces (up to 800 or so separate sets of parameter settings) were executed to find a good set of bands, and to find a good set of thresholds for entanglement swapping at different distances. Each data point represents a single run in which 200 end-to-end Bell pairs of final fidelity 0.98 or better are created, with the exception of a few of the slowest data points, which were terminated early. The throughput is calculated by linear regression to fit a line to the arrival times of the Bell pairs [JAI 91]. Error bars are included but are almost too small to be seen at many data points; they represent the standard deviation of the fitted slope for that run. The coefficient of determination is more than 0.996 for almost every fit except the three data points with the largest error bars in Figure 10.9, for which it is 0.95, 0.80 and 0.78. These fits confirm that despite the stochastic nature of the quantum operations, the mean arrival rate is constant after the initial transient startup latency. Runs of fewer than 200 Bell pairs were found to have unacceptably large variability.

First we analyze the performance of the greedy algorithm, then we present our primary results, comparing the throughput of greedy and banded purification. We backtrack to explain how bands are selected, then we compare several options for setting the fidelity target at each swapping level.

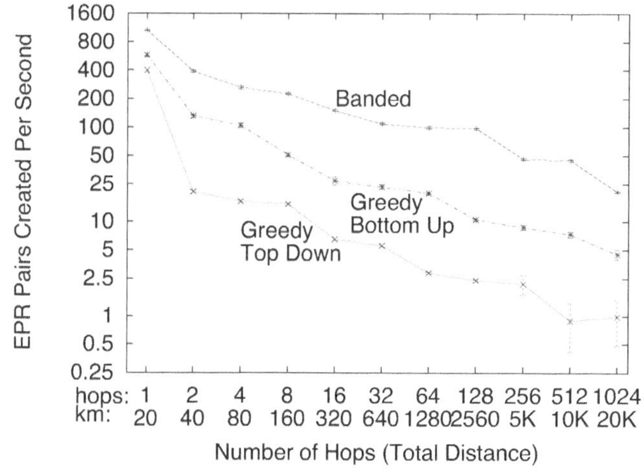

Figure 10.9. *Throughput versus distance for the banded algorithm using five bands, compared to the greedy bottom-up and greedy top-down algorithms. The final fidelity is 0.98*

10.5.2. *Greedy algorithm*

The performance of the greedy top-down algorithm, corresponding to the work of Ladd *et al.*, is the bottom line in Figure 10.9 [LAD 06]. Throughput in end-to-end Bell pairs created per second is plotted against distance. The X axis is labeled with both the number of hops and total distance in kilometers; the rightmost point of 1,024 hops or 20,000 km corresponds roughly to the distance halfway around the world.

For the greedy top-down algorithm, throughput is approximately 21 Bell pairs/s for two hops, and declines to almost exactly 1 Bell pair/s for 1,024 hops. The decline shows a distinct stair-step structure, caused by the discrete nature of purification and our choice to purify until a final fidelity of 0.98 is reached. At a particular length, a certain number of purification steps is required to achieve the final fidelity. As the number of hops increases, the same number of purification steps may continue to serve, until the fidelity drops below the target and an additional round of purification must be added. When this happens, the performance drops by roughly a factor of two, as two high-quality pairs up near the target are required.

The greedy algorithm sorts the Bell pairs by fidelity, and pairs them starting with the two highest-fidelity pairs. We discovered that pairing beginning from the bottom of the list, which we term greedy bottom-up, increases performance by a factor of 3–8, as the middle curve in Figure 10.9 shows. We attribute this improvement to increased conservatism on the use of the highest-fidelity pair. Beginning at the bottom will bring other pairs up toward the fidelity of the highest pair, perhaps even surpassing it, but first risking the failure of lower-fidelity pairs which have cost less to build.

At the left-hand edge of the graph, the greedy top-down algorithm declines from 400 pairs/s for one hop to 21 for two hops, almost a factor of 20 worse. For this graph, our hardware is assumed to have 100 qubits per station. For one hop, all 100 qubits can directly connect to qubits at the far end. For two hops, the middle station must split the use of its 100 qubits, 50 for the left-hand link and 50 for the right-hand link. The difference is due to more efficient purification pairings as the number of available qubits grows.

10.5.3. *Banded performance v. total distance*

The top line in Figure 10.9 graphs the performance of our banded algorithm. Throughput starts at 1,060 Bell pairs/s for one hop, plateaus at approximately 100 for 32 to 128 hops, then declines to 20 pairs/s for 1,024 hops. Due to the stair-step behavior, the benefit compared to the greedy top-down algorithm varies from a factor of 15 to a factor of 50, with the advantage growing unevenly as distance increases. Compared to the greedy bottom-up algorithm, banded is 2.5–9.3 times better, also increasing unevenly with distance.

Entanglement pumping and symmetric scheduling are not shown in the figure. Entanglement pumping cannot effectively create pairs of fidelity 0.98 with our starting fidelity of 0.638. For the particular configuration shown here, the symmetric algorithm would perform similarly to banding.

An important question is whether band structure changes when the total distance (number of hops) is increased. If the band structure does not change, then we can simulate short lines, and apply the simulation results directly to much longer lines, dramatically reducing the amount of computation time needed in simulations. Likewise, in real-world operational environments, distance-independent system controls would be a boon. Unfortunately, these simulations show that the banding structure does vary somewhat at different distances. The performance for nearby banding structures can be a factor of two worse, meaning that a careful search is necessary for each specific link configuration.

10.5.4. *Finding the bands*

We can theoretically place the boundaries that separate bands at almost any level. To determine a placement that gives good performance, we have performed nearly exhaustive searches over many possibilities, for configurations with 2–6 bands. Figure 10.10 shows a two-band setup. In this figure, we vary the boundary in steps of 0.01, but in most other graphs the steps are 0.02 or 0.04. At the left edge, the division between the two bands is below the initial threshold of 0.638 generated by our physical entanglement process, and at the right edge, the division is above the delivery threshold for our final qubits, resulting in the equivalent of the bottom-up greedy algorithm for the first and last data points. The performance peaks when the band boundary is 0.87–0.89, showing clearly that the operational imperative is protecting the high-fidelity pairs from purifying with low-fidelity pairs.

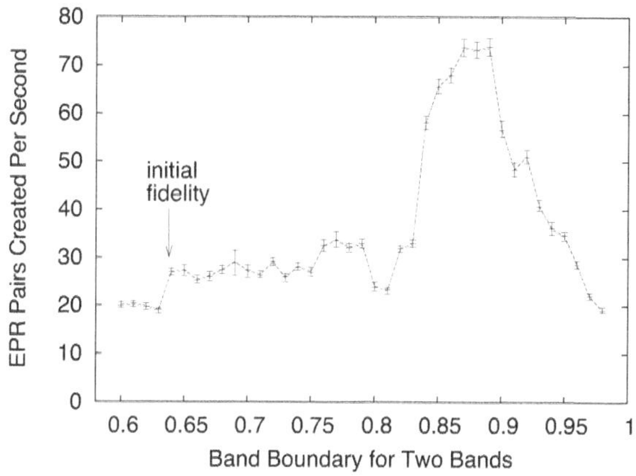

Figure 10.10. *Finding the best band boundary for a 2-band arrangement, for 64 hops of 20 km each*

Increasing the number of bands gives a smooth increase in performance for up to five bands, which perform nearly 50% better than two bands. Figure 10.11 shows the increase in performance for increasing numbers of bands. Moving from one band (equivalent to greedy bottom up) to two increases performance by more than a factor of three. The performance has saturated with six bands; it is not clearly better than five bands, because the behavior has essentially been constrained to that of a symmetric tree. For more than two bands, the number of simulation runs to cover the space increases geometrically, so the granularity of our boundary steps is somewhat larger. For example, for three bands, we tried all combinations of boundaries with the lower bound varying from 0.60 to 0.95, and the upper boundary varying from 0.80 to 0.99, in steps of 0.02.

10.5.5. *Varying swapping thresholds*

Recall the distinction between the purification bands and thresholds at different distances: the former governs purification decisions within PC, whereas the latter governs the promotion of pairs from PC to ESC for entanglement swapping at the next-highest distance. The experiments in the previous subsections were performed with each of the distance thresholds set to 0.98. In this section, we evaluate several possible sets of thresholds that seem like plausible candidates for good configurations:

a. 0.9, 0.9, 0.9, 0.9, 0.9, 0.9, 0.98:
purify only to an intermediate fidelity of 0.9 at distance 1, 2, 4, 8, 16 and 32, then push to the final fidelity of 0.98 at the full distance of 64 hops;

b. 0.98, 0.9, 0.9, 0.9, 0.9, 0.9, 0.98:
purify to fidelity 0.98 at distance 1, then allow the fidelity to slip as far as 0.9 at intermediate distances, before pushing back up to 0.98 at 64 hops; and

c. 0.98, 0.98, 0.98, 0.98, 0.98, 0.98, 0.98:
purify to fidelity 0.98 at distance 1, then maintain that fidelity by purifying as necessary at each distance.

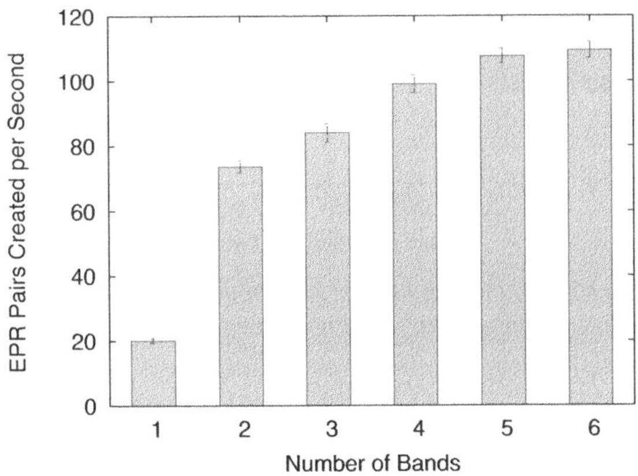

Figure 10.11. *The best band throughput for different numbers of bands, for 64 hops of 20 km each*

Figure 10.12 shows clearly that the preferred method of managing the fidelity of a pair as it hops across the network is case **c**, purifying to the desired level at distance one and maintaining that fidelity at all distances. Case **a** proved to perform so poorly that the simulations were unable to complete. The other two cases are shown in the figure.

These data support the intuitive idea that purifying over short distances will be more efficient than purifying over long distances. Dür *et al.* referred to this approach as maintaining a "working fidelity" [DÜR 99]. They did not report on any alternative schemes, but our data confirm that their approach is correct.

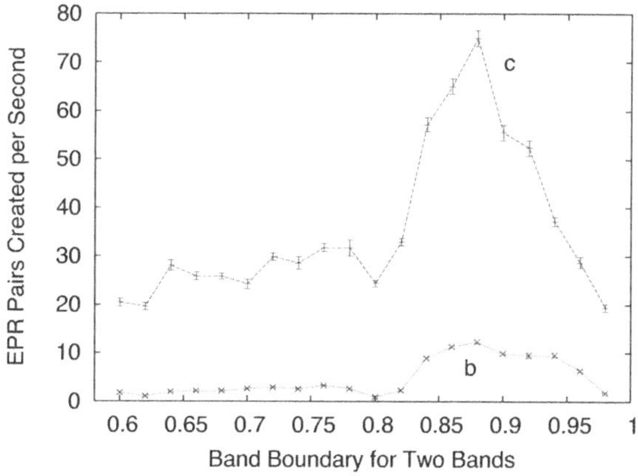

Figure 10.12. *Comparing different distance-swapping thresholds*

Because the curves for **b** and **c** have the same shape, despite radically different distance thresholds, Figure 10.12 also suggests that changing the pattern of fidelity thresholds at different distances is *independent* of the choice of the bands for banded purification. That is, a good choice of bands should remain good regardless of the thresholds at various distances. This fact should allow us to optimize these two parameters independently for a given physical configuration.

10.6. Considerations in the design of a simulator

Because this is the first chapter to present simulation results in detail, it seems appropriate to briefly introduce some "lessons learned" in simulator design. As is natural, our simulation capabilities have evolved over time, reflecting both completion of deferred features and a deeper understanding of important characteristics. Attention to these points when developing a new simulator should result in a code base that can evolve from basic to very complete, capable of reflecting the real world in sufficient detail.

By far the most critical aspects of a repeater are choosing an appropriate abstraction of the physics and correctly modeling the classical messaging needed to accomplish each task. This latter responsibility can be discharged by asking, "What does Software Module A on Node X need in order to make a decision, and *when can it know it?*". The discussion of distributed density matrices in section 8.5 should be helpful here.

The simulations just described were conducted using a third-generation simulator. The first was written directly in Matlab for ease of matrix manipulation, the second, which was written in C++, reflected recognition that a large fraction of the code and execution time are dedicated to managing the messages that are transferred station to station. In our experience, the speed of floating point operations is important, but less important than the decision logic and message management. Both simulators were originally developed by Thaddeus Ladd. The third generation (7,000 lines of C++) is a direct descendant of the second, with extensive code modification by Van Meter.

The simulator faithfully covers the quantum mechanics of the physical interactions and operations (in fact, it is capable of simulating several types of physical layers, though only the qubus scheme results are presented in the earlier text), but in practice, as discussed in section 10.1, what we really care about is the base-level density matrix generated by the link. More complete modularization in software design would suggest that the underlying physics be simulated or measured from experiment separately, and fed into the repeater simulator as a set of parameters. To be useful for simulations of real-world systems, this must include variable link lengths and different fidelities over the same length reflecting differing link qualities, as well. Our third-generation simulator remained a discrete-time event simulator, in which all simulator nodes moved in lockstep, constraining links to all be nominally the same length. For example, the simulations earlier assumed 20 km links, making one time step of 100 μs. Varying per-link channel loss was added to the simulator later.

Memory lifetime and gate infidelity must be supported and be configurable, and require somewhat more active code than the simple parameters for the base Bell pair, as seen in the discussion of decoherence in section 8.2.3. In general, abstraction of the noise model for individual interactions and a simple hardware model (as in section 10.1) will be adequate for most purposes, and will simplify the simulator.

Likewise, the purification scheduling should be configurable, and it may involve a significant set of tuning parameters of its own. Indeed, modularity adequate for easily installing new purification mechanisms will be a benefit.

The simplest simulations use only a power of two number of hops, which is convenient, and avoids some hard design work in e.g. node addressing, but this affects the behavior of purification and swapping, and it will not reflect the real-world operations. Entanglement swapping with a non-power-of-two number of hops is discussed in Chapter 14.

A full network simulator will incorporate the ability to define an arbitrary graph for the network topology. This will necessitate the addition of minimal node naming, path selection (routing), and link multiplexing functionality. We will see some of these issues and simulation results taking them into account in Part 4.

Following a presentation on quantum network simulations, a question that is often asked is, "Why do you trust the simulator? How have you validated it?". The underlying physical operations described in Chapter 8 are well understood theoretically, and the experiments generally match the theory, though experimentalists fight a difficult battle against new sources of noise and other physical imperfections. Simulations that closely hew to observed behavior are on steady ground there. The messaging sequences and behavior of larger networks, in contrast, model systems that do not exist in the real world and hence cannot be fully validated. They must be vetted as carefully as possible through formal models of events and messages, as in the state machines presented throughout this book; it bears repeating that the modeling of distributed management of density matrices and management of buffer memories are critical.

10.7. Conclusion

We have just finished discussing the simulation of repeaters, but the components for repeater operation have been demonstrated in the laboratory, and it seems that complete experimental proof of principle lies just around the corner. Systems adequate for the real-world deployment remain some distance away, however.

The purify-and-swap approach presented in this chapter is the canonical model of quantum repeaters, the first developed and still the reference to which new proposals are compared. Lengthening entanglement via swapping is a fundamental capability, and underlies many quantum communication schemes, if sometimes in a not-so-obvious fashion. Purification is the simplest form of error management, testing propositions about a quantum state and discarding those states that fail the test. It has the advantage that it requires very minimal resources and hardware capabilities, with one rather substantial drawback: the need for long-lived memories, much longer than the in-flight time of a photon from one node to the next. This need, and the accompanying wasted time in which memory is buffered awaiting communication from another node, drove the hunt for architectures with more direct coupling of end-to-end qubits and re-examination of transportation akin to the hop-by-hop nature of the Internet, which we discarded in section 10.2.1. We will see the first of these alternatives in Chapter 11, and the second in Chapter 12.

Chapter 11

Quantum Error Correction-Based Repeaters

Purification as described in Chapter 9 can be viewed as a form of error *detection*, with error states being discarded. Of course, classical computing systems depend heavily on error *correction*, and the same will be true for quantum computers. An important advance in repeater design has been the introduction of error correction-based connection architectures. The two principal ideas are the Calderbank-Shor-Steane (CSS) code-based approach [JIA 09] and the surface code-based approach [FOW 10]. In this chapter, we will first give a very brief introduction to quantum error correction and then look at these two connection architectures.

The principal advantages of these approaches at the application level are that they generate very high fidelity connections (high enough for true distributed, digital computation), and will integrate smoothly with error-corrected systems at both ends. At the network level, both schemes will simplify operations compared to purify-and-swap. After the path setup, essentially all of the communications can be done with the nearest neighbors only, rather than the complex set of overlapping connections and layering required for nested purification, as in Figure 10.5.

The primary disadvantage is that the resource demands are very high for potentially usable configurations. One prospective advantage is that memory lifetimes can be very low (a multiple of the local gate times, rather than a multiple of the single-hop or end-to-end round trip time). However, this only holds under certain conditions, as we will see; more likely configurations require memory lifetimes proportional to single-hop round trip time (RTT), with the memory errors accounted for in the error rate allowable before application of the error correction.

11.1. Quantum error correction

QEC operates on the same principle as classical error correction: redundancy is introduced into the data state (the encoding process), then the encoded data are sent through a channel (for communications theory purposes, either a memory over time or a communications channel can be treated the same). At the far end of the channel, the data are examined for errors by calculating the *syndrome* values, which, with high probability, will identify if errors occurred and how to correct them. However, two major differences complicate this process in the quantum domain: first, we must learn how to extract the error syndromes without causing a superposition or entangled state to collapse; second, the gates that are used for calculating the syndromes can introduce errors into the state, or propagate errors from place to place. Detailed discussion of QEC is beyond the scope of this book; readers are encouraged to study tutorials on this important topic [DEV 13, RAU 12, TER 13, GRA 09].

The stabilizer formalism discussed in section 6.1.1 is very useful for describing the syndrome extraction process, and the nature of states that have or do not have errors in them [GOT 97]. We may, for example, use a circuit that determines if three qubits are in the $+1$ or -1 eigenstate of the XXX stabilizer. Stabilizers found in the $+1$ eigenstate are not in error, whereas those in the -1 eigenstate are in error, and must be corrected. The details of the error correction mechanisms and the possible residual logical error mechanisms are beyond the scope of this book, but the error correcting capability of the code is at least $\lfloor (d-1)/2 \rfloor$. Any group of physical errors capable of causing error correction to miscorrect the state must be at least that large, and must be arranged in specific patterns on the lattice.

In classical computation, many logical bits are typically encoded in a larger block of physical bits. In quantum computation, for reasons we do not need to go into, typically only a single logical qubit is encoded into a group of physical qubits, using e.g. nine physical qubits to hold one logical qubit [SHO 95], or as few as five [LAF 96]. One important, very broad class of error correction schemes is known as *Calderbank-Shor-Steane codes*, which are used in this chapter [CAL 96, STE 96].

If a simple QEC encoding does not provide strong enough error correction, the code can be concatenated, with each encoded logical qubit serving the role of a component qubit in another level, using e.g. 9^x or 5^x qubits for x levels of encoding [KNI 96]. If the physical error rate is below a level known as the *threshold*, then each additional level of encoding gives an exponential suppression of residual logical error rates.

An error correcting code will have a threshold error rate in the physical components; if the physical error rate is below the threshold, then applying error correction will improve the fidelity of the logical qubit, but if the physical error rate is above the threshold, then applying error correction actually makes the logical state worse rather than better. If memory is imperfect and the operations for measuring syndromes and applying corrections are perfect, the threshold may be several percent. More typically, analysis is simplified by assuming that memory, gate and measurement operations all have the same error rate, resulting in a threshold of e.g. 10^{-3} or lower. For our purposes here, we will not delve into the detailed calculations, but recognizing that transmission through a channel will have a different error process than local gates on memory in a repeater is important.

Quantum codes are often described using the notation $[[n, k, d]]$, where n is the number of physical (or lower-level logical) qubits in a block, k is the number of logical qubits encoded in the state, and d is the Hamming distance of the code. The code is capable of correcting any set of physical errors $\leq \lfloor (d-1)/2 \rfloor$.

11.1.1. *Steane code*

The $[[7, 1, 3]]$ code developed by Andrew Steane was one of the first quantum error codes [STE 96]. With distance $d = 3$, it is capable of correcting any single-qubit error. Encoding one logical qubit in seven physical qubits requires six stabilizers,

$$\begin{aligned} S_1 &= X_1 X_2 X_3 X_4 \\ S_2 &= X_1 X_2 X_5 X_6 \\ S_3 &= X_1 X_3 X_5 X_7 \\ S_4 &= Z_1 Z_2 Z_3 Z_4 \\ S_5 &= Z_1 Z_2 Z_5 Z_6 \\ S_6 &= Z_1 Z_3 Z_5 Z_7 \end{aligned} \qquad [11.1]$$

where X_i represents the X basis of qubit i, $1 \leq i \leq 7$. Direct measurement of any of the qubits is not done; a circuit is used to extract the collective parity of the four qubits to determine the ± 1 eigenvalue of the four. Because two-qubit gates can propagate errors, to prevent contamination of the data qubits, a specially prepared four-qubit ancilla state is used in the measurement process. The circuit for syndrome extraction for stabilizer S_i is shown in Figure 11.1.

11.1.2. *Surface code*

In the two-dimensional surface code, a 2D square lattice of qubits is used. There are several variants of codes building on the core idea that a logical qubit can be robustly encoded in the relative parity of a chain of qubits extending from one boundary to another. The variant receiving the most attention at the moment allows the lattice to encode more than one logical qubit [RAU 07a], but here we will need only the form in which a large lattice holds only a single logical qubit [BRA 98, DEN 02, HOR 12, KIT 03].

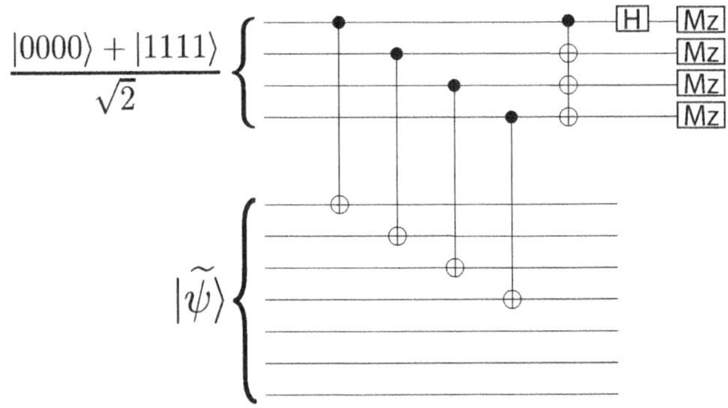

Figure 11.1. *Circuit for extracting the S_i syndrome for the Steane $[[7,1,3]]$ code*

A square lattice consisting of $d \times d$ lattice cells uses $\sim 4d^2$ physical qubits, with a qubit at each vertex, in the middle of each edge, and the middle of each square. About half of the qubits are used to hold the quantum state, and the other half are used as syndrome qubits, briefly entangled with the long-term quantum state and then measured to give us error information. The lattice and stabilizer circuits are shown in Figure 11.2.

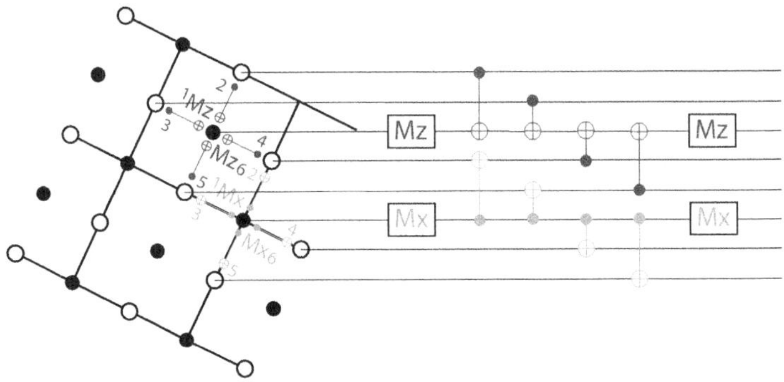

Figure 11.2. *The qubit lattice and syndrome extraction circuit for the 2D surface code. Open circles are the data lattice, filled circles are syndrome qubits. Mz and Mx are measurement in the Z and X basis, respectively*

The lattice is laid out so that when there are x physical qubits in the quantum state, there are $x - 1$ independent stabilizers. As each stabilizer constrains the state, we are left with one degree of freedom, which is used as the logical qubit.

Each syndrome cycle requires four CNOT gates and a measurement for each lattice square or vertex. Each syndrome qubit tells us whether the neighboring four data qubits are in the $+1$ or -1 eigenstate of a stabilizer. Depending on the details of the circuit, it will give us either the XXXX or $ZZZZ$ stabilizer value.

11.1.3. *An early communication proposal*

In an early paper discussing concatenated codes, Knill and Laflamme used the similarity of memory and transmission channels to analyze the possibility of teleporting encoded states from location to location [KNI 96]. Using the 5-qubit code as example, they found that the tolerable error probabilities for transmission were less than 10^{-2}, provided that the local gate errors used in the error detection and correction were less than 5×10^{-5}. Naturally, this approach uses 5^x transmitted quantum optical states for an x-level encoding. Unfortunately, the 1% channel error threshold *includes* loss of the photon or optical state carrier as well as errors in the state, and no optical system comes anywhere close to that level – or indeed *can* come close to that level when losses through any sort of long channel or optical coupler are considered, as we discussed in Chapter 8.

The use of QEC in repeaters stood at this level until the introduction of the CSS repeaters by Jiang, Taylor, Nemoto, Munro, Van Meter and Lukin (JTNMVL), in 2009 [JIA 09]. In this chapter, we first present this approach, followed by another QEC-based scheme by Fowler, Wang, Hill, Ladd, Van Meter and Hollenberg (FWHLVH) [FOW 10] that uses the *surface code* approach to quantum error correction [RAU 07a].

11.2. CSS repeaters

The JTNMVL scheme can be divided into three steps: first, logically encoded, high-fidelity Bell pairs spanning each hop are created; second, a Bell state measurement (BSM, section 2.5.2) at the *logical* level is done on the two logical qubits at each intermediate repeater; third, Pauli frame correction based on the results of the logical Bell measurements completes the operation. Note that this runs as a single, coordinated end-to-end operation, with no need for the multi-hop recursive nesting of the purify-and-swap repeaters in Chapter 10. This sequence is summarized in Figure 11.3.

The link-level high-fidelity logical Bell pairs are created by generating, at each repeater, (1) an encoded logical qubit in memory for each link the repeater will be using and (2) a number of medium-high fidelity, physical Bell pairs along each link. If Alice is at the left end of a link and Bob is at the right end, Alice creates the logical state $|\tilde{+}\rangle$, and Bob creates the logical state $|\tilde{0}\rangle$ in memory. We will then entangle Alice and Bob's logical qubits using a logical CNOT,

$$|\tilde{+}\rangle_A |\tilde{0}\rangle_B \xrightarrow{\text{CNOT}} |\tilde{+}\rangle_A |\tilde{+}\rangle_B = |\widetilde{\Psi^+}\rangle^{(AB)}. \qquad [11.2]$$

This is achieved using the teleported CNOT gate described in section 4.4, which requires a four-qubit ancilla state $|\chi\rangle$ (see equation [4.3]). This is the operation that consumes the base-level Bell pairs created by the physical link.

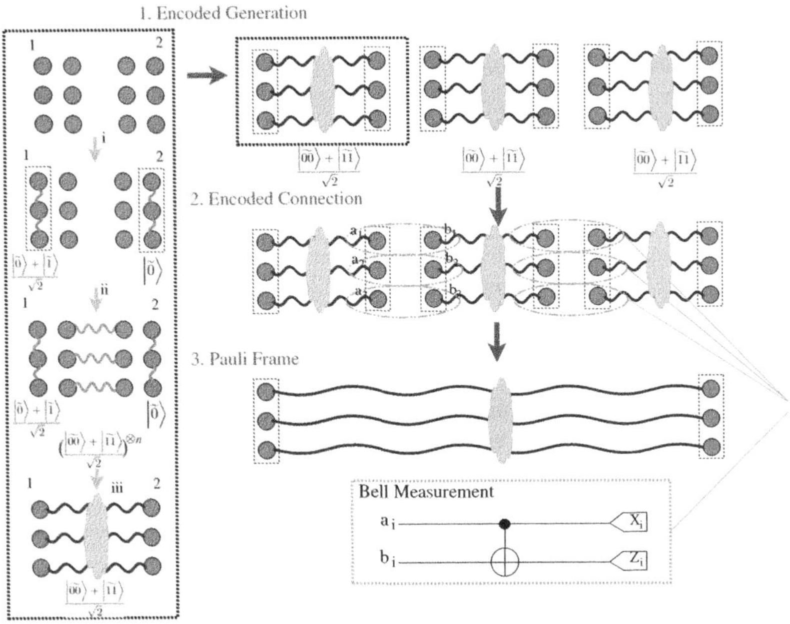

Figure 11.3. *The three major steps in CSS repeaters (adapted from [JIA 09])*

The second step, the Bell state measurement, requires a logical CNOT between the Bell pair on the left and the Bell pair on the right, followed by a Hadamard gate on the logical control qubit of the CNOT gate, then measurement of both logical qubits in the computational basis. It is identical to the Bell state measurement used in entanglement swapping for the purify and swap repeater schemes from Chapter 10, except that the operation is performed on logical qubits instead of physical.

For the [[7,1,3]] code and some other codes, the CNOT is transversal, meaning that seven physical CNOT gates connecting the qubits on the left to those on the right one-by-one executes a CNOT gate on the logical qubits. Similarly, the Hadamard gate and measurement can be performed by executing the same operations on the physical qubits. The operation is therefore straightforward.

Each of those logical Bell state measurements results in two classical bits that must ultimately be communicated to one or both end points of the repeater chain. The BSM splices the two shorter Bell pairs into one longer Bell pair, but *which* Bell pair has been fabricated is random. The BSM results tell us which Bell pair we have, and the results are used to correct the Pauli frame of the spliced long-distance Bell pair to give us the actual desired Bell pair, by applying either X or Z gates. However, those communications and the corrections can be deferred essentially indefinitely, limited only by the application use of the final, resulting end-to-end Bell pair. In some cases, especially QKD, the corrections can even be applied classically, long after the application qubits have been measured.

Because the CSS repeater scheme does not require nested purification, the end-to-end throughput is largely independent of the number of hops, or the total distance (modulo the usual networking concerns of starting up or shutting down a connection, dealing with contention for resources, etc.; we will cover some of these issues in Chapter 13). However, as even the encoded BSM is an imperfect operation building on imperfect Bell pairs, the fidelity does decline as the connection lengthens. The detailed derivation is too complex to discuss here, but the authors found that the end-to-end fidelity is

$$F \approx (1 - Q)^{2L} \qquad [11.3]$$

where L is the number of hops and Q is the probability of error in one encoded logical qubit,

$$Q \approx \binom{n}{t} q^t, \qquad [11.4]$$

where q is the residual infidelity in the physical Bell pairs (possibly after purification) and $t = (d + 1)/2$, the number of errors that can be corrected in the logical state. These approximations only hold for small q and small Q.

11.2.1. *Protocols*

The first major step, entangled logical Bell pair creation, consists of several tasks:

– logical qubit creation in either $|0\rangle$ or $|+\rangle$ (local to each node);

– creation of enough physical $|\chi\rangle$ states for the QEC block size, using the link (possibly in parallel with above);

– assignment of $|\chi\rangle$ state to a position in the code word (must be the same at each end, but a simple algorithm should allow independent, identical decisions);

– execution of the teleported CNOT gates, one per qubit in the encoded block;

– execution of local QEC on the resulting logical state.

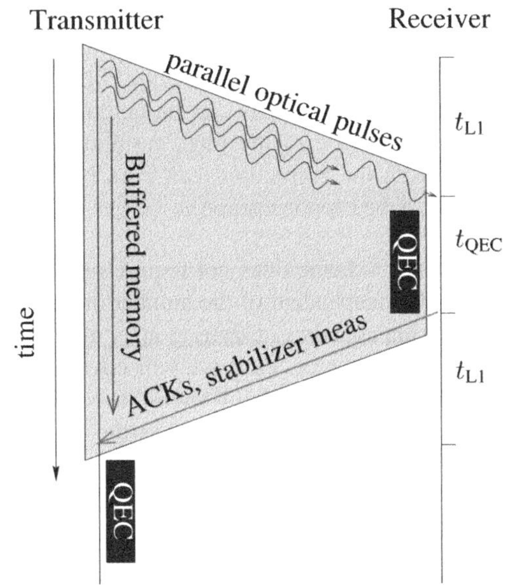

Figure 11.4. *The gray trapezoid represents round-trip processing on a single $M \rightarrow M$ link, including the buffering of memory at the sender, pending the entanglement success/failure acknowledgments*

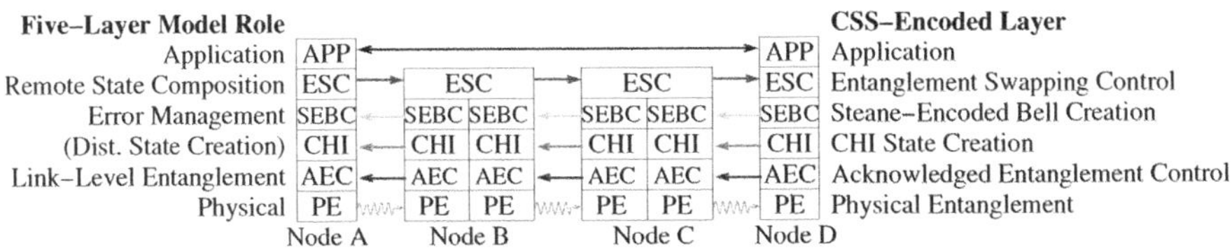

Figure 11.5. *The protocol layering in CSS repeaters*

Figure 11.5 shows the organization of the communication in CSS repeaters. The physical layer (PE) and AEC are the same as in purify-and-swap repeaters. We actually break the first step of CSS repeaters into two new protocols:

– CHI, which creates the distributed $|\chi\rangle$ states;

– SEBC, which creates Steane-encoded Bell pairs.

CHI is our first protocol layer explicitly dedicated to the production of states more complex than Bell pairs, and CHI does not align with the five protocol roles described in Chapter 1. It is very simple, consisting of a single teleportation and some local gates. This functionality could easily be incorporated with SEBC, which executes the actual teleported CNOT gates after assigning a $|\chi\rangle$ state to a specific position in the code word. However, use of $|\chi\rangle$ states appears in a variety of higher-level protocols, so it makes sense to build a reusable protocol here.

To connect the logical states in the second and third steps using the logical BSM and Pauli frame corrections, we can use the same entanglement swapping protocol described in section 10.2.2, with the primary difference being the logical rather than physical operation. Above that will reside some application protocol from Part 2 of the book.

11.2.2. *Operational timing*

The analysis in the papers originally proposing both the CSS and surface code schemes assumed synchronous operation across all links, as illustrated in Figure 11.6. Each gray trapezoid is round-trip processing on a single $M \to M$ link, with qubits awaiting acknowledgment before being used in a QEC state. The horizontal bars are Bell state measurement. Stars indicate the start and end points for complete teleportation of a qubit. The large circles indicate timing for a goal of QKD-like generation of shared, random classical numbers. Open circles mark when qubits may be measured and the memory resources freed; filled circles mark when Pauli frame corrections are complete and the resulting classical numbers can be used. The long diagonal arrow is the propagation of the BSM results used for Pauli frame corrections. The results can be applied as a cumulative correction, passing from node to node where the BSM has been done. Two Z corrections will cancel, as will two X corrections, meaning that the end node ultimately needs to receive only two classical bits, rather than two for each node in the path.

The vertical arrows indicate necessary memory hold time waiting for BSM or final Pauli frame correction. The arrows inside the gray trapezoid in Figure 11.4 are inevitable; we cannot know whether or not an entangling operation succeeds until one RTT later. The arrows outside the trapezoids in Figure 11.6 represent time waiting for another action to become possible. In this case, the arrows sum to one end-to-end round-trip time. We would like to alter this value if possible.

The assumption of synchronous operation is one of the most important differences with the later Munro *et al.* quasi-async design discussed in Chapter 12.

While synchronous operation is not strictly necessary, it simplifies the exposition and basic performance analysis. The exact timing of the BSM at each node can in fact be independent, as long as the BSM operations along the chain can be mapped to the correct end-to-end session. If each node supports more than one logical qubit, CSS repeaters are subject to the same concerns about leapfrog mistakes as purify-and-swap repeaters (Figure 10.6), potentially placing practical constraints on the level of asynchrony achievable.

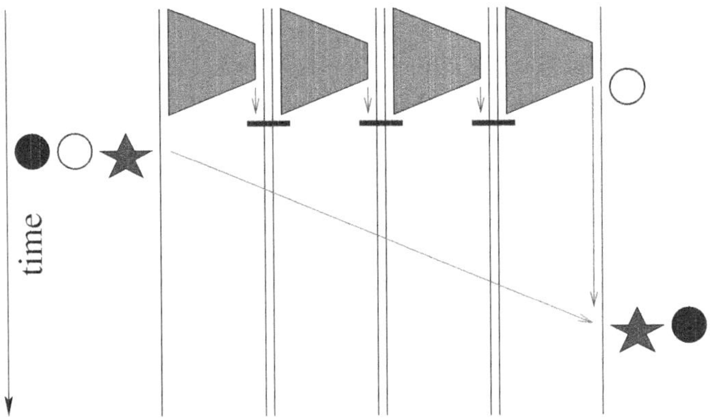

Figure 11.6. *Timing for forward teleportation with flat-timed QEC repeaters and acknowledged link-level entanglement over four hops*

Correct operation requires observing only a few constraints while managing the physical qubits:

– heralded entanglement;

– assignment to the correct end-to-end session;

– if more than one end-to-end (E2E) Bell pair is in process, assignment to the correct application-level Bell pair;

– assignment to the correct *position* within the error correcting code.

If those assignments are tracked correctly, operation is straightforward. End-to-end Bell pairs are delivered directly to upper-layer protocols running at the end points, rather than the further rounds of purification and the attendant difficulties in nested purification. This allows communication for the complete session to be more localized; nodes exchange messages only with their immediate neighbors in the chain, rather than maintaining a complex set of connections with various nodes along the chain.

11.2.3. *Resources and performance*

Consider the circuit in Figure 4.5 for the teleported CNOT used to make the logical Bell pairs in the first step. In addition to the data qubits and the physical Bell pairs, Bob requires three physical qubits and Alice requires two physical qubits per member of the logical state.

The CSS repeaters, if using a single layer of the $[[n,k,d]] = [[7,1,3]]$ code, use a minimum of 13 qubits per network interface: seven for the encoded state, four for an ancilla state used during error syndrome extraction to perform the correction, one for a transceiver qubit and one to purify that transceiver qubit. For the teleported CNOT, the receiver end of a link requires an additional 21 qubits and the sender requires an additional 14 qubits. A minimal configuration, then, is 61 qubits per node: 34 on the receiving end and 27 on the sending.

At this minumum level, performance will be very poor, because the transversal teleported CNOT gates to create the single-hop logically encoded Bell pairs will stall, waiting for the availability of moderate-fidelity Bell pairs. Much better balance will be achieved by increasing the number of transceiver/purification qubits, to provide enough moderately high-fidelity Bell pairs to allow the teleported CNOT gates for an entire encoded word to proceed at the same time.

To this point, we have neglected the physical and link layer. However, this communication session architecture depends on moderately high-fidelity ($F \geq\approx 0.95$) physical Bell pairs. If the physical layer does not provide Bell pairs of this fidelity, we must purify the Bell pairs before performing the QEC. Fortunately, the layered protocol architecture will easily accommodate this. We can slide a purification layer into the stack shown in Figure 11.5, between the AEC and CHI layers.

As we have seen, if the physical link generates a Werner (white noise) state, the first round of purification provides only a small gain in fidelity, but the second round gives a dramatic improvement. Therefore, let us assume here that two rounds of purification are required. We need 7 purified Bell pairs to create the $|\chi\rangle$ states and perform the CNOT gates. Assuming a 100% probability of purification success, this would require 28 physical Bell pairs before purification.

Taking into account the success probability of the first round of purification P_{p1} and the second round P_{p2}, we would like to generate

$$N_b = \frac{4n}{P_{p1}P_{p2}} \qquad [11.5]$$

base Bell pairs per link-level round-trip time, where $n = 7$ for the [[7,1,3]] code. If the base Bell pairs are Werner states with $F = 0.8$ and we perform two purification rounds, the output fidelity will be about 0.96, just barely good enough to run the error correction. $P_{p1}P_{p2}$ will be about 0.56, giving us $N_b = 50$.

If the link-layer entanglement success rate is e.g. 30% (a very high success rate, as envisioned for the qubus technology and its descendants), we would like to have $50/0.3 \approx 167$ transceiver qubits at the sender end of the link in order create enough Bell pairs in the initial round; approximately 50 will be enough at the receiver end of the link. Some qubits are buffered while waiting for purification confirmation; typically approximately 25 will be waiting on first-round confirmation, and approximately 10 will be on second-round confirmation.

In total, a well-balanced repeater with one sending link and one receiving link will want approximately 200 physical qubits on its sending interface and approximately 85 on its receiving interface, or approximately 285 physical qubits in total. This number is highly dependent on the probability of successful entanglement at the physical layer.

This allows a chain of CSS repeaters to generate one high-fidelity, end-to-end Bell pair per link-level round-trip time. If each link is 10 km of optical fiber, RTT is 100 μs, giving a repeat rate of 10 kHz, or 10,000 end-to-end Bell pairs per second.

These values are of course dependent on the underlying physical layer; Liang *et al.* analyzed one particular combination with fewer qubits, a lower entanglement success probability, and a third round of purification, and estimated the logical Bell pair generation time at about 7 ms using 10 km links, allowing a throughput of over 100 end-to-end Bell pairs per second. Of course, as photon transmission time and local gate time are neglible compared to the speed-of-light latency, to a first approximation performance would scale linearly with an increase in available qubits.

11.3. Surface code repeaters

The FWHLVH approach to combining quantum error correction and quantum communication was proposed by Fowler, *et al.* in 2010 [FOW 10], building on the two-dimensional surface code [RAU 07a]. An overview of the scheme is shown in Figures 11.7 and 11.8. In this scheme, we can either move one logical qubit from the left to the right, effectively teleporting it end to end, or we can create a logical Bell pair situated at the two end points, using a technique called *lattice surgery* [HOR 12].

In the surface code, a 2D lattice holds the state of a single logical qubit. In this repeater scheme, each node holds one such lattice. The lattices of a chain of repeaters are briefly coupled, creating one very long, narrow lattice – a single logical qubit whose state spans all of the repeaters. In the surface code, slicing one lattice into two pieces by removing the entangling couplings turns one logical qubit into an entangled pair of qubits of the same state. Thus, we can create a $|\Phi^+\rangle$ Bell pair by creating one large $|+\rangle$ state and splitting it in two. We can also use this approach to propagate a logical qubit already encoded in the surface code from one end of our chain to the other, without the use of an intermediary logical Bell pair.

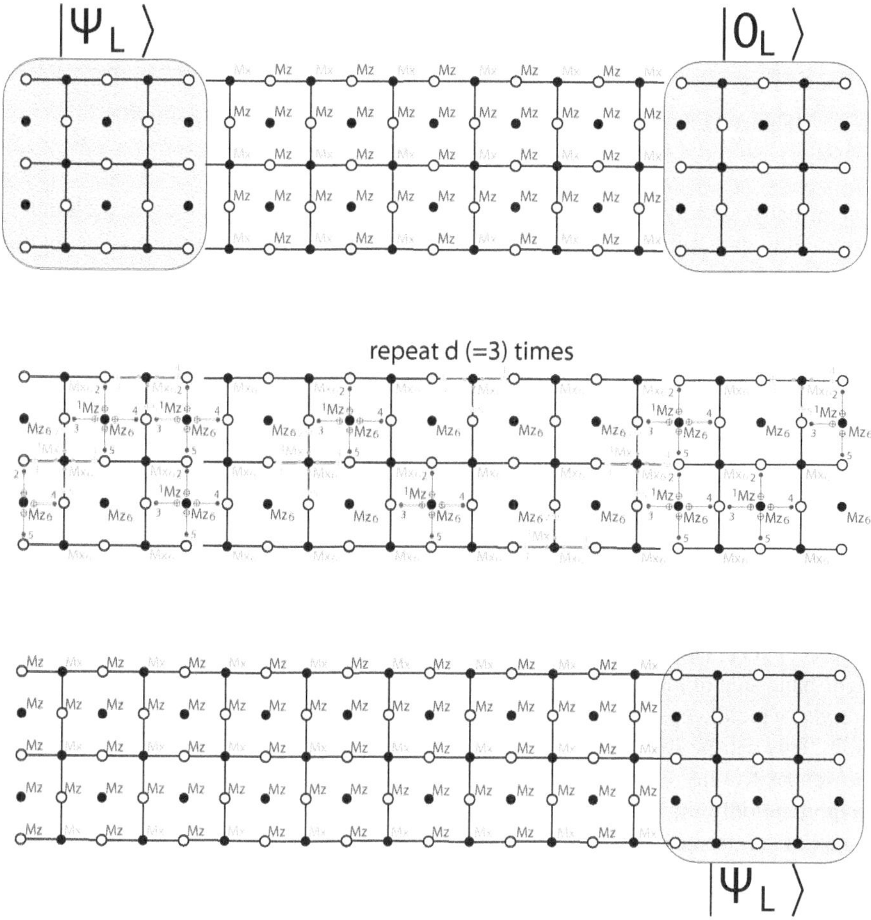

Figure 11.7. *Extending, then contracting, a surface code lattice to move a logical qubit left to right, within a single machine (adapted from [FOW 10])*

In the surface code, half of our physical qubits are part of the logical qubit state, and half are stabilizer, or error syndrome, qubits. For the repeater scheme, a lattice is cut along a lateral, zigzag line of syndrome qubits. Each syndrome qubit we remove is replaced with a Bell pair. This splits the lattice cells, so that one data qubit is colocated with one end of the Bell pair, and the other three data qubits are colocated with the opposite end of the Bell pair.

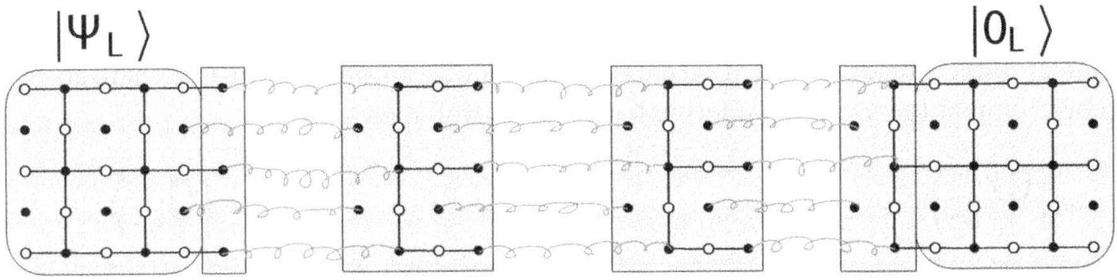

Figure 11.8. *The surface code repeater scheme (adapted from [FOW 10])*

Normally, each syndrome qubit is used in four two-qubit gates, one for each of the surrounding data qubits. In the repeater scheme, those gates are performed using the Bell pair qubits, either as the control or target of the four gates.

Both qubits in the Bell pair are then measured. Final computation of the usual lattice syndrome (e.g. XXXX or $ZZZZ$ stabilizer eigenvalue) requires sharing the measurement results across the physical link. This classical communication can be unidirectional or bidirectional. The lateral number of lattices cells d is our code distance, determining the strength of error correction. This distance needs to be extended in the temporal as well as physical direction, so we repeat the cycle d times, consuming $2d^2$ Bell pairs to make one good connection.

After this is done, we have a single large logical qubit, as in Figure 11.7. The final step is to contract the qubit spanning all of the nodes to a much smaller one held in the receiver. This is done by measuring *all* of the qubits in every node except the receiver (right-hand side of the figure). These classical measurement results are then sent to the receiver, which keeps them and uses them when it wants to measure the final logical qubit. This can be thousands of classical bits for each logical qubit teleported from one sender to one receiver; an alternative to keeping this much classical data is to accumulate only the parity along each horizontal strip in the long lattice, and apply that as a Pauli frame correction to the edge of the remaining lattice at the receiver. This results in a small penalty on the final error rate.

11.3.1. *Protocols*

The protocol stack for the surface code is essentially identical to that of the CSS code repeater, except that CHI and SEBC are replaced with a single protocol layer we can call SURF, for surface code, as in Figure 11.9. We can break down the tasks as follows:

– logical qubit creation in $|0\rangle$ (all of the nodes but one), or $|+\rangle$ (when creating an end-to-end logical Bell pair, done at any one node) or $|\psi\rangle$ (at the "sender", when moving a single logical qubit) (these operations are local to each node);

– creation of $2d^2$ physical Bell pairs on each link, enough to span the lateral distance and repeat d times (link operation);

– assignment of each Bell pair to a lateral and temporal position in the code (identical decisions at each end are trivial, provided they agree on the number and order of Bell pairs created);

– execution of the stabilizer gates and measurement of the Bell pair qubits, repeated d times;

– communication of the measurement results, calculation of any needed corrections.

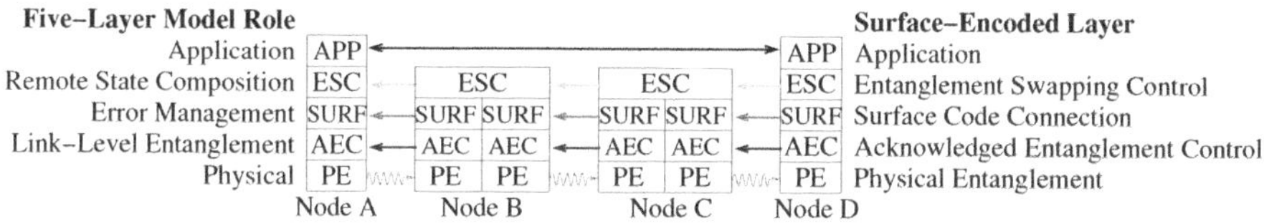

Figure 11.9. *The surface code repeater protocol stack*

Once this is done, the final task is:

– measurement of all of the data qubits in either (1) every node but the left and right (when creating a Bell pair), or (2) every node but the right one (when moving a single qubit), and transmission of the results.

This can all be done with simple unidirectional communication, as in the figure. Strictly speaking, the node-to-node coupling performed here is not entanglement swapping, as we are not performing Bell measurements on the logical qubits and are not really teleporting a qubit from place to place. Rather, a single large quantum state is being formed and manipulated. Thus, the ESC protocol cannot be used as-is; the new SURF protocol fills the role of remote state composition (p. 14) in a new way.

11.3.2. *Operational timing*

As noted above, the scheme for surface code repeaters was originally developed assuming synchronous operation across the chain of repeaters, but this is not strictly necessary. We will see relaxation of this constraint in Chapter 12.

In the basic analysis, it is assumed that Bell pairs are created with high probability of success. If the probability of loss $p_{\text{loss}} < 0.35$, we can use the Bell pairs directly; with an $M \rightarrow M$ link, the sending edge uses its member of the Bell pair immediately to conduct the CNOT gate and its share of the lattice stabilizer measurements. This substantially reduces the demand for long-lifetime memory. If the probability of loss (or unsuccessful entanglement) is greater than 35%, then we must resort to link-level, acknowledged creation of Bell pairs before running the surface code. In this case, the timing pattern will be essentially identical to that in Figure 11.6.

Because this procedure consumes *all* of the resources at every node, coordination of access to all of the nodes is critical; the simplest approach is to pre-establish our right to use a path through the network. In Chapter 13, we will see different schemes for multiplexing of individual links for purify-and-swap repeaters, but for error correction-based repeaters, particularly the surface code, it seems that allocation of a circuit through the network is preferred.

It is possible to modify the timing such that each node performs its share of the work somewhat asynchronously. We will see schemes for this in the next chapter, in the context of CSS-encoded repeaters, but the concepts there can relatively easily be applied to the surface code repeaters as well.

11.3.3. *Resources and performance*

The lattice at each repeater is $3(2d-1)$ qubits in size for code distance d. The lateral dimension d becomes substantial for high-fidelity connections, up to about three hundred, requiring up to about 2,000 physical qubits.

With d transceiver qubits and a high probability of entanglement success, the cycle time can be very rapid; four CNOT gates and a measurement operation run one cycle of the error correction. If T_g is the CNOT time and T_m is the measurement time, d cycles requires only $(4T_g + T_m)d$ seconds to perform. If physical gates require a few nanoseconds, even with $d \sim 300$, end-to-end communication can approach MHz rates.

With direct operation and 35% loss of photonic qubits, raising d by 30 will reduce the residual error rate by a factor of 10. Working with higher loss requires using pre-heralded Bell pairs, but a code distance of $d = 30$ will give a very low residual error rate, at the expense of far more resources.

Likewise, the required fidelity is fairly high, at 96%. If the physical Bell pairs are of lower fidelity, single-hop purification is required before running the surface code. Naturally, this requires longer lifetime memory, and impacts throughput of the system.

As with the CSS codes, if we begin with Werner states of $F = 0.8$ and purify twice before attaching to the surface code, we want

$$N_b = \frac{2d-1}{P_{p1}P_{p2}} \qquad [11.6]$$

base Bell pairs for each of the d rounds of stabilizer measurements needed, requiring N_b receiver qubits and N_b/P_b transmitter qubits, if P_b is the probability of successful entanglement of base-level Bell pairs. If $d = 30$, $P_b = 0.3$ and $P_{p1}P_{p2} = 0.56$, then $N_b \approx 105$ receiver qubits and $N_b/P_b \approx 350$ transmitter qubits would support the 30 qubits in the middle of each node's lattice. An additional 100 or so are required on each interface to buffer states during purification. The total for a two-interface repeater for these parameters would be nearly 700 physical qubits. These resources are linear in d.

As described here, the final fidelity of the logical qubits is far higher than that achieved using the normal purify-and-swap approach. However, it bears noting that these operations are already encoded in the topmost level of error correction in the system, whereas a common fate for the physical Bell pairs of purify-and-swap will be a further layer of error correction handled directly by the end nodes.

Using acknowledged Bell pairs, the performance will decline dramatically. $d \sim 300$ will require some 1.8×10^5 Bell pairs on each link for *each* end-to-end transmission, a time that could take many seconds. However, this would give an infidelity of perhaps 10^{-6}, a level at which not just QKD but long-running distributed computations can be performed.

11.4. Conclusion

The two quantum error correction-based schemes presented here represent a significant advance in our capabilities, potentially altering the demands made of the underlying physical layer, especially memory time, and may simplify the network architecture by reducing the number of nodes that each repeater must communicate with.

While these approaches may eliminate some round-trip communication delays, they demand more physical resources and higher base-pair fidelity, and they are not robust against high photon loss. Therefore, a typical operation mode likely involves building on top of link-level acknowledged purification.

A significant advantage of both of these schemes is that they may interface well with large quantum computers by directly using logical qubits encoded in the system, avoiding changes to the encoding applied to a logical qubit, and especially excessive use of "bare" physical qubits, reducing the residual logical error rate.

Because of their rather complete occupation of a given node, these approaches will present different challenges in resource management (e.g. link multiplexing and buffer memory allocation) than purify-and-swap repeaters. Extending to multi-NIC repeaters, as is necessary for complex networks, will likely require pre-establishment of the end-to-end path. Although the principles used to make engineering design decisions are the same across all of these communication session architectures (including the ones presented in the next chapter), we will not explore all of the network functionality in detail for each scheme; extending the material in Chapters 13 and 14 is left as a research topic for the reader.

Chapter 12

Finessing the Key Limitations

For any quantum repeater scheme, we face several key constraints that drive our engineering decisions. Local gate infidelities and imperfect entangling operations prevent us from propagating states indefinitely without error. Probabilistic operations limit our ability to reliably predict when entanglement will be ready for use, due to physical layer signal loss in all schemes, and because of probabilistic operations such as purification in some schemes. Thus, coordinated distributed decision-making becomes hard. This would be only a minor annoying source of inefficiency, if not for the limited memory lifetimes of most qubit technologies and the limited memory capacities we can currently build.

In section 10.2.2, we asked the question, *when can we guarantee* that a node knows enough to make a decision about the next action that will be *consistent* with decisions made by other nodes? The original Dür-Briegel-Cirac-Zoller purify-and-swap approach presented in Chapter 10 requires round-trip classical communication, not just between the end points but between a complex overlapping set of nodes, as shown in Figure 10.5. By using QEC for the full end-to-end encoding, as we have just seen in Chapter 11, we can simplify the communication to round-trip on the link and unidirectional multi-hop flow of information toward the destination. Waits are reduced, memory lifetime becomes less of an issue and local gate errors are protected against. However, the resource demands are very high for maintaining the full end-to-end encoded state, and the issue of low entanglement success rate P_b is only partially resolved. Operation was presumed to be synchronous across the chain of repeaters, at least for the purposes of making the analysis fairly tractable, although we hinted repeatedly that this constraint can be relaxed.

In this chapter, we examine two approaches that attempt to finesse some of these constraints, then provide a comparison of the schemes we have presented in the last two chapters. We will evaluate the timing and resulting consumption of three types of memory: transmitters, receivers and buffer.

Building on the QEC repeater schemes seen in Chapter 11, Bill Munro and the group of Kae Nemoto have created a quasi-asynchronous architecture that cleverly combines error correction with techniques for dealing with high loss in the quantum channel [MUN 10]. This scheme allows nearly one-way propagation of the classical control signals and explicitly addresses timing to smooth propagation of the information (both classical and quantum), maximize pipelining and deal with links of different latency.

We will discuss the quasi-asynchronous scheme and an extension they refer to as "memoryless", allowing repeaters to operate more effectively without long memory lifetimes [MUN 12]. As discussed in section 8.1.1, several of the types of quantum optical states can actually represent more than one qubit using a single photon. This allows those bits to "share fate"; if the photon is lost, all of the qubits are lost. Conversely, if the photon is received, all of the encoded qubits are received. Munro *et al.* realized that this feature can be used with a clever encoding to improve our transmission capabilities and circumvent a shortcoming of the original quasi-asynchronous approach. The memoryless scheme is the first communication session architecture that approximates the true send-and-forget, hop-by-hop nature of packet-switched networks such as the Internet.

12.1. Quasi-asynchronous

The new, key insight in this scheme is that we can essentially predict successful entanglement of one of the qubits at the receiving end, and use that qubit in enabling a more direct hop-by-hop communication session architecture. A large number of transmitters can all target a single receiver, sending photon pulses in a train. The receiver gates in the first photon pulse, testing for successful entanglement. If it fails, the second pulse is gated in. After a pulse succeeds, the rest of the incoming pulses are discarded, and the successful pulse identifier is acknowledged to the sender. With a large enough train of pulses, the probability of the receiver's single qubit becoming entangled with *some* qubit at the sender can be made arbitrarily high. If the entanglement success is P_b, the probability of failure is $(1 - P_b)^x$ with an x-pulse train. Munro *et al.* named this the *fusilier* scheme.

The number of pulses that would be successfully received with a train of x pulses is $x \cdot P_b$, so if $x > 1/P_b$, successful pulses often will be discarded, which would be wasteful. This is easily resolved by adding more receivers. In fact, earlier simulations used just such a scheme, but did not properly balance transmitters and receivers, instead using the same number of each [LAD 06, VAN 09, MUN 08]. Thus, at this level, the quasi-asynchronous approach can be viewed as a variant of AEC.

Operation consists of three steps like those described in section 11.2: application of error correction-related operations to raw Bell pairs, entanglement swapping and Pauli frame correction. The final Pauli frame reconciliation, which can be done classically and can be deferred essentially indefinitely, is less important operationally than determining which Bell pairs/qubits to use for the quantum operations.

The biggest difference between quasi-asynchronous and the CSS repeaters is operational timing. Transmitters are triggered by a classical pulse flowing left to right, as in Figure 12.1. The state to be sent (either half of a Bell pair, or a single, valuable data qubit) moves left to right, in the same direction as the light propagates in each of the links, which we call *forward* propagation. As in Figures 11.4 and 11.6, each gray trapezoid is round-trip processing on a single $M \rightarrow M$ link. All of the optical pulses are transmitted at the same time, at the upper left-hand corner, each entangled with a memory qubit at the sender. Those memories must be held safely until we determine if the entangling pulse succeeds at the right-hand edge of the link. That memory buffering is the vertical edge on the left of the trapezoid (see the arrow in Figure 11.4, not shown here; see page 227 for a more complete description of the graphical notation).

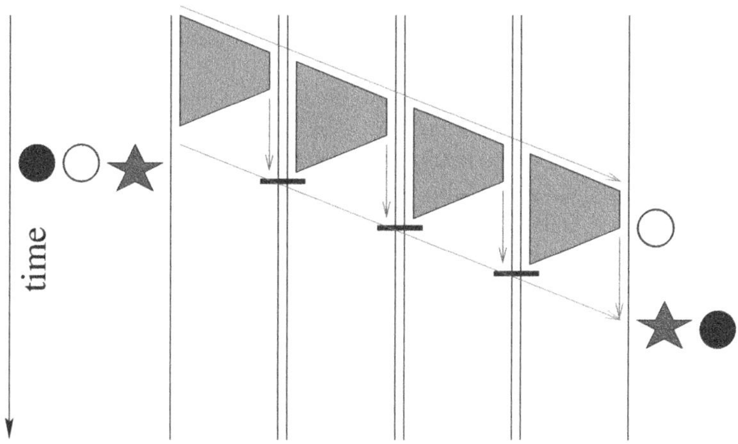

Figure 12.1. *Timing for forward propagation in quasi-asynchronous repeaters. Vertical arrows indicate necessary memory hold time waiting for BSM or final Pauli frame correction, and here sum to one end-to-end round-trip time*

Quasi-asynchronous operation requires relatively high-fidelity base-level Bell pairs, $F >\sim 0.95$, including link-level RTT memory decoherence (i.e. measured along the bottom edge of the trapezoid). High-fidelity local gates with error rates on the order of 0.1%, including for entanglement swapping, are required. Even at this level, error correction can be difficult to run effectively, so higher-quality components are of course desirable.

The advantage of the scheme is that, given enough transmitter memories, we can work with arbitrarily low success probability, while retaining the strengths of QEC and avoiding the multiple RTT delays of long-distance purification.

The system can be configured to operate in one of two modes: as a replacement for purification of physical Bell pairs using only one-way communication, or in full QEC mode. In full QEC mode, the system is very similar to the JTNMVL scheme, except for more explicit attention to timing and a few optimizations.

12.1.1. *Purification replacement operation*

The purification replacement approach uses one-way purification (1-EPP). In section 9.5, we introduced 1-EPP as a variant of standard purification. Here, we use a more complex, and powerful, error correction code, applying the equivalent of the QEC *decoding* circuit at both ends of the link. To encode a qubit using QEC, we begin with a single qubit prepared in an arbitrary state, $|\psi\rangle = \alpha|0\rangle + \beta|1\rangle$. This qubit is then encoded into the larger logical state $\left|\widetilde{\psi}\right\rangle$ using a series of gates. The decoding step is the reverse, taking the logically encoded state and executing a related circuit, measuring the stabilizers for the QEC code and applying corrections, leaving a single physical qubit behind.

For a quasi-asynchronous repeater operation, this decoding step is done at both ends of the link, resulting in the best possible physical Bell pair. Before applying the correction operations, we must reconcile the stabilizer measurement results, because they depend on both ends of the Bell pair. This can be done at one end, without requiring two-way classical communication; it is a one-way entanglement purification protocol. A simple 1-EPP for correcting bit flip errors is shown in Figure 12.2.

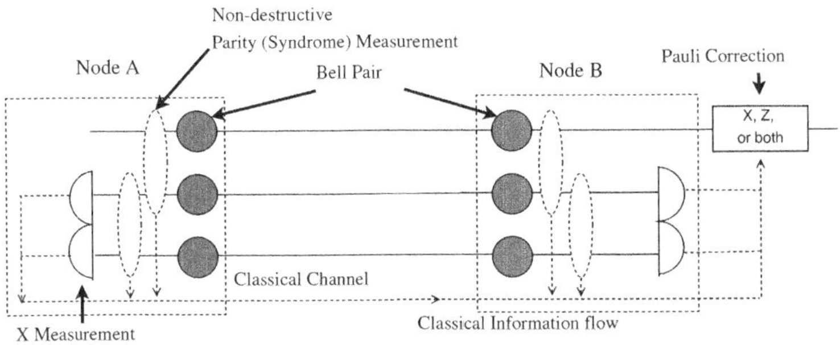

Figure 12.2. *The simplest unidirectional, QEC-based purification protocol for bit flip errors (adapted from [MUN 10])*

Note that the timing of the operations is not specified in this figure; the only visible timing constraint is the left-to-right classical information arrow. A second, implicit constraint is the creation of the Bell pairs and their assignment by the two nodes to the correct role within the QEC code. The transmitter, of course, cannot do its share of the work until it knows which physical Bell pairs have been successfully created, which occurs after arrival of the acknowledgments (ACKs) from the receiver, as in Figure 8.7.

The receiver can assign Bell pairs to roles within the QEC state, then perform the stabilizer measurements, moments after receiving enough Bell pairs. In quasi-asynchronous operation, the receiver can actually perform the stabilizer measurements first, before the sender. However, the corrections must await the arrival of the stabilizer measurement results from the transmitter so that the final error correction is complete after three one-way latency times on the link. The QEC-related messaging flows only left to right, following the diagonal arrow along the bottom edge of the trapezoids in Figure 12.1.

After this decoding step, we are left with a purified physical Bell pair on each link. Entanglement swapping is then performed, splicing multiple links into a longer Bell pair. This follows the diagonal arrow, immediately on the heels of the final QEC operations.

The operation is finally complete when the diagonal arrow reaches the destination. The latency is the end-to-end one-way transmission time, plus small amounts for local operation times, plus one round-trip link time. For QKD-like operations, the received qubit is measured upon reception, marked with the open circle. Final reconciliation using the BSM results must await the arrival of the arrow. For complete teleportation of a qubit, marked with a filled circle, the operation likewise must await the arrow.

Note that in this scheme, purification is done only over a single hop, leaving the buffer memory vulnerable to decoherence for some time, and to gate errors in the entanglement swapping. However, eliminating the need for multi-hop purification reduces buffer memory demands in both capacity and memory lifetime, giving the approach good balance in light of current technological capabilities. It is a strong candidate for deployment in early metropolitan repeater networks, though is unlikely to be used for longer distances.

12.1.2. *QEC-based operation*

To make this truly scalable over large numbers of hops, we cannot use the purification replacement method. Instead of actually decoding the received state down to a single physical qubit, we keep states in the QEC encoding. The entanglement swapping is done using the QEC-encoded logical qubit, as in

JTNMVL. Any time that the logical qubit spends waiting, QEC is applied, allowing us to protect against memory errors. This scheme requires the use of additional memory at each repeater, to hold a logical $|\tilde{0}\rangle$ state used to measure stabilizers while minimizing damage to the important data state.

Vertical arrows indicate necessary memory hold time waiting for BSM or final Pauli frame correction. For the purification replacement, this is the total lifetime that bare, unprotected qubits must retain high fidelity. For the full QEC mode, this is the time that QEC must be applied to the logical state the receiver is holding. Here, the arrows sum to one end-to-end round-trip time. This is identical to the total buffer memory hold time for the JTNMVL scheme in Figure 11.6.

The buffer time is important, but the primary factor in our resource consumption is the busy set of transmitter qubits on the left edge of each trapezoid. All of the resources along the path in the parallelogram defined by the upper and lower diagonal arrows are dedicated to the production of a single end-to-end Bell pair. This turns out to be identical to the JTNMVL repeater scheme: all of the transmitters are busy for exactly one link round-trip time. The difference is only when that period starts and ends for the different nodes.

12.1.3. *Timing variants*

In addition to the forward propagation timing for uniform links in Figure 11.6, Munro *et al.* proposed two other timing regimes. We will examine their "long pole" and "ridge fold", or "butterfly", approaches, then examine two that are original to this book, reverse propagation and "valley fold" timing.

Long pole is a chain of repeaters where one or more links is of higher latency than the others, helping to extend the applicability of repeaters to real-world situations. For quasi-asynchronous operation, all of the links in the path are synced to the slowest link, as shown in Figure 12.3. This timing allows the diagonal arrow, representing QEC information, to flow uninterrupted end to end. However, it leaves many of the resources in the middle of the chain idle for part of the cycle.

The "ridge fold" timing is shown in Figure 12.4. This configuration is targeted specifically at creating a Bell pair in the middle of the network, then propagating it toward the two end points of the communication session. The timing of the entanglement swapping is still constrained to the locations of the horizontal boxes. The memory hold time marked with vertical arrows represents the case where the left edge wishes to use the end-to-end Bell pair to teleport a qubit to the right edge. The total is the sum of the link RTTs of all links except the middle two, plus the one-way E2E latency; or, approximately 1.5 E2E RTT for a large number of hops. Thus, the total resource consumption, in terms of memory buffer time, is larger than for the flat-timed or forward-propagation schemes.

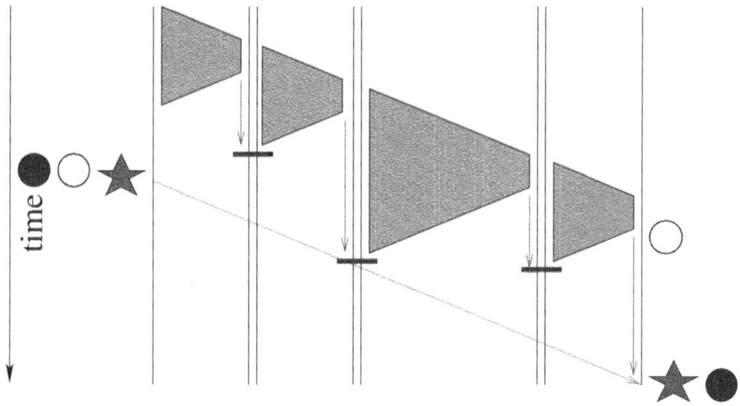

Figure 12.3. *Timing for forward propagation in quasi-asynchronous repeaters with uneven link latencies, clocked to the "long pole". The memory hold time marked with vertical arrows is two times RTT of the long pole link plus the RTT of every other link* except *the first*

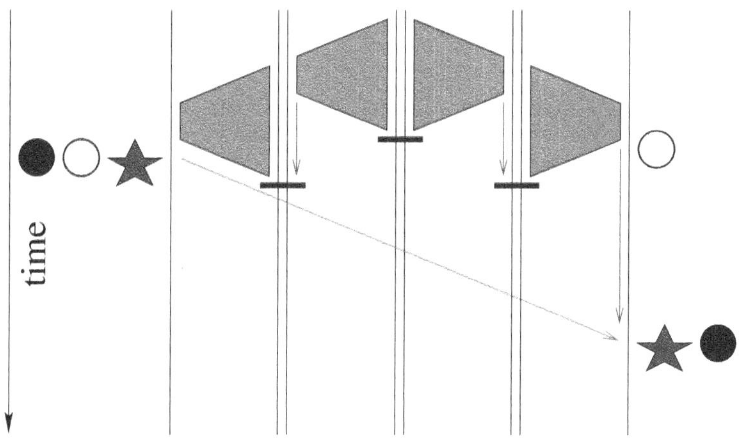

Figure 12.4. *Timing for "ridge fold" (butterfly) propagation in quasi-asynchronous repeaters, with operation initiated from the midpoint of the path*

Figure 12.5 shows the reverse timing, with links transmitting right to left but 1-EPP or QEC operating left to right. The memory hold time here is exactly one E2E RTT, all at the destination. Forcing the destination to buffer the state, rather than consuming buffer memory at nodes in the middle of the path, will have limited impact on an individual connection but may improve performance of a large network by freeing resources in the middle of the network for other users. Note the location of the open circle at the destination; some classical operations using the data or Clifford group operations on the qubit could be performed *before* execution at the transmitting end, subject to Pauli frame corrections one full E2E RTT later.

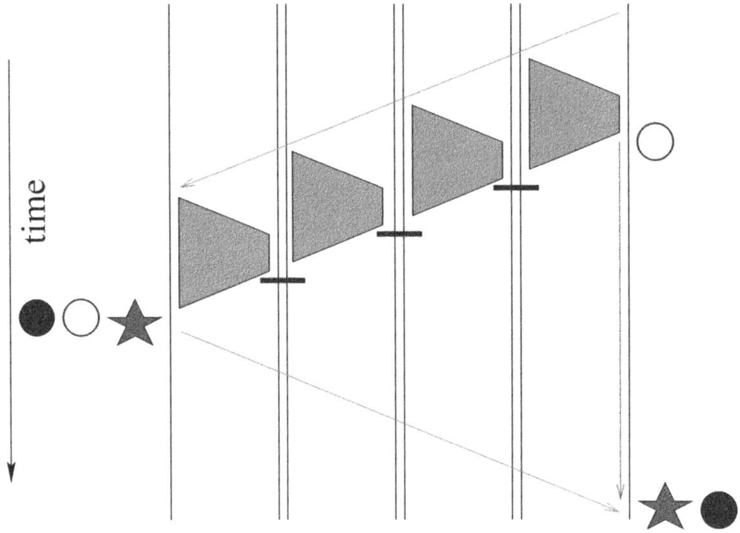

Figure 12.5. *Timing for reverse propagation in quasi-asynchronous repeaters, with operation initiated by a classical reverse trigger pulse upper diagonal arrow. The memory hold time is one E2E RTT, all at the destination*

Our final, and preferred, configuration is the "valley fold" timing of Figure 12.6, the inverted form of ridge operation. Pulses initiating a connection are fired from both ends toward the middle at the same time. Memory hold time is exactly the one-way E2E latency, provably optimal and almost one full E2E RTT less than the ridge fold. Here, as in reverse propagation, all of the buffering is done at the destination.

We have discussed the orientation of links as if we can choose which way light propagates in each one. In practice, we will often have to deal with cases where a path consists of some links with one orientation and some with the other. The analysis techniques used in this chapter should help us identify a timing regime that suits any such path.

12.2. Memoryless

The need for the elaborate structures we have described so far, whether purify-and-swap or QEC-based, has been driven by two major factors: the low fidelity of entangled base Bell pairs, and the low probability of successful capture of a photonic pulse. The various approaches have given hope that the fidelity issues can be resolved, but have more or less punted on the photon reception problem, working with link-level round-trip acknowledgment, as in the gray trapezoids in the figures in Chapters 10, 11 and 12. Munro *et al.* have proposed a new link-level encoding of information in photons that may circumvent this problem.

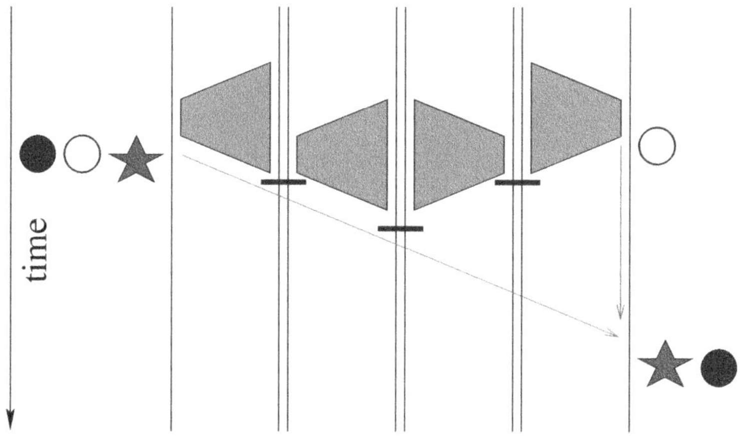

Figure 12.6. *Timing for "valley fold" (inverted butterfly) propagation in quasi-asynchronous repeaters. Memory hold time is exactly the one-way E2E latency*

In the "memoryless" scheme, a single photonic qubit is copied m times onto separate photonic pulses, creating an m-qubit GHZ state,

$$|\psi\rangle_{\text{GHZ}} = \frac{|0\rangle^{\otimes m} + |1\rangle^{\otimes m}}{\sqrt{2}} = \frac{|000...\rangle + |111...\rangle}{\sqrt{2}}. \quad [12.1]$$

Groups like this are created n times, so that we have n groups of m qubits each. A CNOT gate is performed once between each GHZ group and a local memory qubit, before sending the GHZ pulses. This circuit is shown in Figure 12.7 for $m = 3, n = 4$.

All mn pulses are sent down the channel toward the destination, but of course many of them are lost. In general, loss of an entangled qubit would leave us with some mixed state, but with this coding, we can compensate for the missing qubits if an appropriate subset of the pulses are received correctly. Our goal is to receive *all* of the pulses in one GHZ group, and *at least one* pulse from *each* of the remaining groups. In effect, we retain the GHZ group for which all pulses are received as our true qubit variable. Each incomplete group is in a mixed state, but we can undo the effect of each of the incomplete groups on the total state if we have one or more of its components.

We measure the qubit(s) we do receive, and if necessary correct the Pauli frame of the remaining quantum state.

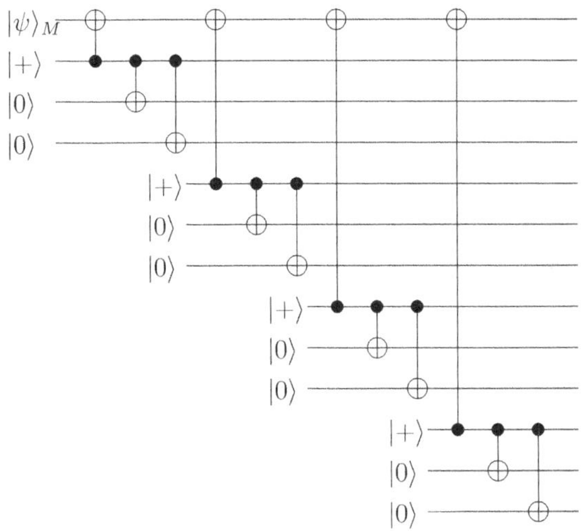

Figure 12.7. *Photon and memory circuit to create the states to be sent for the "memoryless" link, assuming a $M \rightarrow M$ link*

The probability of receiving all of the photonic states in a group is, of course, P_b^m, and the probability of receiving zero of them is $(1 - P_b)^m$, where P_b is the probability of successful entanglement. If $P_b > 0.5$, we can manipulate m and n to give our desired probability of success. Munro *et al.* calculate that $P_b = 0.67$ requires $m = 7$ and $n = 111$ to achieve a failure rate of less than 10^{-3}. With a 1 GHz pulse rate, the 777 pulses at this setting would take 777 ns, giving a qubit transmission rate of 1.29 MHz.

So far, this is just a clever link-level scheme, reducing our need for buffering memory while awaiting acknowledgment of success or failure. If $|\psi\rangle_M$ in the figure is initialized to $|0\rangle$, in fact, it will create link-level Bell pairs for use as in previous approaches. However, with the nearly guaranteed reception of a state, we can replace the $|0\rangle$ with an arbitrary valuable qubit variable, $|\psi\rangle_M = \alpha|0\rangle + \beta|1\rangle$. After the circuit as shown, apply a Hadamard gate to the memory qubit and measure it, then transmit the resulting classical bit, and we now have the ability to transfer a qubit in a hop-by-hop fashion down a line of repeaters without requiring long-lifetime memory. The ridge fold approach of Figure 12.4 becomes a true direct, bidirectional distribution of a Bell pair with the components arriving at the end nodes more or less simultaneously, without the need for the gray trapezoid link-level acknowledgments in the figure.

In section 10.2.1, we discarded being as an Internet-like hop-by-hop forwarding session architecture impractical. The memoryless scheme trades a scarce resource, long-lived quantum memory, for a plentiful resource, photonic pulses, and brings the hop-by-hop approach closer to feasible. However, the twin hurdles of a moderately high P_b and low gate error rate in the encoding and entanglement swapping circuits must be cleared.

Values of $P_b < 0.5$ can be compensated for by using a complex encoding in which one photonic pulse carries more than one physical qubit. In section 8.1.1, we saw that a photon has a number of characteristics that can encode data, including polarization, time bin, physical path and orbital angular momentum. Using these characteristics carefully can alter the relative probability of total failure events versus success events. Encoding 31 separate qubits in one photonic pulse (requiring that the detector be able to distinguish some 2 billion separate states upon reception) and using one pulse in place of the GHZ group would allow $n = 245$ pulses to successfully carry our qubit through the link with 10^{-3} error probability when P_b is as low as 0.2. Dealing with gate errors will still require us to apply an error correction code.

12.3. Summary: comparing quantum communication approaches

A near-ideal technology would have long quantum memory lifetime, high-fidelity local operations, a high probability of entanglement success and high-fidelity coupling. Such a technology would allow us to teleport states hop by hop with no concerns. Purify-and-swap was developed because such a physical quantum link does not exist. However, purify-and-swap has drawbacks: the need for round-trip, end-to-end communication limits throughput, demands long memory lifetimes and results in a complex classical communication suite. The quest to better match available technological capabilities and improve performance has driven the development of several new approaches to the vertical layering and horizontal distributed communication interaction, which we have discussed over the previous two chapters. The relationship between communication session architecture and technological requirements is summarized in Tables 12.1 and 12.2.

Some new approaches affect only the internals of a node, and hence deployment will be invisible to other nodes. Measurement-based repeaters [ZWE 12] can be considered a new implementation of purify-and-swap, and a carefully engineered protocol stack would allow use as a drop-in replacement for the individual nodes. Conversely, memoryless is a new link architecture whose benefits are realized only when the entire protocol stack is optimized. Encoded link, surface code and quasi-asynchronous are not specific to a particular link layer, and may as a group be able to support the same upper-layer protocols, including ESC.

	Requirements				
	Transmitters		# Receivers	Buffer	
Approach	# Per node	Dwell time	per node	# Per node	Cumulative buffer time
Hop-by-hop teleportation					
AEC (ACKed link)	small	Link RTT	small	small	≈ 0
Purify & Swap [BRI 98]					
AEC	small	Link RTT	small	$\propto$ log(# of hops)	$\propto$ log(# of hops) $\times$ E2E RTT
(distributed)					
Encoded link (flat timed) [JIA 09], AEC	$\propto (n/P_b)$	Link RTT	$\propto n$	$\propto n$	$\propto$ $n\times$ E2E RTT (weighted toward destination)
Surface code (flat timed) [FOW 10]					
direct transmit	$\propto (d/P_b)$	short	$\approx d$	$\propto d$	$\propto d\times$E2E one-way latency
AEC	$\propto (d/P_b)$	Link RTT	$\approx d$	$\propto d$	$\propto d \times 1\times$ E2E RTT
Quasi-asynchronous [MUN 10]					
AEC, 1-EPP					
Forward propagation	$\propto (n/P_b)$	Link RTT	$\propto n$	1	$1\times$ E2E RTT (distributed)
Reverse propagation	$\propto (n/P_b)$	Link RTT	$\propto n$	1	$1\times$ E2E RTT (all at destination)
Ridge fold	$\propto (n/P_b)$	Link RTT	$\propto n$	1	$1.5\times$ E2E RTT
(weighted toward destination)					
Valley fold	$\propto (n/P_b)$	Link RTT	$\propto n$	1	$0.5\times$ E2E RTT (all at destination)
AEC, QEC encoded					
Forward propagation	$\propto (n/P_b)$	Link RTT	$\propto n$	$\propto n$	$\propto n \times 1\times$ E2E RTT (distributed)
Reverse propagation	$\propto (n/P_b)$	Link RTT	$\propto n$	$\propto n$	$\propto n \times 1\times$ E2E RTT
(all at destination)					
Ridge fold	$\propto (n/P_b)$	Link RTT	$\propto n$	$\propto n$	$\propto n \times 1.5\times$ E2E RTT
(weighted toward destination)					
Valley fold	$\propto (n/P_b)$	Link RTT	$\propto n$	$\propto n$	$\propto n \times 0.5\times$ E2E RTT
(all at destination)					
Memoryless [MUN 12]	small	short	Tens?	small	≈ 0

Table 12.1. *Memory requirements for several quantum repeater communication session architectures. Dwell time for receivers is not listed because it is very short in all approaches. RTT is round-trip time, and E2E is end to end. Code distance for error correcting codes is d, and code block size is n. Entanglement success probability is P_b*

	Requirements		
Approach	Local Operation Fidelity	Entanglement	
		Success Probability	Fidelity
Hop-by-hop teleportation			
AEC (ACKed link)	Very high	Low	Very high
Purify & Swap			
AEC	High	Low	Low
Encoded link (flat timed)			
AEC	High enough for QEC	Low	Fairly high
Surface code (flat timed)			
direct	High enough for QEC	High	High
AEC	High enough for QEC	Low	High
Quasi-asynchronous			
AEC, 1-EPP			
Forward propagation	Very high	Low	Fairly high
Reverse propagation	Very high	Low	Fairly high
Ridge fold	Very high	Low	Fairly high
Valley fold	Very high	Low	Fairly high
AEC, QEC encoded			
Forward propagation	High enough for QEC	Low	Fairly high
Reverse propagation	High enough for QEC	Low	Fairly high
Ridge fold	High enough for QEC	Low	Fairly high
Valley fold	High enough for QEC	Low	Fairly high
Memoryless	Very high	High	High

Table 12.2. *Gate and entanglement requirements for several quantum repeater communication session architectures*

The memory table lists the number of transmitter, receiver and buffer memories, and their required lifetimes, in qualitative terms. The times mentioned are for high-quality preservation of the memory, with errors of a few percent, at most; meeting these values will necessitate T_1 and T_2 lifetimes of one order of magnitude longer to keep error rates approximately 10% or two orders of magnitude longer to keep error rates approximately 1%. Round-trip times for 100 km are approximately 1 ms, and intercontinental times can exceed 300 ms. Van Meter *et al.* found that the latency of starting up a purify-and-swap connection is 2–8 times the E2E RTT [VAN 09]. The necessary T_1 and T_2 values for 1% errors over intercontinental distances can thus reach into the low hundreds of seconds, far beyond projected technical capabilities for physical qubits in the near future. Even metro networks will require memory lifetimes on the order of a second.

If our memory lifetime is too low, we can compensate by using QEC internal to the repeater nodes without modification to the externally visible behavior, or by re-engineering the protocol stack to avoid round-trip delays. Over the previous two chapters, we have discussed several schemes taking the latter approach. Schemes using error correction allow us to eliminate the multi-hop purification, but in general still require an acknowledged link layer. We gain one to two orders of magnitude, moving into the realm of reachable lifetimes for ion trap and possibly NV centers in diamond, but remaining well out of reach for semiconductor memories in the immediate future.

Qubit lifetime is the major constraint, but the availability of sufficient buffer memory is also a significant problem. The earliest purify-and-swap proposals required a few tens of qubits per node, proportional to the log of the number of repeater hops in a network's longest path. Although this suggests a scalable solution, it exceeds current experimental capabilities. An adapted version uses the minimum two qubits per node [CHI 05]. Encoded link and surface code, which depend on QEC, require orders of magnitude more memory than purify and swap.

The memoryless approach takes advantage of a clever encoding to avoid storing qubits in memory, but requires a complex decoding process. We must be able to receive each of the qubits into a separate memory at least long enough to perform this decoding. With mn qubits in flight in various forms, this may be a formidable challenge, especially when using OAM.

As we saw in Table 8.2, the fidelity of entangled states that can be created in the laboratory covers a wide range, and it is improving rapidly. If the base-level-generated entanglement is of high fidelity, all of these schemes will work well. Purify-and-swap is the only scheme that works well with low-fidelity entanglement, and of course, it also benefits when entanglement fidelity is high, reducing the number of rounds of purification, and hence round-trip delays.

12.4. Conclusion

Ultimately, our question is how to select a communication session architecture with the flexibility to work within current technological constraints and to grow with both technological evolution (push) and market growth (pull). Metropolitan-area networks seem to be within reach using either purify-and-swap or quasi-asynchronous 1-EPP. Which of these is used will depend on the details of the implementation technology; quasi-1-EPP will be more sensitive to local gate errors, for example. Whether it is easier to add more transceiver and buffer memories or to increase memory lifetime on the order of a second will be a major factor. Strides in memory lifetime are steady, and interfaces to the optical components are improving, but the additional complexity of optical switching for transceiver multiplexing may be the more stringent limitation. Any of the more complete uses of QEC in the protocol itself are likely to be a number of years behind, but promise far higher fidelities when achievable.

The market pull for quantum networks currently rests on the shoulders of QKD. We discussed the rationale for QKD in section 5.1. QKD can operate with Bell pairs of relatively low end-to-end fidelity but the secret key bit rate improves rapidly with increasing fidelity, so it meshes well with the needs of the technology providers. However, in networks of a modest number of nodes, repeater-based QKD must compete with the simpler trusted relay model.

This concludes our discussion of the quantum repeater communication session architectures, and Part 3. In the next, and final, part of the book, we come at last to *networks* of quantum repeaters.

PART 4

Networks of Repeaters

Chapter 13

Resource Management and Multiplexing

Up to this point in the book, most of the concepts presented were developed by a large community of researchers, and my contribution has been small. The state machine-based perspective on protocols we have developed and published over the last few years is unique; the resource analysis of applications is new to this book; and the study of purification scheduling published several years ago is an original contribution to the engineering of repeaters, but these are only a small fraction of the work in the field. In this part of the book, we turn to material that is largely original to my research group, as few researchers in repeaters have considered in detail the complex issues in moving beyond a chain of repeaters.

The previous three chapters dealt with the *communication session* architectures: how to manage messaging so that the desired end-to-end operation (generally, entanglement creation) is performed. The next four chapters will extend these notions to a more complete *network* architecture. They are based on work by Luciano Aparicio, Hiroshi Esaki, Clare Horsman, Thaddeus Ladd, Bill Munro, Kae Nemoto, Takahiko Satoh, Joe Touch and by the author of the current book [APA 11a, APA 11b, VAN 11, VAN 13b].

The first issue we must address is multiplexing: how can resources (principally quantum memory and access to quantum communication channels) be shared among competing communication sessions, as we discussed in section 3.3.4? Research on repeaters has generally studied a single connection in isolation, assuming that the physical links in the path are completely dedicated to building end-to-end Bell pairs, ignoring any other activity. In reality, in a complex network, more than one conversation is taking place at the same time.

In this chapter, we will examine the applicability of several known multiplexing disciplines to purify-and-swap quantum repeater networks: straightforward circuit switching, TDM, statistical multiplexing (e.g. the equivalent of packet switching), and a buffer memory allocation and sharing mechanism. Although it is also possible to discuss frequency (wavelength) multiplexing, in fact the channel itself is of less interest here than the memories (transceivers and buffer), which are presumed to be in short supply due both to the difficulty of fabricating them and the need to retain buffered state in the middle of the quantum repeater network for longer periods than in a classical network.

The different multiplexing schemes were studied in order to recommend a mechanism for sharing resources in a multi-user network, and ultimately to be able to predict the performance of a given network under certain traffic patterns. To recommend a scheme, we want to know how the performance changes as traffic changes, beginning with the aggregate performance of the network. We also want to be assured that short-distance communication sessions are not able to "shut out" long-distance communication sessions and prevent them from making forward progress.

13.1. Simulated network and traffic

13.1.1. *Network topology and simulator*

Aparicio and Van Meter simulated a small network, with five competing traffic communication sessions on 12 links among 13 nodes, as shown in Figure 13.1 [APA 11a]. This may be the first quantum repeater network simulated using competing communication sessions and examining their behavior. Each link is a 20 km instance of the weak nonlinearity type discussed in Chapter 8, with a base Bell pair density matrix

$$\rho = 0.633\left|\Phi^+\right\rangle\left\langle\Phi^+\right| + 0.244\left|\Psi^+\right\rangle\left\langle\Psi^+\right| + 0.061\left|\Phi^-\right\rangle\left\langle\Phi^-\right| + 0.061\left|\Psi^-\right\rangle\left\langle\Psi^-\right|. \qquad [13.1]$$

Each transmitting interface has 50 qubits of buffer memory, and each receiving interface has 32 qubits. Node E, being on the receiving end of five links, has 160 qubits in total, while F, being on the transmitting end of five links, has a total 250 qubits. Node J has one receive and three transmit interfaces, for a total 182 qubits. (Note that the grey lines represent communication sessions, rather than physical links; all four of the communication sessions crossing between E and F are sharing the same physical link and a single memory buffer.)

In the simulations presented here, memory and gates are assumed to be perfect.

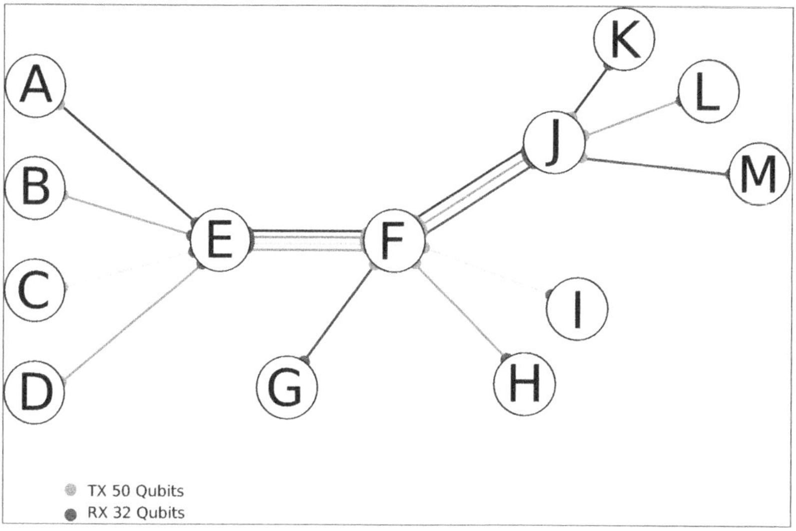

Figure 13.1. *Simulated network. Each line in a shade of gray represents one communication session. The four-hop AK communication session is our primary communication session, and the others are enabled and disabled in various cases to test the impact of multiplexing schemes on that primary communication session*

These simulations were run as extensions to OmNet++ 4.0, a C++-based network simulator, using protocols modeled after the state machine-based approach discussed in the prior chapters. Additional details of the simulation setup are shown in Table 13.1.

Hardware	Properties
Fiber length	20 Km
Fiber attenuation	0.17 dB/km
Transmitter qubits (TX)	50
Receiver qubits (RX)	32
End-to-end fidelity target	0.98
Contested links purification threshold	0.98
Uncontested links purification threshold	0.99
Base-pair fidelity	0.633
Entanglement success probability	0.36

Table 13.1. *Hardware and software configuration for multiplexing simulations*

13.1.2. *Traffic load*

The traffic load is the set of communication sessions initiated by applications running on nodes in our network. We will call each session a *flow*. Each flow is assigned a path through the network, possibly using the mechanism described in the next chapter. We assume the path is fixed for a given session. When the path of more than one flow passes through the same link at the same time, we have competition, or *contention*, for the resources of that link.

The five flows in Figure 13.1 are enabled and disabled in different simulation scenarios, to test the behavior of each multiplexing discipline in different circumstances. The five scenarios are listed in Table 13.2. The first scenario is the baseline, where we only activate one flow at a time. This is the maximum possible for each path. In the second scenario, we enabled the flows AK and CI only, with competition for shared resources only on one link (between E & F). In the third scenario, we enabled flows AK and BL, with competition for resources on two links (between E & F and F & J). In the fourth scenario, we enabled flows AK, CI and DH, where two flows compete on one link with AK (between E & F). Finally, we studied all the flows together, where there are several competing flows in different parts of the network.

Scenario	Flows	Comments
1	(each alone)	Baseline case, equivalent to circuit switching
2	AK+CI	Two flows competing on one link
3	AK+BL	Two flows competing on two links
4	AK+CI+DH	Three flows competing on one link
5	AK+BL+CI+DH+GM	Several competing flows in different parts of the network

Table 13.2. *Traffic flows*

13.1.3. *Adjusting link target fidelity*

We use a target end-to-end fidelity of 0.98, as in the simulations presented in prior chapters. This level is high enough for the applications discussed earlier, including running quantum error correction on teleported logical states, given local gates of adequate fidelity. This is also used as our working fidelity (section 10.3.1) on a dedicated chain of hops.

However, when the use of the network is not uniform, using the same working fidelity throughout the entire network results in portions sitting idle while a flow waits for its turn to use a shared resource, as discussed in section 3.3.4. Aparicio, Van Meter and Esaki have shown that setting a higher fidelity threshold for those under-used resources can result in a net gain in performance [APA 11b].

Consider scenario 2 from the table, in which the connections are AK and CI. Contention exists only for the EF link, so naturally the other links have less work to do while the sessions wait for access to EF. With a working fidelity of 0.98 on each link, entanglement swapping will result in a noticeable drop in fidelity. However, if the CE and FI links were perfect, $F = 1.0$, the entanglement swapping would give a final output fidelity identical to the EF Bell pair fidelity, assuming perfect swapping. Thus, rather than leaving CE and FI idle, we can work to improve the fidelity of their Bell pairs and reduce the fidelity penalty of swapping, potentially improving the throughput of Bell pairs above our desired end-to-end fidelity. The simulations presented here set the purification threshold for the 10 uncontested links (AE, BE, JK, etc.) to 0.99, whereas the contested links EF and FJ are set to 0.98.

13.2. Simulations

13.2.1. *Circuit switching: upper and lower throughput bounds*

First we studied each flow separately using circuit switching and measured the throughput that the network can provide when no other traffic is present. This multiplexing scheme naturally provides the best performance possible for the active session on the chosen path, and is used here as a reference to compare the other schemes when we try to accommodate additional flows. Table 13.3 shows the measured throughput for our baseline scenario for each flow when no other traffic is present. From this table, we can arrive at upper and lower bounds for the throughput for our multiplexing disciplines.

Flow	Number of hops	Throughput
AK	4	64
BL	4	65
CI	3	133
DH	3	135
GM	3	124
sum	-	521

Table 13.3. *Maximum traffic per flow using circuit switching. Throughput is in end-to-end Bell pairs per second of fidelity* $F \geq 0.98$

Using circuit switching, only one flow on each link is enabled. In this particular network topology, with this traffic pattern, the GM flow can operate concurrently with either the DH or CI flow, but any other pair of flows conflicts on either the EF or FJ links, and must wait.

The throughput of a single flow is the lower bound for other multiplexing schemes; any scheme that gives an aggregate throughput below this level while attempting to share resources would be worse than the simplest circuit switching approach. That is, as can be the case with humans, dividing attention among multiple tasks results in less effective work on all of them.

The sum of the flows serves as an upper bound on the throughput for any multiplexing scheme. If it were possible to run all of the sessions at the same time with no negative impact on any of them, the performance would reach this number. Presumably, this is only achievable when the sets of resources desired by each session are completely disjoint.

13.2.2. *Other multiplexing disciplines*

In addition to circuit switching, we simulated TDM, statistical multiplexing and buffer memory multiplexing. We looked for both throughput and fairness. We also examined the time to complete a fixed workload using each discipline.

In TDM, time is divided into a number of *time slots* and each flow is assigned one, so for each simulation, we have the same number of time slots as the number of flows. The measurements are shown in Figure 13.2, compared to the throughput of an ideal impossible-to-achieve case of totally uncontested access to resources for all flows (the "SUM" line in Table 13.3).

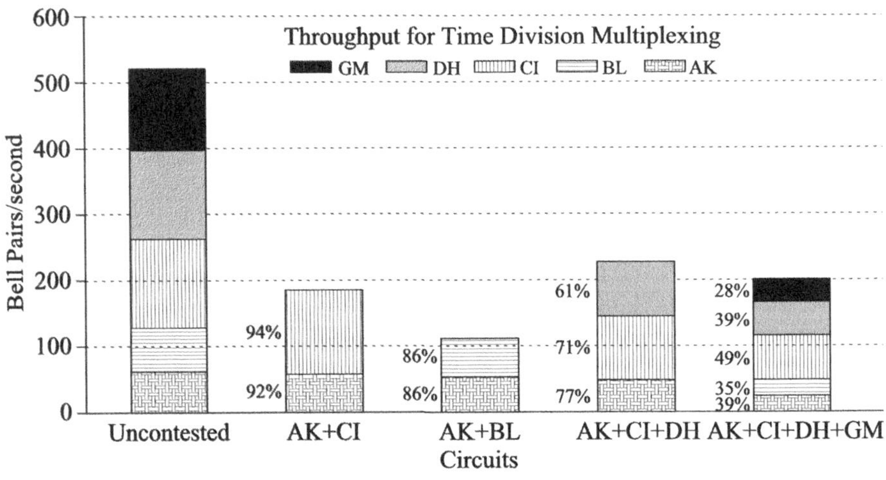

Figure 13.2. *Throughput of TDM compared to uncontested flows*

Figure 13.3 shows the performance of our five scenarios using statistical multiplexing, again compared to the uncontested case.

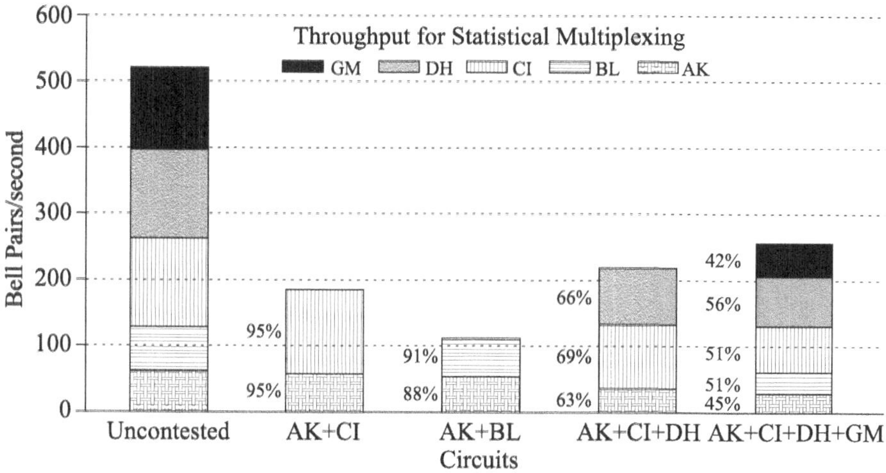

Figure 13.3. *Throughput of statistical multiplexing compared to five uncontested flows (left)*

Figure 13.4 shows the throughput for buffer space multiplexing compared with the uncontested scenario. We divide the shared resources by the number of flows in transit. For the first scenario AK+CI, there are two flows competing for access to link EF, so we assign half the resources for each flow (16 qubits in station E and 25 qubits in station F). The allocation of memory for each scenario is summarized in Table 13.4.

Scenario	AK(EF)	AK(FJ)	BL(EF)	BL(FJ)	CI(EF)	DH(EF)	GM(FJ)
2	16, 25	–	–	–	16, 25	–	–
3	16, 25	25, 16	16, 25	25, 16	–	–	–
4	11, 17	–	–	–	11, 17	10, 16	–
5	8, 13	17, 11	8, 13	17, 11	8, 12	8, 12	16, 10

Table 13.4. *Buffer memory allocations for network interfaces on shared links EF and FJ in buffer multiplexing simulations*

Two flows using one shared link can sometimes exceed the performance of a single flow. This counter-intuitive behavior arises because the unshared resources can continue to improve beyond their minimum required fidelity threshold, making more efficient use of the shared resources when they do gain access. This can be seen clearly in the AK+CI and AK+BL scenarios, where two flows *each* achieve 86%–95% of their ideal performance. The behavior of a single flow, then, under at least some circumstances, can be said to be only minimally affected by the presence of a second flow using one or two of the same links. The addition of a third flow (AK+CI+DH) raises the total throughput again, but begins to have significant impact on the performance of each flow relative to the ideal case. Knowledge of this behavior can be used to guide the design of a network topology, if the expected traffic pattern is understood.

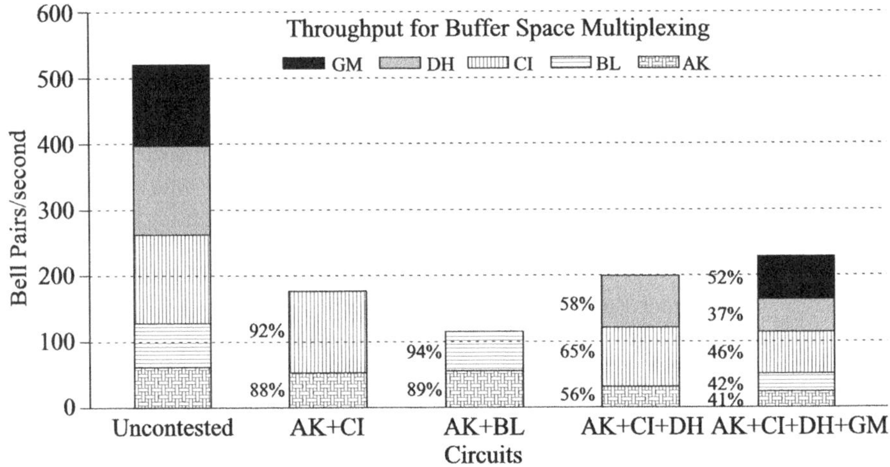

Figure 13.4. *Throughput of buffer space multiplexing compared to uncontested flows*

The total throughput of all five flows is highest for the statistical multiplexing case, achieving 257 Bell pairs per second, compared to 228 per second for buffer space multiplexing and 201 per second for TDM. Statistical multiplexing substantially outperforms the other two schemes (by 13% and 28%, respectively), though the specific numbers are traffic- and network specific. For the scenarios with fewer flows, the performance advantage was smaller, only a few percent. Further confirmation of this advantage by simulating additional networks and traffic patterns (especially larger, more complex networks) is needed.

The above analyses assess the steady-state throughput of our flows. Let us briefly compare these schemes for a variant with a fixed amount of work by assuming that all five flows in Table 13.2 issue their initial requests to use the network at the same time and run until 100 Bell pairs have been created.

For the circuit-switched case, first we would run the AK flow to completion, then the BL flow. Next, the CI and GM flows can run at the same time, because they use independent parts of the network. Fourth and last would come the DH flow. This would take approximately $100/64$ (for the AK flow) $+$ $100/65$ (for the BL flow) $+$ $100/133$ (for the CI and GM flows simultaneously) $+$ $100/135$ (for the DH flow) $=$ 4.6 s.

In contrast, in statistical multiplexing all five flows begin work at the same time. Using a slightly more complex calculation to take into account that the three-hop CI, DH and GM flows would complete more quickly than the four-hop AK and BL flows, we estimate that all 500 qubits could be teleported in 2.7 s, or $1.7\times$ as fast. TDM and buffer space multiplexing require more detailed simulation to produce reasonable estimates of their performance under changing workloads, which we defer.

Although statistical multiplexing has the highest throughput, we might suspect that with no control on resource use, it is potentially susceptible to being unfair to some flows. In particular, we are concerned that long-distance flows, which naturally react more slowly to changing conditions than shorter flows, may be penalized more than short ones as the dynamic network state changes.

We evaluated the fairness of the multiplexing schemes using Jain's fairness measure for resource allocation (equation [3.1]) for the five flow scenarios for each multiplexing method. A fairness of $\mathcal{J} = 1.0$ indicates perfectly even distribution of resources among the users, whereas $\mathcal{J} = 1/5$ would indicate that one user had acquired all of the resources, shutting out all other users. Because the maximum capability of each flow differs even when given uncontested access to all links, we applied the measure to the set of throughputs normalized to each flow's circuit-switched throughput, using the percentages shown in Figures 13.3, 13.2 and 13.4.

Statistical multiplexing, with a range of 42%–56% of maximum, has a nearly perfect fairness of $\mathcal{J} = 0.99$. Buffer space multiplexing likewise comes in with $\mathcal{J} = 0.99$. TDM, despite the relatively large spread from 28% to 49%, also has an excellent fairness of $\mathcal{J} = 0.97$. In particular, the four-hop flows fall in the middle of the group in terms of performance degradation, giving us no reason to infer that long-distance flows are penalized more heavily, although further confirmation with longer flows is desirable. From these values, we conclude that all three multiplexing schemes share contested resources fairly.

13.3. Conclusion

The simulations discussed here are only a first step toward a definitive answer on multiplexing for quantum repeaters. Larger networks need to be examined, and of course, other session architectures besides purify-and-swap are very likely to behave differently. The dynamic assignment of different working fidelities for different sub-paths through the network, discussed in section 13.1.3, will present operational difficulties, which bear further investigation.

Statistical multiplexing is simpler to implement than any scheme requiring explicit resource management, whether it is circuit switching, TDM or buffer space multiplexing. Both the software implementation and the network protocols have fewer requirements using statistical multiplexing, reducing implementation and deployment cost. Given its performance, as an isolated engineering decision, we would lead toward statistical multiplexing as our choice.

However, multiplexing is only one facet of a complete architecture. Other factors, such as path selection (addressed in Chapter 14) and the arrangements necessary to identify swapping points and control purification may lead us toward circuit switching.

Chapter 14

Routing

In the previous chapter, we discussed how communication sessions behave when competing for link resources in a network of shared links. We asserted that the path through the network for each session was fixed, but did not discuss how such a path is chosen. In a complex network of heterogeneous links, selection of a path between the two communicating nodes is an important problem. Here, we examine a form of Dijkstra's algorithm adapted for repeater networks [VAN 13b].

The delicacy of quantum states makes a practical path selection algorithm imperative, but classical notions of resource utilization are not directly applicable, rendering known path selection mechanisms inadequate. To adapt Dijkstra's algorithm for quantum repeater networks that generate entangled Bell pairs, we quantify the key differences and define a link cost metric, *seconds per Bell pair* of a particular fidelity. Simulations that include both the physical interactions and the extensive classical messaging confirm that Dijkstra's algorithm works well in a quantum context. Simulating about 300 heterogeneous paths, comparing our path cost and the total work along the path gives a coefficient of determination of 0.88 or better.

14.1. Introduction

In section 3.3.9, we introduced the concept of routing. The emerging field of quantum communication has, to date, experimentally demonstrated the basic principles of entangled quantum networking [CHO 07, KIM 08, REI 06, TAS 10, ZHA 03], and laid the theoretical foundations of creating long-distance high-quality entanglement [BEN 93, BRI 98, LLO 04], but topologically has considered primarily channels and linear networks, leaving us with an urgent need for a path selection mechanism as quantum networks develop.

The first, theoretical studies on entangled quantum networks focused primarily on an abstract model consisting of a linear chain of repeaters, with a power of two number of hops of identical length and quality [BRI 98, DÜR 07]. Recent work [FOW 10, JIA 09, MUN 10] has targeted more realistic chains of repeaters, relaxing those constraints. Here, we analyze the behavior of more complex network topologies, as in Figure 14.1. In a network of heterogeneous links and irregular topology, path selection affects both the performance of individual connections and global network load. In this chapter, we study only the purify-and-swap session architecture. Path selection for QEC-based repeaters may be different in some respects, but the principles outlined here can easily be applied to a simulator for the other types.

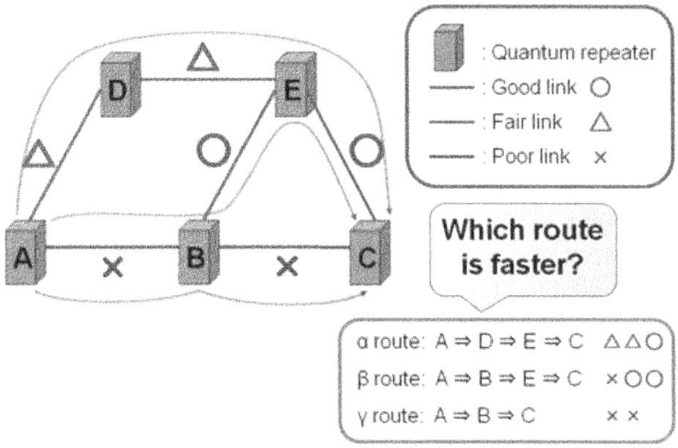

Figure 14.1. *Path selection is a critical problem in all networks; in quantum networks, due to the delicacy of quantum information, it may determine whether or not a connection can be made successfully*

In this chapter, we apply Dijkstra's algorithm (section 3.3.9) for ranking candidate paths to quantum repeater networks. We evaluate our algorithm for its ability to select a path that both maximizes the throughput of end-to-end connections and minimizes global work for this class of repeater.

Our proposed link cost is the inverse of the throughput of the link, measured in *Bell pairs* per second of a particular *fidelity*. Other candidates for link cost, including lower-level metrics such as the number of laser pulses and quantum measurement operations, are found to be useful for evaluating the total work actually consumed on a path, but are poor metrics for prioritizing a link because they reflect the physical link characteristics but not the system factors that are equally important influences on end-to-end performance.

The results of three sets of simulations of various paths using four different qualities of links are presented. The first set consists of 46 paths that vary in length from one to nine hops, whereas the second set covers 256 link combinations in four-hop paths, both using a target fidelity of $F \geq 0.98$. The third data set replicates the first 46 paths, but for a target fidelity of $F \geq 0.90$, which is too low for some distributed quantum computations but high enough for successful quantum key distribution [BRA 13].

Across the first two data sets, the coefficient of determination is 0.88 or better between the path cost and the total work performed (counted as the number of *quantum measurements* performed along the whole path), supporting our choice of link cost and the effectiveness of Dijkstra for this type of quantum network. Comparing the results of pairs of simulations, the path with the lower cost also has higher throughput in more than 80% of all tested cases. We demonstrate that, in direct analogy to classical networks, the performance of a quantum path will be limited by the throughput of the *bottleneck link*, while total work is a function of both the path length and the quality of all the links.

To build a complete argument, we discuss the differences between quantum and classical networks and the difficulties encountered (section 14.2). After defining the problems in path selection and proposing several solutions (section 14.3), we evaluate those solutions via simulations that answer a series of specific questions about the behavior of quantum repeater networks (section 14.4).

14.2. Difficulties: differences between quantum and classical networks

So far, we have discussed quantum communication technologies and outlined where we apply classical techniques to the quantum problems. However, we have not specifically articulated the fundamental differences that make the merger of classical and quantum networking concepts less than straightforward. There are both theoretical results and practical reasons for believing that the answers to our questions require thought, rather than simply asserting that classical and quantum networks can use the same solutions. Indeed, Di Franco and Ballester have studied the end-to-end fidelity of paths in quantum repeater networks, and shown that under some circumstances, the final fidelity is not a simple calculation based on the hop-by-hop input fidelities, potentially affecting routing in repeater networks [DI 12].

The differences stem from several sources: the engineering difficulties of creating and protecting quantum states and the real-time decay of quantum information; the impact of (probabilistic) photon loss; and the fundamentally probabilistic nature of some quantum operations. These issues manifest themselves both locally and globally, requiring in some cases additional classical messages to be exchanged, further exacerbating the problems.

For a classical path consisting of a series of identical links, throughput (at least in the ideal, sustained case) can match the throughput of a link, independent of the number of hops. However, the maximum throughput of a symmetric chain of purify-and-swap repeaters declines polynomially with length, due to the need for additional purification [DÜR 07]. Thus, it *may* be desirable to assign a path cost that grows more than additively.

When operated to deliver Bell pairs above a specified fidelity threshold (typically related to the fidelity threshold for executing quantum error correction), simulations show a stair-step decline in throughput versus hop count [VAN 09]. The total work performed grows linearly as the throughput holds steady, then increases suddenly when a "stair" is crossed (see Figures 10.9 in Chapter 10 and 14.2 and 14.4 in section 14.4). Likewise, dealing with non-power-of-two numbers of hops will affect throughput and total work performed in hard-to-predict fashion, due to changes in the swapping and purification patterns.

14.3. Problems and solutions

Our goal for this chapter is to develop and analyze a routing algorithm for heterogeneous quantum repeater networks, as shown in Figure 14.1. Our metric for success, therefore, is agreement between the prospective functioning of a network (represented by detailed simulations) and an easy-to-calculate algorithmic cost: does our algorithm allow us to make effective choices?

To develop a routing algorithm, we require a definition for the cost to use a link, a function to calculate a path cost based on a set of link costs, and a goal for the algorithm (e.g. a metric for deciding if the algorithm meets our needs). More precisely, we set as our problems:

– *PS.1* – choose a goal for the routing algorithm;

– *PS.2* – for a quantum link, identify the characteristics of interest for routing, and reduce them to one number or a small set of numbers that represent the link cost; and

– *PS.3* – for a path (an ordered set of links, with associated costs), define a function that gives a path cost.

To solve these problems, we evaluate the following potential solutions:

– *PS.1: Goal* – as the goal for the routing algorithm itself, we choose *minimizing work along the path*, with attention to the secondary goal of selecting the highest-throughput single path between the defined communication endpoints, measured in Bell pairs per second of the target fidelity. We propose a goal for the *system* of delivering Bell pairs useful for teleportation, with a target fidelity of $F \geq 0.98$ or $F \geq 0.90$. The phenomena presented here are independent of the exact value chosen, but this value will be adequate for various uses.

Assessing the work for a particular path is not a simple problem. Intuitively, we want our measure of work to reflect use of some scarce resource. We evaluate two candidates: *total measurement operations* and *total pulse count*.

Without some reasonable idea of global traffic (which we do not yet have), evaluation of global success is difficult, so as a first step we are evaluating the correlation between total work on the path, achieved connection throughput and our definition of path cost.

– *PS.2: Link cost* – we are exploring several link cost definitions. In OSPF, the link cost is nominally unitless. The network operator is free to assign any value for any reason. In practice, one commonly used link cost definition is the inverse bandwidth, equating to "transceiver time", normalized to ten nanoseconds (i.e. cost of a 100-Mbps hop is $C = 1.0$). Simply assessing the clock speed at which pulses can be emitted is clearly an inadequate link cost metric in a system where the fidelity of the output state is important and the probability of success depends on the characteristics of the link. Before execution of our experiments, differing intuitions led one project member to suggest Bell pair generation time and another member to suggest number of measurements; eventually we had a list of five candidates:

– *Loss* – the loss in the channel, in decibels;

– *InvTrans* – the inverse of the transmittance of the channel ($1/T$, where the transmittance T is the percentage of photons received through the path);

– *Pulse* – the number of laser pulses used to create an entangled Bell pair of a high fidelity over a single hop, corresponding to the number of uses of the transmitter (each fixed-time);

– *Meas* – the number of measurement operations used to create an entangled Bell pair of a high fidelity over a single hop (this differs from the above because (1) some pulses are discarded rather than accepted when receiving qubit resources are busy, and (2) measurements are used in entanglement swapping and purification);

– *BellGenT* – the inverse of the throughput of the single link, when run as a single-hop system, measured in seconds per Bell pair.

This list can be divided into two groups: the first two candidates are simply physical characteristics of the link that can be measured easily, whereas the last three require simulation or monitoring of the link to determine. The first two differ by a logarithmic factor; InvTrans corresponds to addition, whereas Loss in dB corresponds to multiplication of cost when placed in a Dijkstra context. While at first glance it may appear desirable to have such an easily determined link cost, by tying the cost so directly to the physical mechanism, the definition may not transfer well when links of heterogeneous physical technologies are involved.

Pulse and Meas seem to correspond most closely to the "transceiver time" definition used in classical networks, but BellGenT incorporates system factors and may give more accurate estimates for little additional complexity. (Although the technical details are very different, these could be considered roughly analogous to the raw transceiver rate, the throughput of a flow-controlled link and the throughput of a reliable link-layer protocol, in terms of the increasing functionality present.)

– *PS.3: Path cost function* – the principal hypothesis of this chapter is that Dijkstra's algorithm can be used as is, with an appropriate choice of link cost. More formally,

$$C_{path} = \sum_i c_i, i \in \{P\}, \qquad [14.1]$$

where $\{P\}$ is the set of links in a path and c_i is the cost for link i. When it is necessary to distinguish among the link cost candidates, we will refer to them as Dijkstra/BellGenT and similarly.

14.4. Simulation and results

Our goal is to determine the range of conditions under which Dijkstra correctly selects the lowest-work path and highest-throughput path, and when it selects some reasonable approximation. We also wish to articulate the conditions under which lowest work and highest throughput are not the same path and when the algorithm does not pick the lowest-work path. We wish to examine whether we consider those to be acceptable cases, or if the algorithm is "failing". This can be achieved by creating a set of candidate paths and comparing the ordering established by Dijkstra/BellGenT and the ordering according to simulation of the whole path.

In this section, we begin by asking a set of questions about the behavior of heterogeneous paths (14.4.1). Next, we describe our simulator, proposed hardware configuration and single-hop simulation results, which both set the Dijkstra parameters and allow us to evaluate our link cost candidates (14.4.2). After enumerating a set of interesting path candidates (14.4.3), we can answer our questions (14.4.4) and use those answers to solve the research problems posed in section 14.3 (14.4.5).

14.4.1. *The behavior questions*

We use the simulator to assess the behavior of systems that are too complex to solve analytically and cannot yet be built and measured directly. Such simulation results will help to guide the development of actual hardware. We can pose a series of specific questions that will help us understand how heterogeneous paths will behave:

1) How does the *number* of hops in a path affect throughput and total work? As a specific case, do the throughputs of $2^n - 1$, 2^n, and $2^n + 1$-hop paths vary?

2) How does the number of *weak* links matter? Does the introduction of a single weak link become a *bottleneck*, as in classical networks? Does adding a second or third weak link further reduce throughput?

3) How does the *position* of weak links in the chain affect throughput? Does a weak link at the beginning, in the middle or at the end differ?

4) Under what circumstances will the path cost mis-order candidate paths with respect to throughput or total work?

For classical systems, we know the answers: although cost increases, (1) the length of the path (in hops) has no effect on the (theoretical) throughput, and (2) the (theoretical) throughput of a path is capped by the throughput of the bottleneck link(s). Question (3) is answered in the negative: ordering does not (or should not) matter, and ordering of the path segments should always agree with ordering of the full paths. Question (4) is open-ended, but there are known cases where a lower-performance path can be selected in the service of a larger global goal. The answers to these questions are presented in section 14.4.5.

14.4.2. *Simulated hardware and link costs*

All of the simulations presented in this chapter model the qubus physical entanglement mechanism (section 8.1.1). The simulator used is the same as in work by Van Meter, Ladd, Munro, Nemoto, van Loock and Yamamoto, with extensions to support the heterogeneous paths [VAN 09, LAD 06, MUN 08]. The simulator was developed for modeling the quantum-level behavior of a cavity QED system for qubus and two other physical layer candidates, and uses the purification mechanism of Dehaene *et al.* [DEH 03]. The necessary classical messaging is carefully modeled. The simulator consists of approximately 11,000 commented lines of C++, and the production runs of the simulations presented here consumed about 100 h of CPU time on 2.2 GHz AMD Opteron CPUs.

The simulator is based on well-understood physical equations and materials whose behavior and relationships have been experimentally validated at the lowest levels, although the complete mechanisms have not yet been demonstrated together; as experience with larger-scale quantum networks develops, it is necessary to continue tracking the agreement of theory, simulation and experiment. We believe the prospective agreement between a real-world network and our simulations will be more than adequate for the purposes of this chapter, but a detailed analysis is beyond the scope of this book.

Figure 14.2 shows the results of simulating a single qubus hop, with parameters as in Table 14.1; additional details of the hardware configuration are the same as in [VAN 09, LAD 06, MUN 08]. The original qubus mechanism is very sensitive to loss, but works well in low-loss situations. This form of qubus repeater fails to work for losses greater than about 5.5 dB from transmitter qubit to receiver qubit, limiting hop length to about 30 km over high-quality optical fiber at telecom wavelengths. Other types of physical link, including variants of the qubus mechanism [MUN 08], will work over longer distances and with higher fidelity. By choosing to simulate basic qubus links, we can see very clearly the impact of purification and low-fidelity entanglement. As future work, we plan to confirm the behavior of Dijkstra with other physical link types.

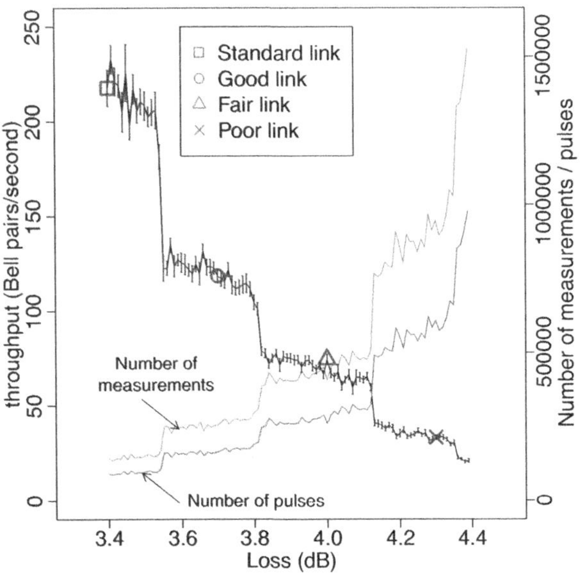

Figure 14.2. *Single-hop simulated throughput versus loss using the parameters in Table 14.1, used to define linkcost. The four link types chosen for simulation of more complex paths are marked on the throughput curve (left axis) with the corresponding symbols. The stair-step behavior is due to the increasing number of purification rounds as the initial fidelity declines due to increasing loss. The details of this curve are very specific to the qubus technology we simulate, but the principles are general. In these simulations, final fidelity is* $F \geq 0.98$

For the multi-hop simulations, we choose four specific points as example links, as shown in Tables 14.2 and 14.3. The four link types chosen are marked in the figure with the corresponding symbols. The terms "standard", "good", "fair" and "poor" are relative to this simulation only, not indicative of all possible physical quantum link types. As can be seen from this table and Figure 14.2, even small differences in loss have a large and uneven impact on throughput, suggesting that not only is the utility of the proposed link costs Loss and InvTrans rather technology-specific, but even in the isolated case of qubus they may be poor metrics.

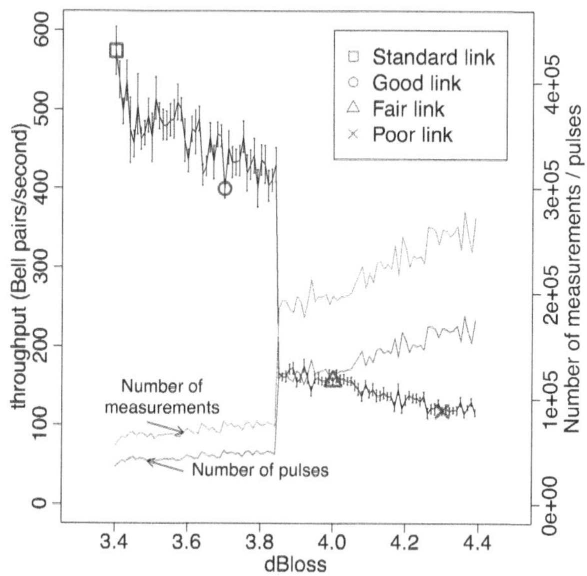

Figure 14.3. *Single-hop simulations of the same conditions as Figure 14.2, except that purification is only done until reaching a final fidelity of* $F \geq 0.90$*, reducing the number of rounds and resulting in the differing stair-step pattern*

Links with the same physical loss but different numbers of qubits at the transmitter and receiver will behave differently, as shown in Table 14.4. Although the throughput varies by a factor of two for different configurations, the number of pulses and number of measurements to teleport 200 qubits does not vary significantly except for the clearly misconfigured 100/25 case. For the conditions simulated, approximately 450 pulses and 700 measurements (including basic entanglement, swapping and purification) are required to teleport one qubit. Although these two measures are arguably more direct representations of cost, this important throughput difference affects our notion of a preferred path.

Qubus Quantum Repeater	
Number of qubits per repeater link connection	25 transmitter, 25 receiver
Number of qubits teleported (length of simulation)	200 qubits
Final target fidelity	$F \geq 0.98$
Optical Fiber	
Length	20 km
Signal loss	3.4∼4.4 dB/20 km

Table 14.1. *Link cost simulation parameters*

Link	Loss	InvTrans
Standard (□)	3.4dB	2.19
Good (◯)	3.7dB	2.34
Fair (△)	4.0dB	2.51
Poor (×)	4.3dB	2.69

Table 14.2. *The link configurations and two candidates for link cost. These are physical characteristics of the channel*

Reasoning about the behavior of only a single hop and confirming the results via simulation allows us easily to eliminate two of our prospective link cost candidates (Loss and InvTrans). Two others (Pulse and Meas) remain good measures of total work, but the results shown in Table 14.4 cast doubt on their viability as candidates for link cost when the goal is to achieve high throughput. Thus, *we settle on BellGenT as our link cost metric*.

14.4.3. *Simulated path candidates*

The candidate paths we choose to examine are designed to answer the questions in section 14.4.1. We simulated 46 paths of one to nine hops in various patterns, as well as all $4^4 = 256$ four-hop combinations for the four chosen link types. The 46 paths are enumerated along the bottom of Figure 14.4. We have simulated many more paths with a variety of link conditions over the course of this experiment, notably very long, homogeneous paths (up to 2,048 hops). The findings from other simulations do not contradict the results presented in this chapter, which were chosen to clearly show the effects of interest.

Link	Pulse	/pair	norm	Meas	/pair	norm	T'put	BGT
Standard (□)	90441	452	1	140519	702	1	217.7	1
Good (○)	163628	818	1.80	254691	1237	1.76	118.4	1.83
Fair (△)	258852	1294	2.86	404117	2020	2.87	74.3	2.93
Poor (×)	606278	3031	6.70	945247	4276	6.72	33.1	6.57

Table 14.3. *The link configurations and three additional candidates for link cost. These are the results of single-hop simulations. Simulation parameters are as in Table 14.1. "/pair" is per pair, and "norm" is a normalized cost. BGT is BellGenT*

Xmtr	Rcvr	Throughput	Pulses	Meas
25	25	237 ± 8	80587	125905
25	50	213 ± 8	92071	143582
50	25	436 ± 14	99708	142723
50	50	456 ± 14	89533	139506
100	25	462 ± 22	199240	138543
100	50	956 ± 31	92221	137828
100	100	984 ± 32	88694	138145

Table 14.4. *Throughput depends on the number of available qubits at both transmitter and receiver. Due to the round-trip latency in the Entanglement Control (EC) protocol for acknowledging successful entanglement, having more qubits at the transmitter boosts throughput. Simulations are for a single hop with parameters as in Table 14.1. Throughput and confidence interval (std. dev.) are determined by a linear fit to Bell pair completion times*

14.4.4. *Answering our behavior questions*

Figures 14.4, 14.5 and 14.6 show the simulation results for the paths. In Figure 14.4, the throughput for each specific path can be seen, as well as the two measures of total work, Pulse and Meas. In Figure 14.5, throughput is plotted against the Dijkstra-calculated path cost using BellGenT as our cost metric, and in Figure 14.6 the total work measures are plotted against the calculated path cost. Figures 14.7 and 14.8 plot the results for all 256 four-hop paths simulated.

Perhaps the most important factor to note about the behavior we see is the discrete nature of changes to throughput, due to the discrete nature of purification and entanglement swapping and our choice to establish a particular threshold for final acceptance of an end-to-end Bell pair. In many circumstances, minor changes force an additional round of purification, generally causing a 50% reduction in throughput.

As the *number of hops* (*Q. 1*) increases, our results show a stair-step phenomenon in throughput, but not necessarily at powers of two: the decline of fidelity resulting in additional rounds of purification does not move in concert with entanglement swapping. For Standard □ paths, we see steps at the 2nd and 7th hops.

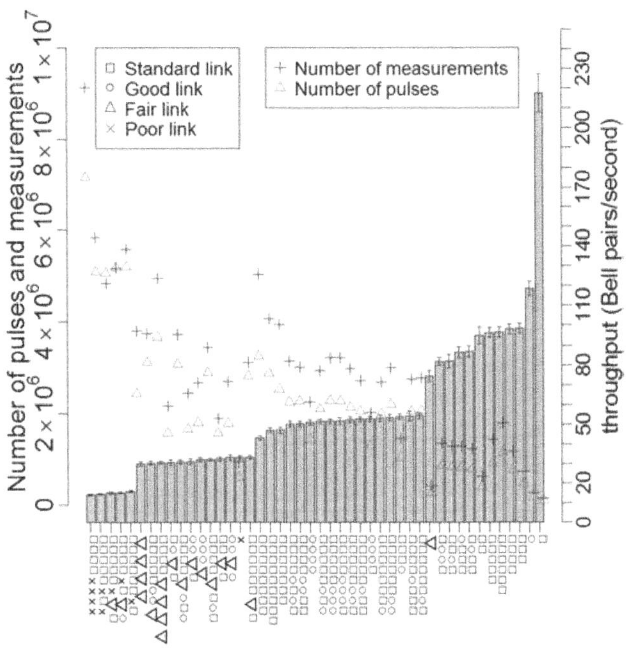

Figure 14.4. *Total number of pulses (△) and measurements (+) for forty-six of our candidate paths (right scale), for end-to-end fidelity* $F \geq 0.98$. *The paths vary in length from one to nine hops. They are ordered left to right according to ascending throughput, plotted using bars (left scale). The legend below the graph shows the individual path configurations; □, ○, △ and × represent our standard, good, fair, and poor links, respectively. The stair-step behavior reflects increasing numbers of rounds of purification*

Examining the symbols in Figure 14.5 shows clearly the existence of a *bottleneck link* phenomenon (*Q. 2*): with one low-quality link in the path, the quality of the other links is almost irrelevant, as can be seen by the clustering of each type of data point (e.g. △ at approximately 30 Bell pairs/s in Figure 14.5 and the concentration of × links at the left edge of Figure 14.4). The various paths can largely, though not entirely, be grouped according to the throughput of the slowest link in the path. The most eye-catching anomalies in Figure 14.5 are the single ○, △ and × marks above and to the right of their respective clusters. These are the single-hop paths of the corresponding link type, indicating that we do not have a pure bottleneck phenomenon. The bottleneck plus the polynomial decline in performance as the number of hops grows work in combination determine to the path performance. In most cases for the longer paths, adding a second link of the same quality as the bottleneck link does not result in a statistically significant reduction in throughput, but optimization of the path becomes more difficult. In a few cases, our optimization fails to find an acceptable pattern, and an additional round of purification becomes necessary. In particular, four-hop paths with ◯ as the bottleneck link(s) are split into two plateaus at approximately 80 Bell pairs/s and 50 Bell pairs/s. However, the total work shows strong correlation even for these cases.

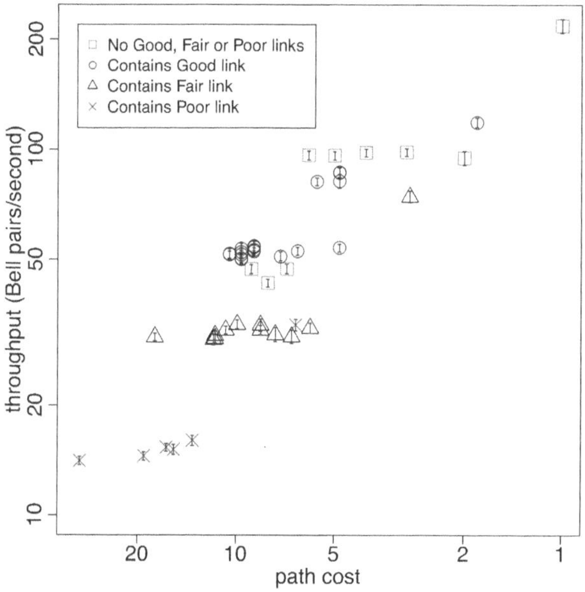

Figure 14.5. *Throughput ($F \geq 0.98$) versus BellGenT path cost for forty-six of our candidate paths. Each path is represented by the symbol for the weakest type of link in the path. The clustering of each type of data point shows clearly that throughput is limited by the bottleneck link. The length of the vertical bar (mostly contained within each symbol) shows the std. dev. of the throughput*

Despite this general bottleneck behavior, the *weak link position* (*Q. 3*) is a more subtle one. Among our simulations, we found a single case where the bottleneck position produced different results. With one Good link and three Standard ones, having the Good link at the left end of the path resulted in 53 Bell pairs/s, whereas the other three paths produced 82 to 87 Bell pairs/s. The low-throughput case required an extra round of purification before one entanglement swapping operation. As the path hovers near a threshold demanding an extra round of purification, optimization of the path usage becomes more difficult; three paths successfully did so, whereas the fourth did not do it. Thus, we must answer that the weak link position *may* have an impact on throughput, depending on our ability to effectively use the path.

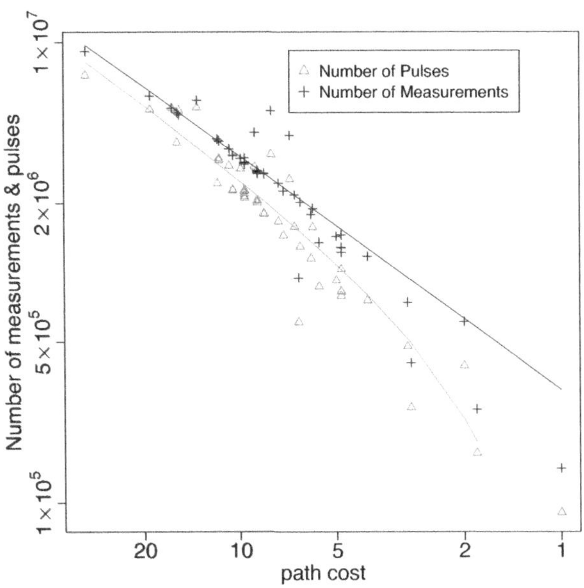

Figure 14.6. *Total work ($F \geq 0.98$), measured in Pulses (△) and Measurements (+), versus BellGenT path cost for forty-six of our candidate paths. The coefficient of determination of each linear fit is 0.88, showing that our path cost is a strong predictor of total work*

Dijkstra/BellGenT occasionally *mis-orders* (*Q. 4*) pairs of path candidates with respect to throughput. Comparing the 256 four-hop paths we simulated, there are 32,640 possible pairs, of which 1,230 had the same path cost. Of the pairs with different costs, 82.6% of the time Dijkstra/BellGenT chooses the higher-throughput path of the pair (the "correct" choice), and 17.4% of the time it chooses the lower-throughput path (the "incorrect" choice). In only 5% of those mis-ordering cases was the difference in throughput more than 10%. For the 46 variable-length paths, the rate of "correct" choices was similar, i.e. 81.6%, but the impact of those choices is higher, with a throughput penalty of 25% or more in almost half of the "incorrect" cases.

The most remarkable case of incorrect ordering we found is the pair of paths □×□□□□□□ (throughput: 16.0 ± 0.51 cost: 13.57) and △△△△□□□□ (throughput: 30.7 ± 0.77 cost: 15.72). The higher-cost path has twice the throughput of the lower-cost one. Examining Figure 14.4 shows that the total work (in pulses or measurements) is quite similar. This isolated case would suggest that the bottleneck poor link might warrant a link cost even higher than 6.57; however, in all of the other cases we examined, it has worked well.

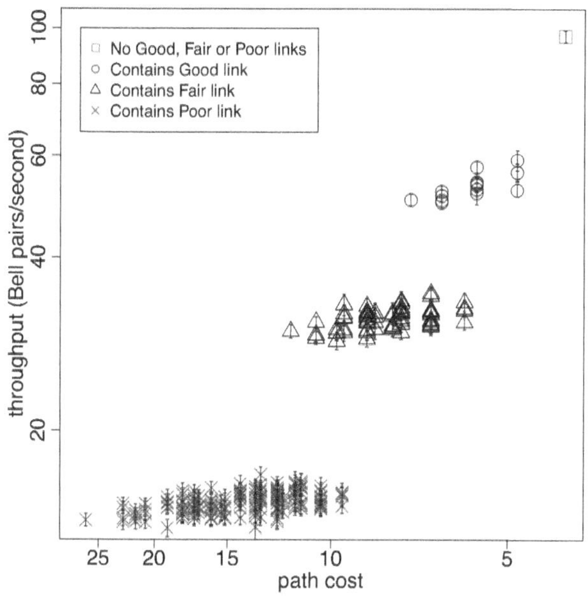

Figure 14.7. *Throughput ($F \geq 0.98$) versus BellGenT path cost for all 256 four-hop candidate paths. Each path is represented by the symbol for the weakest type of link in the path. The clustering of each type of data point shows clearly that throughput is limited by the bottleneck link*

The performance and work are both dominated by purification costs. The use of a given path must be optimized as a whole; our current solution is brute force, trying a large number of possibilities. In a large majority of cases, optimization of that process for a path is straightforward and robust. However, a noticeable minority of cases require delicate adjustments to the entanglement swapping settings. Further automated optimization of this process may result in both better performance and agreement between path cost and work.

For total work, we find a coefficient of determination of 0.88 for both pulses and measurements for the 46 variable-length paths. For the 256 four-hop paths, we find a coefficient of 0.81 for pulses and 0.99 for measurements.

Figures 14.9 to 14.11 show data for a lower end-to-end target fidelity of $F \geq 0.90$, corresponding to the $F \geq 0.98$ data in Figures 14.4 to 14.6. In this data set, we find a coefficient of determination of 0.77 between number of pulses and our cost function, and a coefficient of determination of 0.81 between the number of measurements and our cost function. These values are lower than for the $F \geq 0.98$ tests. However, when comparing two paths, we find that the Dijkstra/BellGenT cost will select the higher-throughput path correctly 88% of the time, more often for $F \geq 0.90$ than for $F \geq 0.98$.

Thus, we conclude that the validity of our results holds across a range of fidelities and behaviors.

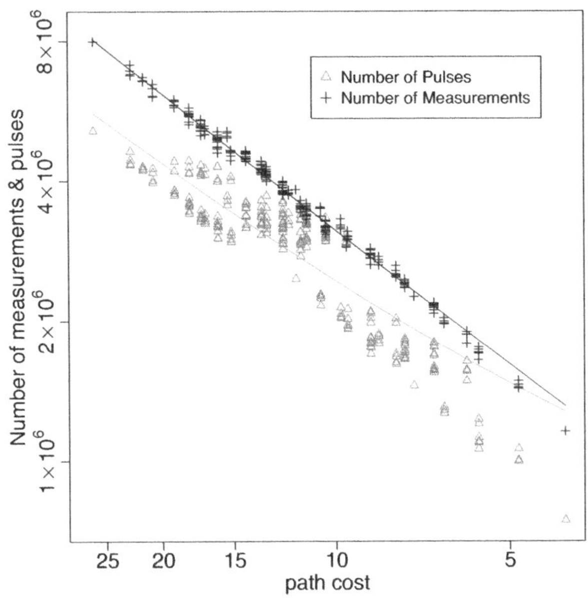

Figure 14.8. *Total work to achieve output fidelity $F \geq 0.98$, measured in pulses ($\triangle$) and measurements ($+$), versus BellGenT path cost for all 256 four-hop candidate paths, with linear fits. The coefficient of determination for the number of pulses is 0.81, and for the number of measurements is 0.99*

We can summarize the behavior as follows: performance is determined by the number of rounds of purification used anywhere on the path; the number of rounds is dominated by the bottleneck link. However, work is spread across the entire path in rough proportion to relative link quality.

14.4.5. *Solving our problems*

With the simulator results and the answers to our questions in hand, we are now prepared to assess the solutions we proposed in section 14.3, and determine whether Dijkstra/BellGenT meets our goal of developing an acceptable routing algorithm for quantum repeater networks.

We have seen that the highest-throughput path is also strongly correlated to measures of work, including total pulse count and number of measurements. In some pathological cases, number of measurements is a better metric, and across a broad range of cases it matches well with our chosen link cost below. Our solution to *Path Selection Problem PS.1* is therefore: *to minimize total work along the path, using number of measurements as the metric.*

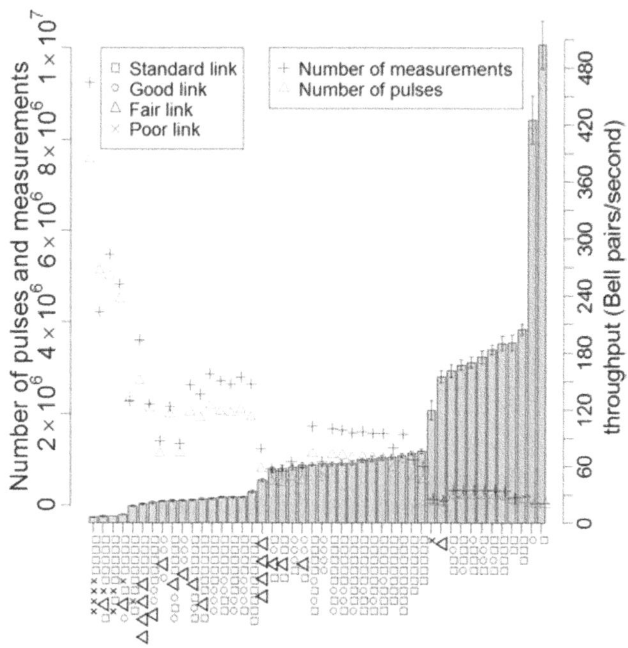

Figure 14.9. *Simulations for $F \geq 0.90$, plotted as in Figure 14.4*

We saw in section 14.4.2 that defining BellGenT as our link cost metric is likely to better suit our purposes than Pulse, Meas, Loss or InvTrans, allowing us to propose a solution to *Problem PS.2*: *seconds per Bell pair.*

Our simulation results presented in the previous subsections indicate clearly that we can usually, if not always, correctly predict the highest-throughput, lowest-work path from among a set of candidates. Dijkstra/BellGenT solves *Problem PS.3* well: *path cost is the scalar sum of link costs.*

Intuitively, we can see why this works well, as confirmed by the simulations: BellGenT is directly related to the number of purification rounds required, and hence the amount of work on a link. Despite the concerns we expressed earlier about the nonlinear amount of work as the number of hops grows, the predictive ability for ordering paths remains strong. Thus, we can assert that even as the use of complex repeater networks evolves, and various traffic patterns arise, Dijkstra/BellGenT likely will remain an effective choice.

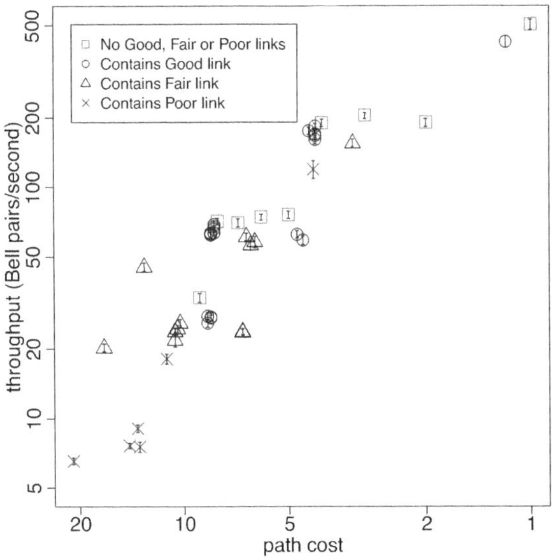

Figure 14.10. *Throughput ($F \geq 0.90$) versus BellGenT path cost for forty-six of our candidate paths*

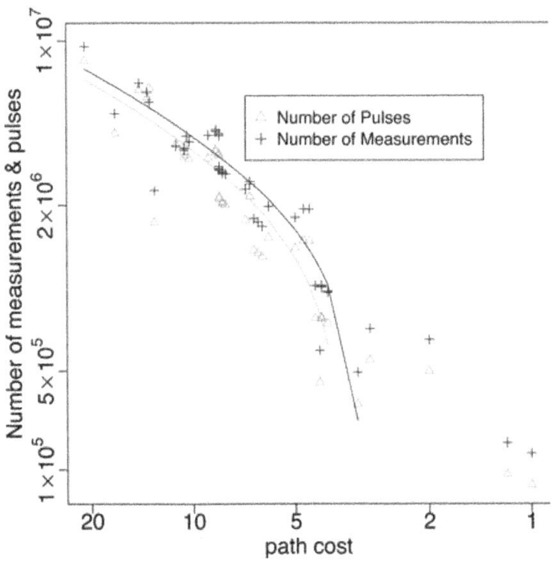

Figure 14.11. *Total work to achieve output fidelity $F \geq 0.90$, measured in Pulses ($\triangle$) and Measurements ($+$), versus BellGenT path cost for forty-six of our candidate paths*

14.5. Conclusion

Real-world deployment of networks of quantum repeaters will inevitably be physically heterogeneous, with complex topologies and high- and low-quality links and many possible paths through the networks, rather than an idealized, homogeneous power-of-two number-of-hops.

In this chapter, we have investigated critical problems in the use of purify-and-swap repeater networks. The focus has been on the path selection and the need for a routing algorithm. An acceptable routing algorithm must be an easy-to-calculate metric that reliably chooses a reasonable, if not optimal, path.

The results show that, despite many important differences, quantum repeater networks behave similarly to classical networks in useful ways, but the classical principles cannot be applied without thought. Via detailed physical simulation of both the physics and the classical messaging protocols, we have investigated several variants and explored the range of conditions under which these principles apply.

We can predict the *throughput* of a connection based primarily on the bottleneck link in the path, while the *total work*, in number of operations performed (pulses or measurements), increases with the addition of other, non-bottleneck links, much as in a classical network path. Applying a form of Dijkstra's algorithm with the inverse throughput of each hop as the link cost (Dijkstra/BellGenT) results in strong correlation between our easily calculated path cost and actual throughput, and between cost and total work. This is achieved with reasonable computational effort, allowing us to recommend the use of Dijkstra.

These results are in contrast to, but not direct contradiction of, Di Franco and Ballester's conclusion [DI 12]. Di Franco and Ballester examined a form of Dijkstra's algorithm in which the goal is to maximize the end-to-end fidelity outcome of entanglement swapping along a path in a static graph of links. The links do not just have varying fidelities, but potentially different forms of infidelity. They demonstrate that the best end-to-end path is not necessarily a composition of the best choice for each subsequence of that path. That is, in a path $A \to ...B \to ...D$, the best AB path does not necessarily appear in the best AD path. However, with some reasonable constraints on the density matrix (the "shape" of the impure quantum state), we have found that in practice these cases are likely to be rare, in part due to the relatively uniform Bell pairs that result from purification.

Moreover, we have established an operational focus on delivering the largest number of Bell pairs at or above a threshold in unit time, that is, the highest throughput. In addition, we find that implementation details such as the availability of memory and the stochastic generation of Bell pairs matter. Thus, although the

Di Franco finding stands as an important theoretical result, our results show that Dijkstra works well in practice.

More work is required to confirm the general behavior of Dijkstra with other physical-layer repeater types, adding single-photon [JIA 07a] and low-photon number [MUN 08] to the qubus systems explored here. The simulator is capable of modeling the important factor of finite quantum memory lifetimes [HAR 07], but the sheer additional combinatoric complexity that would have come from including variation of this parameter in both the simulations and algorithmic arguments is substantial. To keep the argument straightforward, perfect memory was assumed.

Perhaps, the most important open question is whether these results apply to error-corrected, rather than purified, repeater networks [FOW 10, JIA 09, MUN 10]. Demonstrating the applicability of Dijkstra to both purify-and-swap and error-corrected repeater networks would be a strong indicator of the universality of the results.

Chapter 15

Quantum Recursive Network Architecture

Finally, in this chapter we will present an architecture for a true quantum Internet, as it might be deployed in the real world, growing and evolving over time and involving many participants and technologies. Internet-scale quantum repeater internetworks will be heterogeneous in physical technology, repeater functionality and management. The classical control necessary to use the internetwork will therefore face similar issues as Internet data transmission.

In 2011, Van Meter, Touch and Horsman introduced the concept of a Quantum Recursive Network Architecture (QRNA), developed from the emerging classical concept of recursive networks, extending recursive mechanisms from a focus on data forwarding to a more general distributed computing request framework. Recursion abstracts independent transit networks as single relay nodes, unifies software layering and virtualizes the addresses of resources to improve information hiding and resource management. The architecture is useful for building arbitrary distributed states, including fundamental distributed states such as Bell pairs and GHZ, W and cluster states.

QRNA's organizing principle is recursion, and in conjunction with developing our recursive techniques we will construct the state machines for each of the roles that must be filled. Perhaps the most radical difference from classical networks arises from the need to extend message semantics. The classical messages carry an implicit request: please forward me toward my destination. Messages in our QRNA will carry requests more explicitly: please build this state for me, and dispose of it like so once it is built.

After reviewing the concepts of network architecture we covered in Chapters 1 and 3, we describe the request structures that make the recursive architecture possible (section 15.2), then show how this concept makes real-world deployment of truly large-scale heterogeneous networks practical (section 15.3).

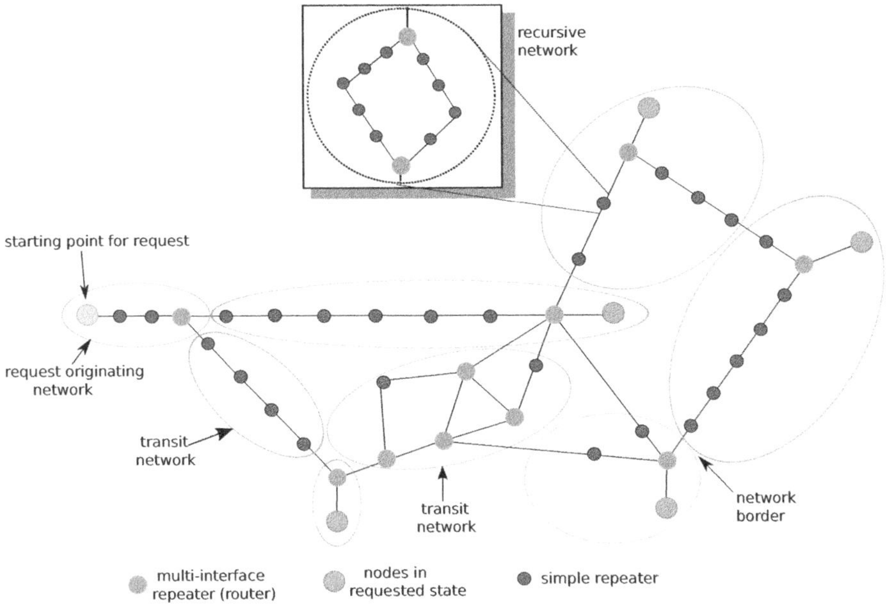

Figure 15.1. *Large-scale quantum repeater networks will consist of many nodes. Many long-line connections between routing-capable vertices ("routers") will consist of multiple-hop chains of simpler repeaters. In a recursive network, each node in the figure can actually represent a complete network itself that provides the services of a single node*

15.1. Review: network architecture

It is important to return to our networking roots, and see how we are doing relative to some of the principles, problems and ideas we laid out earlier in the book. Do the communication session architectures we have covered align with these ideas? What about the network topics discussed in the last couple of chapters? We will also take the opportunity to preview some of the value of QRNA, as developed in this chapter.

In Chapter 1, we introduced four foci for decisions in designing a network architecture: (1) the semantics of the messages that are exchanged; (2) the naming system used, including the form of identifiers used to name elements of, and participants in, our network, the range of these items, and how and where the names may be modified or translated; (3) the paths used to enable communication between nodes not directly connected; and (4) a means of managing the resources in the network. In the last two chapters, we addressed the third and fourth points; they are isolated relatively easily from other aspects of the architecture. This leaves us with the first two issues to address, and our solution will form the core of QRNA.

Chapter 3 presented several challenges in scaling networks to Internet scale and beyond, and introduced ten design patterns commonly used in Internet-related systems. The challenges are (1) heterogeneity, especially of deployed technologies and local conditions; (2) sheer scale, affecting routing and naming in particular; (3) dealing with the out of date information about current network conditions (e.g. routing or congestion) and the success or failure of requested operations; (4) meeting the needs of participating organizations, such as privacy, desired traffic transit policies and autonomous management; and (5) dealing with misbehaving nodes on the network, whether the misbehavior is deliberate or accidental. The communication session architectures addressed in Chapters 10 to 12 are aimed primarily at solving the problems of decoherence and probabilistic success in the context of issue (3). QRNA will address issues (1), (2) and (4). Any complete network architecture must support at least one of the session architectures in issue (3); QRNA, in fact, aims to allow any of the session architectures to be deployed although the exact relationship to demands on network nodes remains an open question. Issue (5) is, so far, altogether unaddressed in quantum networks.

The 10 design patterns introduced in Chapter 3 are hardly exhaustive, but they have been helpful over the previous two chapters, and all figure into the design of QRNA. Hierarchy is necessary to achieve scalability in routing and naming systems; calculating the optimal path through a network of a billion devices is out of the question for current computing systems.

Layering is a natural means of dividing functionality, and the associated modularity allows us to replace individual functions more or less independently. The layering we have presented in this book, such as in Figure 1.4, is not exactly equivalent to the layering of classical systems. In fact, two of my own collaborators have argued that the error management and state composition functions are better described as *roles* rather than layers, though (with apologies to them) I have retained the simpler terminology of layers for this book.

A key, related factor in the technical success of the Internet has been the "narrow waist" of the protocol stack, with the Internet Protocol itself serving as the waist. QRNA aims to similarly enable replacement of individual modules through a unified messaging system, although a true waist will be less obvious in a recursive architecture.

The generic notion of multiplexing the use of resources provided us with a framework for the major issue of resource management, and we assessed the viability of several schemes in Chapter 13. This issue is largely independent of the QRNA messages and semantics discussed in this chapter, but the results we found are encouraging for statistical multiplexing. However, at that time, we did not assess the real-time decoherence of qubits, and we also noted that other issues such as request management might push us toward a circuit-switched architecture. QRNA strives to relax the need to tie down a specific path and all of its resources through the judicious use of recursion, but the details remain to be worked out.

We discussed smart and dumb networks, with the distinction being how much application domain-specific functionality is built into nodes in the interior of the network and the resulting restrictions on deploying the new services. Certainly, each of the error management schemes we have discussed so far requires fairly sophisticated nodes, but all in the service of creating Bell pairs, which are useful, generic entangled states. It should be relatively easy to deploy a variety of services on top of this basic functionality. QRNA provides a measure of programmability alongside the data movement or state construction requests, and so raises both the intelligence and flexibility at the same time.

Distributed management and autonomy have not appeared directly in the prior chapters, but are a major goal of the use of recursion, and hence of QRNA. This also allows us to effectively deploy the distributed path selection algorithms, not just in a two-level, interior/exterior arrangement as in today's Internet, but also in a recursive fashion.

We used the state machine approach to protocol design extensively in our designs for the link management, purification and remote state composition. It will not figure prominently in the visible portions of the QRNA design, but the requests we will describe here likely will be issued and absorbed by such state-driven code.

The network itself is in constant flux as nodes join and leave and links go up and down, and other management changes are made. The weak consistency and soft failure design approaches in IP-related networks are not incorporated directly into the models we have presented so far; hence, their robustness can be called into question. QRNA attempts to partially rectify this by abstracting networks, allowing a portion of a path to be relocated transparently within a network.

And, of course, the notions of overlay networks, virtualization of topology and explicit recursion are the heart of QRNA.

15.2. Recursive quantum requests

Recursion is a natural model for quantum repeater networks because purification, entanglement swapping and Calderbank-Shor-Steane (CSS) error correction [CAL 96, STE 96] or surface code error correction [FOW 09, RAU 07a, RAU 07b, WAN 10] build on mixed entangled states and produce other mixed entangled states, working toward a common goal of a high-fidelity, wide area-distributed quantum state. The similarity of the interfaces on the top and bottom of a given protocol layer simplifies recursion, allowing more or less arbitrary composition of protocol stacks.

In a large network (millions to billions of nodes spread across many countries and organizations), direct management of the network as a single, synchronous, shared, centrally managed system is impractical, and even optimization of smaller portions of the network becomes a computationally intractable combinatoric problem. Applying recursion takes away much of this complexity and allows us to effectively manage the larger set of resources. Each protocol layer, node or network needs to recognize and be able to reach only a small subset of the entire network's resources, and hides much of the underlying complexity to allow its own clients to operate in a smaller subspace.

In order for recursion to be effective, we must have a well-defined request-response model that allows us to combine protocol layers. Before the requests and responses can be defined, we must have some idea of the processing necessary, we must be able to name the distributed entangled states themselves and we need to understand the resources comprising the states. The next three sections address these issues.

15.2.1. *Processing in recursive networks*

Classical networking uses recursion to represent topology hiding, but we can also consider the entire network architecture as recursive as well [DAY 08a, TOU 06]. As an architectural principle, recursive networking explains layering of protocols (and their modular software architecture), name resolution, routing and forwarding as more than just artifacts of the current Internet [TOU 10].

Consider the steps of classical recursive networking, shown in Figure 15.2. When a packet is received by a node, the packet is implicitly requesting that the node forwards the packet on toward its destination. This algorithm is executed by the node to perform the forwarding. The `process()` step may alter the packet, including turning a single packet into more than one. The `FOREACH` loop passes the packet(s) down the protocol stack, as in Figure 3.3.

QRNA adopts an outline similar to Figure 15.2, with different semantics. In quantum networks, `data` contains a computation request using virtual identifiers for resources, and the `process()` step represents the local operations that are performed in a repeater toward fulfillment of that request, such as the entanglement swapping that happens when Bell pairs are spliced to form a longer pair [DÜR 99, VAN 09]. The output of `process()`, `newdata`, may be more than one request. The `map()` function may modify the addresses in a given request. In our architecture, the node identifiers do not change within a forwarding path, but requests may be retargeted from a network destination to a node destination, as we describe in section 15.3. The corresponding concept is shown in layered communication, as supported by recursive networking, in Figure 3.3.

```
deliver(data, src, dst) {
  process(data) -> newdata
  WHILE (here != dst) {
    found = FALSE
    FOREACH (lowerlayer) {
      map(src,dst,lowerlayer) -> newsrc, newdst
      IF (deliver(newdata, newsrc, newdst)
          == TRUE) {
        found = TRUE
      }
    }
    IF (found == FALSE) {
      /* if you get here, you failed to deliver
         the data */
      FAIL
    }
  }
  /* if you get here, you're at the
     destination */
  RETURN TRUE
}
```

Figure 15.2. *The algorithm for recursive resolution and forwarding, adapted from Ref. [TOU 06]. This algorithm is executed at each node as it receives data to be delivered.* `src` *and* `dst` *are the source and destination addresses*

Beyond this basic structure for forwarding, in QRNA the requests themselves become recursive, and must be carried explicitly through the classical network. Next, we turn to the structure of these requests.

15.2.2. *Naming a state*

Over the course of the lifetime of a quantum network, many entangled quantum states will be created and consumed. Repeaters will make both independent and coordinated decisions about which states to purify, swap, error correct, forward, buffer and discard, as they build states that satisfy users' requests. This is, of course, intimately tied up with the distributed density matrix management (section 8.5) and path management.

To communicate successfully about these states, nodes must be able to *name* states using a namespace that other repeaters will understand: "do operation U on this particular state we share." To construct such requests unambiguously, the qubits within the states must also be named.

The simplest naming scheme for a particular qubit is the tuple (N, A), where N is the node name and A is the physical qubit address within the node. However, there are three key problems with this scheme:

– each node is entitled to move the logical state of a qubit from one physical qubit to another *without notifying other nodes*;

– physical qubits are reused after being freed; and

– the node issuing the original request may need to refer to the qubit by name even before physical resources for it are actually allocated (e.g. a request for a gate to be executed may be issued at the same time as the initial entangling pulse).

These factors mean that physical address is a constraining and unreliable identifier for the quantum states that are our true subject of interest. All of these problems can be solved by allowing the original requesting node to assign a *virtual address* or other abstract identifier for the qubit; the node (or network) housing the physical resource is responsible for maintaining the mapping of virtual to physical resource. That mapping information is private to the node and need not be disclosed or coordinated with other nodes. To ensure that the virtual address assigned by the requester is unique, the full address tuple must include the requesting node and the actual request identifier.

The naming scheme must be prepared for names to shift as operations proceed. Multiple quantum states often merge to become a single state. Purification, entanglement swapping and error correction all result in such mergers flowing up the protocol stack, and result in multiple requests moving down the protocol stack. Names for states and qubits may be remapped when crossing boundaries. Names for nodes, when visible, do not generally need to change, but requests moving from the outside of a network to the inside may become more specific at the boundary.

Each boundary in the system, whether a software boundary between modules or a hardware boundary between nodes or networks, represents a point at which resource names and requests may change. Logically, these boundaries represent points where these mappings and requests must be maintained, although in implementation this may vary.

15.2.3. *Defining quantum requests*

As classical distributed computation proceeds, applications running on several nodes request that the network subsystem send and receive messages or, using higher-level constructs, synchronize the state of distributed copies of shared data structures [COU 05, LYN 96]. In the quantum world, as we have seen, requests are more about managing a quantum state than explicit sending and receiving of messages, though that state may in fact be in motion from node to node.

Fundamentally, the conversation between nodes, or between the application and other software subsystems within a node, is about answering a series of questions: *What state* do we want? *Where* do we want it (which nodes should it span)? And, because quantum systems are imperfect, *how good* does it have to be? Finally, to deal with high latency in networks and fragile quantum data, and to support our recursive architecture, *what should be done* with the state once it is completed?

The interface to the network subsystem must allow the requester to specify the desired state $|\psi_S\rangle$, while the network will actually return

$$\rho = \mathrm{Tr}_{AB}|\Psi\rangle\langle\Psi| \text{ where } |\Psi\rangle = |\psi'_S\rangle \,\widetilde{\otimes}\, |\psi_{A+B}\rangle \qquad [15.1]$$

where $|\psi'_S\rangle$ spans the set of state qubits, $|\psi_{A+B}\rangle$ is the set of ancillae (defined but unused qubits, for this state) plus the bath (the environment) and $\widetilde{\otimes}$ indicates that what we get in the real world is only an approximation of a separable state. The aim is to have

$$\rho \approx |\psi_S\rangle\langle\psi_S| \qquad [15.2]$$

within certain tolerances. The request must therefore also specify these tolerances on the state: a minimum fidelity and a maximum entanglement with the ancillae and bath. Thus, a density matrix should be viewed as *no-less-than* for the element(s) corresponding to the desired state, and *no-more-than* for the elements corresponding to undesired states.

Both the fidelity $F = \langle\psi_S|\,\rho\,|\psi_S\rangle$ and entropy $S = -\mathrm{Tr}(\rho\log\rho)$ appear in the request to constrain the returned state ρ to be near the desired state. The fidelity is to ensure closeness to $|\psi_S\rangle$; the constraint on the entropy of ρ allows the system to filter out returned states that may be non-trivially entangled with other nodes in the system. In the limit of $F \rightarrow 1$, the entropy becomes unnecessary, but for fidelities bounded farther away from 1, the entropy becomes a useful tool. We assume that the repeater nodes make repeated use of the same physical resources and sometimes swap data qubits with ancillae, which if done imperfectly leaves behind some residual entanglement between qubits that should not be entangled. Further reuse of those ancillae can therefore further entangle data qubits in an undesired fashion. Because both the qubits on which $|\psi_S\rangle$ are defined and the ancilla qubits may be entangled with the environment (i.e. in a mixed state), the state of ρ alone cannot determine if any node qubits are entangled with any of the ancillae. Limiting the entropy of ρ serves to limit the possible entanglement with ancilla qubits by limiting all external entanglement.

In addition to these properties, the requester must specify the desired logical or physical encoding of the quantum state. An application will request an absolute encoding, whereas each layer in the protocol stack provides a relative encoding (discussed further below), with the entire stack to provide the absolute encoding.

The tuple specifying a request for a state is

$$T = (ID, |\psi_S\rangle, F, S, ((N_i, A_i)), E_A), \quad [15.3]$$

where ID is the transaction identifier assigned by the requester, F is the minimum acceptable fidelity of ρ with $|\psi_S\rangle$, and S is the maximum acceptable entropy of ρ. $((N_i, A_i))$ is the set of nodes that are requested to comprise the state and the virtual addresses A_i that are to be used for the qubits, and E_A specifies the absolute quantum error correction encoding. $|\psi_S\rangle$ is the desired pure state; the exact encoding of the description of the requested state does not have to be specified at this point, but can take numerous forms, including state vector, stabilizer and circuit descriptions. (Density matrix is of course possible, but given that we are specifying a pure state, would be overkill.)

Requests may also be for *actions* to be executed on specific states, in which case the tuple is

$$T = (ID, C, F, S, ((N_i, A_i)), E_A), \quad [15.4]$$

where C is a circuit that may include both unitary and measurement operations.

The return value of a request is the tuple

$$R = (ID, \rho) \quad [15.5]$$

where ρ is the density matrix of the delivered state for request ID. The set of resources represented by ρ is specified by the basis $((N_i, A_i))$, the tuple of tuples including node (or network) identifiers N_i and the virtual addresses A_i included in the original request.

Benjamin *et al.* described a brokered approach to building large-scale graph states from smaller ones, tailored to a specific hardware implementation [BEN 06]. QRNA provides a framework for abstracting and generalizing this process, including support for cost functions that will allow intelligent decisions for constructing the sub-graphs.

15.3. Implementing recursion in quantum networks

15.3.1. *Satisfying quantum requests*

Requests naturally originate at applications running on specific nodes and are processed through a series of software protocol modules that implement the layers of the protocol stack, with carefully defined interfaces between the layers. Each layer in the protocol stack has access to a set of resources it can use to satisfy requests: it knows about a certain set of network nodes (or, more scalably, how to *find out* relevant information about a set of network nodes), can ask for certain states (including entangled states) to be created on that set of nodes and for certain operations to be performed on those qubits, and can use its own internal capabilities. It has exclusive control of a certain set of resources, and may consult with the corresponding layer instances at remote nodes about the best way to satisfy requests. However, it should endeavor to make *independent but coordinated* decisions whenever possible so that the latency penalty for explicit messaging can be avoided.

Each protocol instance has the ability to execute local quantum operations (unitary operations and measurements), as well as compute and communicate classically with other repeater nodes. The instance has no access to distributed quantum states or operations beyond those it currently owns. If additional states are required to complete an operation, they must be requested from protocol layers below or from peers.

Requests are not constrained to be 1:1; a single request from above may be mapped to multiple requests to the layer below. A protocol layer has the right to merge and split states and issue multiple requests to meet its obligations. The ability to *buffer* quantum states, to hold them while waiting for other resources to become available (e.g. other quantum states or answers to classical queries), is generally necessary when coordinating multiple requests.

Protocols that make decisions about how to get from place to place in the network must have access to a *cost function* for specific requests that can be used to make intelligent decisions, discussed next.

15.3.2. *Paths and rendezvous points*

QRNA depends on the ability to find a path through a network. On a modest-sized network, Dijkstra's algorithm can be applied to select a path through the network, though on a larger internetwork, a multi-layer scheme becomes necessary. As in the Internet, a path will be calculated locally based on the distributed information exchanged through the network dynamically. The path will consist of node names for nearby portions of the path, but network names (e.g. AS numbers) for portions that are farther away, assisted by the recursive nature of QRNA. Whether the path itself, once calculated, is visible outside of the software on a given node is an open question, and ultimately is related to resource management (e.g. circuit reservation). If so, of course an identifier for that path must be established.

Purify-and-swap repeaters require the explicit use of named rendezvous points along the path where the entanglement swapping occurs. After the path is selected, the swapping points are optimized on the chosen chain of repeaters, as in Chapter 14. The order of entanglement swapping can be either specified or left unspecified. Again, these rendezvous points must have names, which may be either a node or a network. Using a network as a single rendezvous point requires use of recursion, and will require that the network internally be able to give the appearance of a single node.

Hop-by-hop teleportation and QEC-based repeaters (including the quasi-asynchronous variants) require a similar path selection mechanism, but have no direct need for rendezvous points.

15.4. Example

As an example, consider an application requiring a three-qubit cluster state defined by the circuit in Figure 15.4. The request originates at Node11, with the three qubits requested to be at Node11, Node55 and Node77 in the network in Figure 15.3. The application begins by specifying the state it wants using a tuple as in equation [15.3], then other (system) software running at Node11 creates a global strategy for how to achieve the state, and sends requests to corresponding nodes or networks. The bulk of this work happens in the QRNA equivalent of the `process()` step in Figure 15.2. The nodes that receive the requests in turn will craft their own strategies for the requests they receive. Although the two stages of creating a strategy and choosing where to send the application requests are intertwined, here we will describe them separately for clarity.

The application running on Node11 creates a request of the form

$$\begin{aligned} R_A = & (1, |\psi_A\rangle, F \geq 0.99, S \leq 0.1, \\ & ((\text{Node11}, 1000), (\text{Node55}, 1000), \\ & (\text{Node77}, 1000)), \text{Raw}), \end{aligned} \qquad [15.6]$$

where $|\psi_A\rangle$ is the cluster state created by the circuit in Figure 15.4, 1000 is the virtual address chosen to be used for the qubit requested at each node, and Raw indicates that we are requesting an unencoded state.

To fulfill R_A, the first system software module to process the request (still at Node11) must create a global strategy. The principal decision is whether to create the state in one location and move the qubits via teleportation, or to allocate memory for the qubits at the destination(s), then use teleported gates (as in section 4.4) to execute the circuit in a remote fashion. For this specific circuit, either approach requires three remote operations, each consuming a Bell pair. The exact cost of creating the corresponding Bell pairs will depend on the network topology. For this example, we will assume that the global strategy choice is local creation followed by qubit teleportation.

With the global strategy chosen, the next step is selecting *where* the operations will take place. The routing table at Node11, shown in Table 15.1, contains information on how to get to all destinations on the network. To achieve scalability, the table has more precise information about nearby destinations, and vague information about more remote destinations, achieved using hierarchy and recursion.

Based on a cost function that uses the information in the routing table, Net5 is identified as being close to the "center" of this request. Thus, the strategy module chooses to ask Net5 to create the cluster state, after which Net5 will teleport the qubits to Node11, Node55 and Node77.

As shown in Figure 15.5, the original request (left side of the figure) is broken down into seven separate requests (right side of the figure): one for the state to be created local to Net5 (labeled R_{Net5}), three for Bell pairs to be used for teleportation (labeled $|\Psi\rangle_1$, etc.) and the teleportation operations themselves. In this case, R_{Net5} is the same circuit as in Figure 15.4, with the resources specified as local to Net5 rather than distributed.

Each box in the figure lists the virtual addresses of the qubit resources to be used for that request. The virtual addresses are created when the requests are created, but are not assigned to matching physical resources until the requests are processed at the receiving nodes. Each of these requests must also carry information about fidelity and entropy, with those values chosen to ensure that the delivered final state will meet the originally requested constraints. Based on the routing table in Table 15.1, each of these requests is then sent via the classical network to each node involved; in this case, Node51, as the gateway to Net5, will receive most of the requests. Node51 will then forward the requests onward, or craft its own strategy, as appropriate. Requests can be executed once all dependencies (indicated with arrows in Figure 15.5) are satisfied. The application's request is completed once all of the component requests finish.

Although this example shows only a single layer of recursion, the process may be repeated indefinitely for the physical nodes (as shown in Figure 15.1), or for requests. To achieve adequately high fidelities, the node assigned to process each of these requests may in turn break the request down further into multiple requests for base-level entangled Bell pairs and purification operations. Likewise, for those operations spanning multiple hops, either entanglement swapping or hop-by-hop teleportation can be requested.

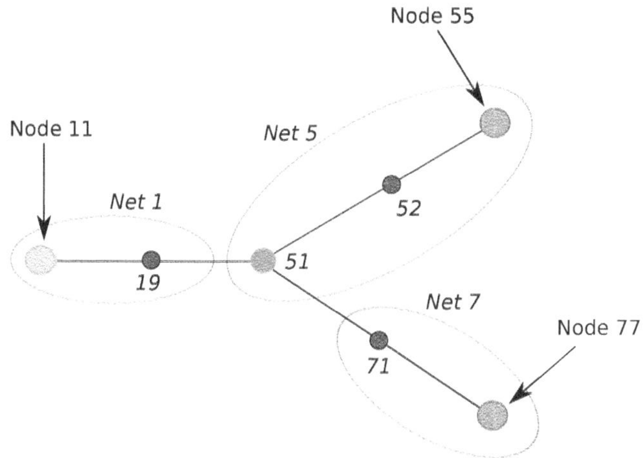

Figure 15.3. *Example of a small-scale internetwork composed of three networks. Our example request is initiated at Node11, and includes Node55 and Node77*

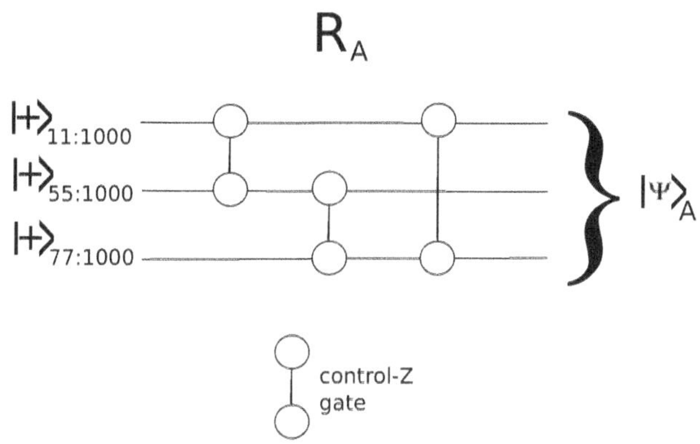

Figure 15.4. *Circuit for the three-qubit cluster state requested at Node11.* 11 : 1000 *and similar are the virtual addresses for the qubits, assigned by Node11*

Destination	Route
Node19	(direct)
Net1	Local
Net5	Node19
Net7	Net5

Table 15.1. *The routing table at Node11 contains information on how to get to all destinations on the network. To achieve scalability, the table has more precise information about nearby destinations, and vague information about more remote destinations, achieved using hierarchy and recursion. Node55 resolves to Net5, and Node77 resolves to Net7, so that independent records are not needed for each node*

Destination	Route
Node52	(direct)
Node55	Node52
Net1	Node19
Net5	(process locally)
Net7	Node71

Table 15.2. *The routing table at Node51*

15.5. Conclusion

The fundamental difference between classical and quantum networks is the services they deliver. Classical networks move data from a source application to one or more destination applications over a distance. Quantum networks may likewise transport data from place to place, but in addition can produce distributed entangled quantum states, connecting two or more quantum applications. This difference requires a new form of interaction between network components. On the Internet, a received packet is implicitly a request: please forward this block of data toward the destination or destinations listed. In our QRNA, rather than such an implicit request, the requester explicitly asks a node or network to participate actively in the creation of a larger state. Thus, rather than simply an information transfer system, a quantum network is a general-purpose distributed quantum computing system.

The problems of truly large-scale quantum repeater networks have much in common with the problems of classical distributed computing: naming and resource management are critical issues, and judicious use of the concepts of hierarchy and recursion provide the right abstraction to keep the systems efficient while the data structures that must be managed at each node remain tractable in size. Dynamic composition of the protocol stacks provides the required flexibility, as well as isolation of responsibility.

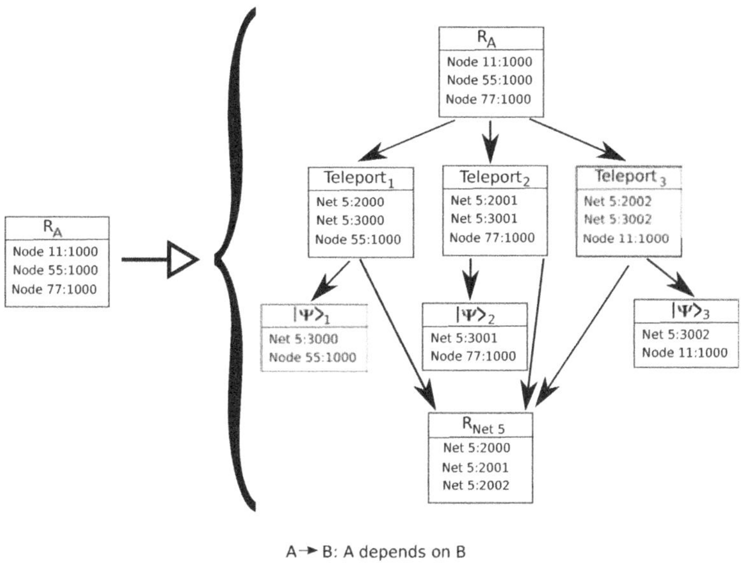

Figure 15.5. *The initial application request R_A is translated into a set of requests for sub-operations before leaving its origin, Node11. Each box lists not the full request tuple, but only the ID of a sub-request and the virtual addresses for the qubits assigned by the request creator*

All of these issues can be addressed through the use of recursive networking. QRNA abstracts subnetworks as individual nodes, allowing technology-independent requests for quantum state creation to be constructed with imperfect knowledge of the total network structure and state, and for those requests to be modified and processed in a recursive fashion as necessary to deliver the end-to-end quantum state required by applications running on quantum computers.

This chapter has assumed that network nodes and repeaters are well-behaved and are not malicious, but in the real world, those assumptions will not hold and the issues of robustness in the sometimes-hostile world will have to be addressed.

Long computations will naturally require not a single distributed state, but a sequence of them; reservations for such longer sequences, especially the real-time requirements, are beyond the scope of the current discussion.

Although we have focused in this book on the creation of a core group of entangled states that are common building blocks for distributed algorithms, the mechanisms generalize quite easily to support direct distributed execution of any quantum algorithm.

This concludes the technical discussion of quantum repeater networks. In Chapter 16, we finish with some thoughts on future directions for quantum networking research, as well as further readings for those whose appetite has been whetted by our discussion so far.

Chapter 16

Coda

Now we have covered quantum networks and internetworks in detail sufficient to allow the reader to find the edge of what is known, or accomplished, and understand how to push that boundary as a researcher; help to transform researchers' accomplishments into valued products as a developer; or decide where, when and how to deploy quantum networking technology as an operator. In this coda, written in early 2014, I would like to suggest some next steps as a community, as well as recommend readings for further study.

16.1. Future development

Some relatively obvious hurdles remain for building and deploying real-world networks. In particular, without significant advances in hardware, all entangled quantum networking will remain a laboratory demonstration, despite the work of theorists to relax the demands made of the hardware. Let us take a quick look at some of the hardware issues, and then speculate about how QRNA can become real.

16.1.1. *Hardware*

Higher fidelity will remain a major pursuit in quantum networking for some time to come, but arguably the existing experiments in Table 8.2 have reached a level at which the base Bell pairs are adequate for some purposes. Likewise, logic gates are not yet good enough for fully general quantum computation, but would work for purification and entanglement swapping. The hardware focus from this point, then, will likely lie elsewhere.

This leaves us with a series of important issues: first and foremost, the probability of successful entanglement is still too poor for production use, primarily due to optical losses and poor coupling of memory elements to the channel. Even relatively recent experiments have achieved success at rates below 10^{-6}. This probability needs to increase by several orders of magnitude for systems to become truly viable. This problem will be exacerbated by the need to couple more than one independently controllable qubit to a fiber or other channel so that we can have many transceiver qubits in a network interface.

A related matter is the need to transfer qubits from place to place inside the repeater, very likely transforming from one optically active physical type of qubit to a different, longer-lived type of memory. Memory lifetime itself is an important matter for quantum computers as well, but is especially critical in repeaters due to the latency of light, as we have seen throughout this book. However, experimentally achievable memory lifetimes in all technologies have been climbing rapidly for the last decade. It is too early to call this a "solved" issue, but it is well on its way to being so.

Of course, the choice of technology for a specific role within a repeater depends on future developments. One concern is that a technology may find favor in the laboratory or even in production quantum computing systems, but be less useful for repeater networks because it is difficult to use in the field, for example, because it requires cryogenic temperatures, constant maintenance attention or hard-to-achieve local conditions such as isolation from stray magnetic fields or radio signals.

However, I am optimistic that the hardware can be built. The advances since I entered this field have been tremendous, and it seems that every week a new step is being reported in the literature.

16.1.2. *Making QRNA real*

A key test of the viability of QRNA will be the implementation of some protocols using the APIs and messages. It is fairly obvious how to extrapolate from the example in section 15.4 to basic creation of end-to-end Bell pairs and other distributed states, but making the system concrete inevitably turns up interesting problems that require us to make difficult engineering decisions. A critical exercise will be to demonstrate that QRNA enables the E91 QKD protocol, and perhaps even the unentangled BB84. QRNA should be useful for the link-level monitoring necessary to maintain the base-level density matrix used in all of the higher-level decisions, although the actual real-time tuning of the link to optimize the fidelity or probability will likely remain beyond the scope of QRNA.

Perhaps one of the most interesting uses to which QRNA could be put in the short term would be the development of a CHSH inequality violation experiment. If QRNA proves to be capable of handling this task when coupled with a single, real physical link, that would be strong evidence that the QRNA model carries semantics that are complete enough.

This coupling of a single experimental link to actual QRNA software will be an important step, but carries us only partway to a functional network. Of course, we would like to be able to run this over multiple hops with fully distributed software, and even across multiple networks. A full demonstration of all of QRNA's capabilities is still a long way in the future.

16.2. Open problems

No list of open problems could possibly be comprehensive, but in this book, we have mentioned a few, and it seems worthwhile to review them. These problems are general, rather than specific to QRNA, and each could easily serve as a solid research topic for a student. This section contrasts with the previous section in that these are more open-ended, and less specifically tied to development of a network for building Bell pairs.

– By far the most pressing problem is the matter of a "killer app" for distributed quantum states. QKD is still the best developed and most important application, but it remains unclear whether it alone is compelling enough to drive an entire industry.

– To that end, the reference frame applications of Chapter 7 are highly intriguing, but a complete analysis of how to run them successfully in the context of a repeater network is still lacking.

– The papers assessing theoretical bounds on distributed quantum problems discussed at the top of Chapter 6 provide some intriguing suggestions about performance. The next step would be to make specific quantum algorithms concrete enough to conduct the same kind of network resource analysis presented for various algorithms in this book.

– The extent to which misbehavior of individual nodes can disrupt the overall operation of the network is a matter of both fundamental concern and engineering practicality.

– More broadly, the issues of security discussed in section 3.1.5 have yet to be addressed.

– We have no idea what the true traffic pattern on a quantum network will be like. In fact, we will not have any firm idea until the networks start to become reality. However, it may be possible to apply some of what we know about Internet traffic patterns [MED 02, SOU 05].

– Although QRNA hopefully provides a strong, flexible framework for internetworking, no one has yet sat down and worked out how to actually accomplish entanglement swapping between qubits of different QEC encodings to determine their error vulnerability, nor studied the connections between purify-and-swap on physical qubits and QEC-encoded qubits.

16.3. Further readings for depth

16.3.1. *Quantum repeaters and QKD*

Bennett *et al.*'s 1996 paper discusses an astonishing range of important ideas in purification and error correction [BEN 96c]. It is a must-read for anyone interested in purification, and hence in quantum repeaters.

Pan's 2003 photonic purification paper demonstrates some of the experimental difficulties well, and how what is done in an experiment does not always correspond exactly to the logic circuit diagrams that theorists draw when developing procedures [PAN 03]. Bernien *et al.*'s 2013 paper demonstrates many of the experimental difficulties with NV diamond [BER 13]. These two might form a good place to start reading experimental papers in more depth.

In this book, we have not delved into the mathematical details of determining a final secret key rate for QKD over quantum repeaters based on known facts about the density matrix of shared Bell pairs, but this is a critical area. Abruzzo *et al.*, from Bruß's research group, have conducted an in-depth analysis [ABR 13]. Some older repeater papers also used secret key bits per second as a performance measure [LAD 06].

16.3.2. *Optics and general quantum physics*

Hecht's *Optics* is a very good textbook in classical optics, and basic knowledge of this field is critical [HEC 02]. However, quantum optics involves some substantial differences from classical optics. Gerry and Knight's book on quantum optics is a good place to understand some of these differences [GER 05].

I learned quantum mechanics as an undergraduate from French and Taylor, which I found to be an excellent, if difficult, text [FRE 79]. Feynman's lectures on physics, of course, provide a wonderful guide to the intuition without skimping on the mathematics [FEY 63].

16.3.3. *Quantum computing*

For general study in quantum computing and quantum information, Nielsen and Chuang (Mike & Ike), originally published in 2000, has yet to be surpassed [NIE 00].

It is *the* standard text in the field. Preskill's lecture notes from an early course on quantum computation are also considered a standard reference, and they are freely available on the Web [PRE 98a]. Kitaev's book is readable and brief; although it is more mathematical and abstract than Mike & Ike, its different perspective is very valuable [KIT 02]. Readers are also referred to both popular articles and books to help grasp the key points at an intuitive level [NIE 03, WIL 99].

For an overview of recent progress in quantum algorithms, Bacon and van Dam and Mosca have published surveys [BAC 10, MOS 09], although recent work on machine learning and related topics post-date these. For a discussion of the size and capabilities of machine necessary to run algorithms at scale, begin with [VAN 13a]. To study the theoretical computational complexity of quantum computation, [BEN 97] is seminal, and Aaronson's PhD thesis is highly readable and comprehensive [AAR 04].

16.4. Further readings for breadth

Despite the breadth of the work covered in this book, we have not come close to exhausting the range of inquiry in distributed quantum computing and communication. Each of these topics would easily warrant a full chapter, or even a book of its own. The readers are encouraged to learn more about them.

16.4.1. *Information theory*

Overall, we have largely ignored the fundamental field of *quantum information theory*. This is not due to a lack of appreciation of the importance of the field; indeed, it is a separate field of its own, with books covering the subject, such as Hayashi's [HAY 06], Wilde's [WIL 13], and Imre and Gyongyosi's [IMR 12]. A large chunk of Nielsen and Chuang is devoted to the area, and is an excellent place to start. Winter and others have substantially extended the field since the publication of Nielsen and Chuang. Smith and Yard derived the startling result that two quantum channels with zero capacity can be bonded together to create one channel with non-zero capacity [SMI 08], a result that is still reverberating through the field some 5 years later.

16.4.2. *Dense coding*

Dense coding is yet another intriguing, fundamental aspect of the behavior of distributed entanglement that came out of the fertile minds of Bennett and Wiesner [BEN 92]. Using this coding, teleporting a group of qubits can be done while seemingly flaunting the rules of ordering operations, as required by relativity.

In reality, of course, the rules still hold; as with basic teleportation, the key is the supporting classical information that must be transmitted. Dense coding was experimentally demonstrated in 1996 by Mattle, Weinfurter, Kwiat and Zeilinger [MAT 96].

16.4.3. *Quantum network coding*

An area that would warrant an extensive analysis in its own right is *quantum network coding*. Building on the concepts in classical network coding [AHL 00], Hayashi, Iwama, Leung, Winter and others have developed a quantum equivalent, with surprising differences from the classical version [HAY 07, IWA 06, LEU 10].

Classical network coding, in its abstract form, shows how multiple conversations propagating across a network in slightly different directions can complete their collective work in fewer uses of unidirectional links in the network than seems intuitively necessary. This is achieved by performing simple calculations on multiple independent messages at nodes in the middle of the network, and inverting those calculations at the destinations. Network coding is especially useful in multicast scenarios and in wireless networks.

The quantum version behaves rather differently, partly due to the incompressibility of quantum information and our inability to destroy information at nodes in the middle of the network without negatively affecting the overall state. The concepts were developed for abstract graphs of entanglement, but recently have been extended by Satoh *et al.*, to work with networks of repeaters [SAT 12].

16.4.4. *Entanglement percolation*

Throughout this book, particularly the last several chapters, we have assumed that the operational approach of a network is to select a single path for each communication session and use that path exclusively, and that multiple sessions will be competing for access to the network at the same time. A rather different abstraction for analyzing networks of complex topology is *entanglement percolation* [ACÍ 07, CUQ 09, CUQ 11, HOL 02, LAP 09, PER 10, PER 13].

In classical percolation theory, links (sometimes called "bonds") or nodes are either present or absent on a defined graph with a certain probability. The questions addressed include the probability of a given pair of nodes being connected by instantiated links, and the probability of the left edge of the network being connected to the right edge. The analyses are often performed on infinite networks (in one or both dimensions) so that behavior in large networks can be examined in the limit.

In quantum percolation, the behavior of a graph of Bell-like entangled pairs is examined, with nodes as the junction points where multiple Bell-like pairs meet. With a *single use* of the entanglement, can a Bell pair connecting two chosen nodes be fabricated? Early papers examining this question assumed pure but incompletely entangled pairs of the form $|\psi\rangle = \alpha_{00}|00\rangle + \alpha_{11}|11\rangle$, with $0 < \alpha_{11} < \alpha_{00}$ and of course $|\alpha_{00}|^2 + |\alpha_{11}|^2 = 1$. Longer distance entanglement can be created by *entanglement concentration*, a purification-like operation on pure states that probabilistically either creates a perfect Bell pair, or fails. More recent papers have addressed the issue of mixed states, including the use of purification, and have found rather different behavior with more stringent limits on range. In the pure state approximation, a pure, perfect Bell state may be generated using percolation, but with mixed states of fidelity $F < 1.0$, it is never possible to construct a Bell pair with $F = 1.0$.

Operational networks will use the repeat-until-success methods discussed in this book, but percolation scenarios are usually presented in terms of a single use of the network. In real-world networks, asynchronous operation and the classical communication to communicate results will dictate much of the behavior of the network.

One area where percolation may offer an insight is the possible use of multiple paths in the generation of a single end-to-end (imperfect) Bell pair in our repeater networks.

Applying these important theoretical results to realistic networks is a critical open problem. For more details, an excellent place to start is the recent review paper by Perseguers *et al.* [PER 13].

16.5. Final thoughts

To date, most research on quantum networking has focused on systems for creating high-fidelity generic entangled states, such as end-to-end Bell pairs [DÜR 99] or larger graph states on model networks [BEN 05b, MUN 05]. The resulting generic states are then used for the remote execution of quantum gates [GOT 99], teleportation of valuable application-level qubits [BEN 93, FUR 98, OLM 09] or the creation of shared classical random bits via measurement (e.g. for QKD [EKE 91, MAR 08]). The more general concept is the creation of arbitrary distributed entangled states; thus, a quantum network is effectively a large-scale distributed quantum computing system.

As I noted earlier, hardware research is progressing well, and I am optimistic that development of deployable hardware is now within reach. The work done over the past several years to improve communication session architectures makes some demands of the hardware, particularly with respect to fidelity adequate for running quantum error correction, but in turn simplifies other aspects, such as reducing the need for long memory lifetimes. With the work in my group and a few other places on quantum network and internetwork architectures, we can see the broad outlines of how all the technologies can come together to be deployed. QKD is a valuable function, but more work on applications will create more compelling use cases and help to drive adoption. As with quantum computing systems, I fully expect that once networks are in the hands of some hackers and engineers, we will see a flowering of so-far undreamed-of applications.

As the book began, we noted that "teleportation" is a magic, evocative word. We have seen that it is also very real and the key component of distributed quantum information. With hard work, some help from Mother Nature and a little luck, soon we will have a quantum Internet and teleportation will be a common event.

Bibliography

[AAR 04] AARONSON S.J., Limits on efficient computation in the physical world, PhD Thesis, University of California-Berkeley, 2004.

[AAR 12] AARONSON S., FARHI E., GOSSET D., *et al.*, "Quantum money", *Communications of the ACM*, vol. 55, no. 8, pp. 84–92, 2012.

[ABR 97] ABRAMS D.S., LLOYD S., "Simulation of many-body Fermi systems on a universal quantum computer", *Physical Review Letters*, vol. 79, pp. 2586–2589, 1997.

[ABR 13] ABRUZZO S., BRATZIK S., BERNARDES N.K., *et al.*, "Quantum repeaters and quantum key distribution: analysis of secret-key rates", *Physical Review A*, vol. 87, Art. no. 052315, 2013.

[ABR 14] ABRUZZO S., KAMPERMANN H., BRUß D., "Finite-range multiplexing enhances quantum key distribution via quantum repeaters", *Physical Review A*, American Physical Society, vol. 89, no. 8, p. 012303, January 2014.

[ACÍ 07] ACÍN A., CIRAC J., LEWENSTEIN M., "Entanglement percolation in quantum networks", *Nature Physics*, vol. 3, pp. 256–259, February 2007.

[ACM 06] ACM, *Computer Architecture News, Proc. 33rd Annual International Symposium on Computer Architecture*, June 2006.

[AHA 93] AHARONOV Y., DAVIDOVICH L., ZAGURY N., "Quantum random walks", *Physical Review A*, vol. 48, no. 2, pp. 1687–1690, 1993.

[AHA 04a] AHARONOV D., VAN DAM W., KEMPE J., *et al.*, "Adiabatic quantum computation is equivalent to standard quantum computation", *Proceedings of 45th Annual IEEE Symposium on Foundations of Computer Science*, IEEE, pp. 42–51, 2004.

[AHA 04b] AHARONOV D., VAN DAM W., KEMPE J., *et al.*, "Adiabatic quantum computation is equivalent to standard quantum computation", *Proceedings of 45th Annual Symposium on Foundations of Computer Science*, ACM, 2004.

[AHA 08] AHARONOV D., BEN-OR M., EBAN E., Interactive proofs for quantum computations, arXiv preprint arXiv:0810.5375, 2008.

[AHL 00] AHLSWEDE R., CAI N., LI S., *et al.*, "Network information flow", *IEEE Transactions on Information Theory*, vol. 46, no. 4, pp. 1204–1216, 2000.

[AHO 03] AHO A.V., SVORE K.M., Compiling quantum circuits using the palindrome transform, available at: http://arXiv.org/quant-ph/0311008, 2003.

[ALL 92] ALLEN L., BEIJERSBERGEN M.W., SPREEUW R.J.C., *et al.*, "Orbital angular momentum of light and the transformation of Laguerre-Gaussian laser modes", *Physical Review A*, vol. 45, pp. 8185–8189, 1992.

[ALL 07] ALLÉAUME R., BOUDA J., BRANCIARD C., *et al.*, "SECOQC white paper on quantum key distribution and cryptography", *quant-ph/0701168*, January 2007.

[ALL 09] ALLÉAUME R., ROUEFF F., DIAMANTI E., *et al.*, "Topological optimization of quantum key distribution networks", *New Journal of Physics*, vol. 11, no. 7, p. 075002, 2009.

[ALT 01] ALTER O., YAMAMOTO Y., *Quantum Measurement of a Single System*, Wiley, 2001.

[ALT 05] ALTEPETER J., JEFFREY E., KWIAT P., *et al.*, "Experimental methods for detecting entanglement", *Physical Review Letters*, vol. 95, Art. no. 033601, 2005.

[AMB 07] AMBAINIS A., CHILDS A., REICHARDT B., "Any AND-OR formula of Size N can be evaluated in time N^{1/2+ o (1)} on a quantum computer", *48th Annual IEEE Symposium on, Foundations of Computer Science (FOCS'07), 2007*, pp. 363–372, 2007.

[APA 11a] APARICIO L., VAN METER R., "Multiplexing schemes for quantum repeater networks", *Proceedings of SPIE*, vol. 8163, Art. no. 816308, 2011.

[APA 11b] APARICIO L., VAN METER R., ESAKI H., "Protocol design for quantum repeater networks", *Proceedings of the 7th Asian Internet Engineering Conference*, November 2011.

[ASP 81] ASPECT A., GRANGIER P., ROGER G., "Experimental tests of realistic local theories via Bell's theorem", *Physical Review Letters*, vol. 47, pp. 460–463, 1981.

[ASP 82] ASPECT A., GRANGIER P., ROGER G., "Experimental realization of Einstein-Podolsky-Rosen-Bohm *Gedankenexperieedingsent*: a new violation of Bell's inequalities", *Physical Review Letters*, vol. 49, pp. 91–94, 1982.

[ASP 99] ASPECT A., "Bell's inequality test: more ideal than ever", *Nature*, vol. 398, no. 6724, pp. 189–190, 1999.

[ASP 03] ASPELMEYER M., JENNEWEIN T., PFENNIGBAUER M., *et al.*, "Long-distance quantum communication with entangled photons using satellites", *IEEE Journal of Selected Topics in Quantum Electronics*, vol. 9, no. 6, pp. 1541–1551, 2003.

[BAC 10] BACON D., VAN DAM W., "Recent progress in quantum algorithms", *Communications of the ACM*, vol. 53, no. 2, pp. 84–93, February 2010.

[BAO 12] BAO X.-H., XU X.-F., LI C.-M., *et al.*, "Quantum teleportation between remote atomic-ensemble quantum memories", *Proceedings of the National Academy of Sciences*, vol. 109, no. 50, pp. 20347–20351, 2012.

[BAR 04] BARRETT M., CHIAVERINI J., SCHAETZ T., *et al.*, "Deterministic quantum teleportation of atomic qubits", *Nature*, vol. 429, no. 6993, pp. 737–739, 2004.

[BAR 12] BARZ S., KASHEFI E., BROADBENT A., *et al.*, "Demonstration of blind quantum computing", *Science*, vol. 335, no. 6066, pp. 303–308, 2012.

[BAS 01] BASU A., RIECKE J., "Stability issues in OSPF routing", *ACM SIGCOMM Computer Communication Review*, vol. 31, ACM, pp. 225–236, 2001.

[BEN 82] BENIOFF P., "Quantum mechanical models of turing machines that dissipate no energy", *Physical Review Letters*, vol. 48, pp. 1581–1585, 1982.

[BEN 84] BENNETT C.H., BRASSARD G., "Quantum cryptography: public key distribution and coin tossing", *Proceedings of IEEE International Conference on Computers, Systems, and Signal Processing*, IEEE, pp. 175–179, December 1984.

[BEN 92] BENNETT C.H., WIESNER S.J., "Communication via one- and two-particle operators on Einstein-Podolsky-Rosen states", *Physical Review Letters*, vol. 69, pp. 2881–2884, 1992.

[BEN 93] BENNETT C.H., BRASSARD G., CRÉPEAU C., *et al.*, "Teleporting an unknown quantum state via dual classical and EPR channels", *Physical Review Letters*, vol. 70, pp. 1895–1899, 1993.

[BEN 96a] BENNETT C., BRASSARD G., POPESCU S., *et al.*, "Purification of noisy entanglement and faithful teleportation via noisy channels", *Physical Review Letters*, vol. 76, no. 5, pp. 722–725, 1996.

[BEN 96b] BENNETT C.H., BERNSTEIN H.J., POPESCU S., *et al.*, "Concentrating partial entanglement by local operations", *Physical Review A*, vol. 53, pp. 2046–2052, 1996.

[BEN 96c] BENNETT C.H., DIVINCENZO D.P., SMOLIN J.A., *et al.*, "Mixed-state entanglement and quantum error correction", *Physical Review A*, vol. 54, no. 5, pp. 3824–3851, 1996.

[BEN 97] BENNETT C.H., BERNSTEIN E., BRASSARD G., *et al.*, "Strengths and weaknesses of quantum computing", *SIAM Journal on Computing*, vol. 26, no. 5, pp. 1510–1523, 1997, available at: http://arXiv.org/quant-ph/9701001.

[BEN 05a] BEN-OR M., HASSIDIM A., "Fast quantum Byzantine agreement", *Proceedings of the 37th annual ACM symposium on Theory of Computing*, ACM, pp. 481–485, 2005.

[BEN 05b] BENJAMIN S.C., EISERT J., STACE T.M., "Optical generation of matter qubit graph states", *New Journal of Physics*, vol. 7, no. 1, p. 194, 2005.

[BEN 06] BENJAMIN S.C., BROWNE D.E., FITZSIMONS J., *et al.*, "Brokered graph-state quantum computation", *New Journal of Physics*, vol. 8, no. 8, Art. no. 141, 2006.

[BER 97] BERNSTEIN E., VAZIRANI U., "Quantum complexity theory", *SIAM Journal on Computing*, vol. 26, no. 5, pp. 1411–1473, 1997.

[BER 13] BERNIEN H., HENSEN B., PFAFF W., *et al.*, "Heralded entanglement between solid-state qubits separated by three metres", *Nature*, vol. 497, no. 7447, pp. 86–90, 2013.

[BLI 04] BLINOV B., MOEHRING D., DUAN L., *et al.*, "Observation of entanglement between a single trapped atom and a single photon", *Nature*, vol. 428, no. 6979, pp. 153–157, 2004.

[BLU 83] BLUM M., "Coin flipping by telephone a protocol for solving impossible problems", *SIGACT News*, vol. 15, no. 1, pp. 23–27, 1983.

[BOT 00] BOTO A.N., KOK P., ABRAMS D.S., *et al.*, "Quantum interferometric optical lithography: exploiting entanglement to beat the diffraction limit", *Physical Review Letters*, vol. 85, no. 13, pp. 2733–2736, 2000.

[BOU 97] BOUWMEESTER D., PAN J.-W., MATTLE K., *et al.*, "Experimental quantum teleportation", *Nature*, vol. 390, pp. 575–579, December 1997.

[BOU 00] BOUWMEESTER D., EKERT A.K., ZEILINGER A., (eds.), *The Physics of Quantum Information*, Springer Berlin, 2000.

[BRA 94] BRASSARD G., SALVAIL L., "Secret-key reconciliation by public discussion", in HELLESETH T., (ed.), *Advances in Cryptology - EUROCRYPT '93*, vol. 765 of *Lecture Notes in Computer Science*, pp. 410–423, Springer Berlin Heidelberg, 1994.

[BRA 98] BRAVYI S., KITAEV A., Quantum codes on a lattice with boundary, Arxiv preprint quant-ph/9811052, 1998.

[BRA 03] BRASSARD G., "Quantum communication complexity", *Foundations of Physics*, vol. 33, no. 11, pp. 1593–1616, 2003.

[BRA 06] BRAVYI S., FATTAL D., GOTTESMAN D., "GHZ extraction yield for multipartite stabilizer states", *Journal of Mathematical Physics*, vol. 47, no. 6, pp. 62106–62106, 2006.

[BRA 13] BRATZIK S., ABRUZZO S., KAMPERMANN H., *et al.*, Quantum repeaters and quantum key distribution: the impact of entanglement distillation on the secret key rate, arXiv:1303.3456v1 [quant-ph], 2013.

[BRI 98] BRIEGEL H.-J., DÜR W., CIRAC J., *et al.*, "Quantum repeaters: the role of imperfect local operations in quantum communication", *Physical Review Letters*, vol. 81, pp. 5932–5935, 1998.

[BRO 08] BROADBENT A., TAPP A., "Can quantum mechanics help distributed computing?", *SIGACT News*, vol. 39, no. 3, pp. 67–76, 2008.

[BRO 09] BROADBENT A., FITZSIMONS J., KASHEFI E., "Universal blind quantum computation", *50th Annual IEEE Symposium on Foundations of Computer Science (FOCS'09)*, IEEE, pp. 517–526, 2009.

[BRO 10] BROWN K.L., MUNRO W.J., KENDON V.M., "Using quantum computers for quantum simulation", *Entropy*, vol. 12, no. 11, pp. 2268–2307, 2010.

[BUH 03] BUHRMAN H., RÖHRIG H., "Distributed quantum computing," *Mathematical Foundations of Computer Science 2003*, Springer-Verlag, pp. 1–20, 2003.

[BUL 09] BULUTA I., NORI F., "Quantum simulators", *Science*, vol. 326, no. 5949, pp. 108–111, 2009.

[BUR 01] BURT E.A., EKSTROM C.R., SWANSON T.B., "Comment on quantum clock synchronization based on shared prior entanglement", *Physical Review Letters*, vol. 87, Art. no. 129801, 2001.

[BYR 06] BYRNES T., YAMAMOTO Y., "Simulating lattice gauge theories on a quantum computer", *Physical Review A*, vol. 73, Art. no. 022328, 2006.

[CAL 96] CALDERBANK A.R., SHOR P.W., "Good quantum error-correcting codes exist.", *Physical Review A*, vol. 54, pp. 1098–1105, 1996.

[CAM 07] CAMPBELL E.T., FITZSIMONS J., BENJAMIN S.C., *et al.*, "Efficient growth of complex graph states via imperfect path erasure", *New Journal of Physics*, vol. 9, no. 6, Art. no. 196, 2007.

[CAR 06] CARLINI A., HOSOYA A., KOIKE T., *et al.*, "Time-optimal quantum evolution", *Physical Review Letters*, vol. 96, Art. no. 060503, 2006.

[CAS 99] CASTRO M., LISKOV B., "Practical Byzantine fault tolerance", *Proceedings of 3rd Symposium on Operating Systems Design and Implementation*, February 1999.

[CER 74] CERF V., KAHN R., "A Protocol for Packet Network Intercommunication", *IEEE Transactions on Communications*, vol. Com-22, no. 5, May 1974.

[CHA 09] CHAILLOUX A., KERENIDIS I., "Optimal quantum strong coin flipping", *50th Annual IEEE Symposium on Foundations of Computer Science (FOCS 09)*, IEEE, pp. 527–533, 2009.

[CHE 10] CHEN T.-Y., WANG J., LIANG H., *et al.*, "Metropolitan all-pass and inter-city quantum communication network", *Optics Express*, vol. 18, no. 26, pp. 27217–27225, December 2010.

[CHI 00] CHILDS A.M., PRESKILL J., RENES J., "Quantum information and precision measurement", *Journal of Modern Optics*, vol. 47, no. 2–3, pp. 155–176, 2000.

[CHI 05] CHILDRESS L., TAYLOR J., SØRENSEN A., *et al.*, "Fault-tolerant quantum repeaters with minimal physical resources and implementations based on single-photon emitters", *Physical Review A*, vol. 72, no. 5, Art. no. 52330, 2005.

[CHI 06] CHILDRESS L., TAYLOR J., SØRENSEN A., *et al.*, "Fault-tolerant quantum communication based on solid-state photon emitters", *Physical Review Letters*, vol. 96, no. 7, Art. no. 70504, 2006.

[CHI 13] CHIEN C.-H., VAN METER R., KUO S.-Y., Fault-tolerant operations for universal blind quantum computation, arXiv:1306.3664 [quant-ph], 2013.

[CHO 07] CHOU C.-W., LAURAT J., DENG H., *et al.*, "Functional quantum nodes for entanglement distribution over scalable quantum networks", *Science*, vol. 316, no. 5829, pp. 1316–1320, 2007.

[CHU 00] CHUANG I., "Quantum algorithm for distributed clock synchronization", *Physical Review Letters*, vol. 85, no. 9, pp. 2006–2009, 2000.

[CIR 97] CIRAC J., ZOLLER P., KIMBLE H., *et al.*, "Quantum state transfer and entanglement distribution among distant nodes in a quantum network", *Physical Review Letters*, vol. 78, no. 16, pp. 3221–3224, 1997.

[CLA 69] CLAUSER J.F., HORNE M.A., SHIMONY A., *et al.*, "Proposed experiment to test local hidden-variable theories", *Physical Review Letters*, vol. 23, pp. 880–884, October 1969.

[CLA 90] CLARK D.D., TENNENHOUSE D.L., "Architectural considerations for a new generation of protocols", *Proceedings of SIGCOMM '90*, pp. 200–208, 1990.

[CLA 09] CLARK C.R., METODI T.S., GASSTER S.D., *et al.*, "Resource requirements for fault-tolerant quantum simulation: the ground state of the transverse Ising model", *Physical Review A*, vol. 79, no. 6, Art. no. 062314, 2009.

[CLA 13] CLADER B.D., JACOBS B.C., SPROUSE C.R., "Preconditioned quantum linear system algorithm", *Physical Review Letters*, American Physical Society, vol. 110, no. 25, pp. 250504, June 2013.

[CLE 97] CLEVE R., BUHRMAN H., "Substituting quantum entanglement for communication", *Physical Review A*, vol. 56, no. 2, pp. 1201–1204, 1997.

[CLE 99] CLEVE R., GOTTESMAN D., LO H.-K., "How to share a quantum secret", *Physical Review Letters*, vol. 83, no. 3, pp. 648–651, 1999.

[COL 07] COLLINS O.A., JENKINS S.D., KUZMICH A., *et al.*, "Multiplexed memory-insensitive quantum repeaters", *Physical Review Letters*, vol. 98, Art. no. 060502, February, 2007.

[COU 05] COULOURIS G., DOLLIMORE J., KINDBERG T., *Distributed Systems: Concepts and Design*, 4th ed., Addison-Wesley, 2005.

[CRÉ 02] CRÉPEAU C., GOTTESMAN D., SMITH A., "Secure multi-party quantum computation", *Proceedings of the 34th Annual ACM Symposium on Theory of Computing, (STOC '02)*, New York, NY, pp. 643–652, 2002.

[CUQ 09] CUQUET M., CALSAMIGLIA J., "Entanglement percolation in quantum complex networks", *Physical Review Letters*, vol. 103, no. 24, p. 240503, 2009.

[CUQ 11] CUQUET M., CALSAMIGLIA J., "Limited-path-length entanglement percolation in quantum complex networks", *Physical Review A*, vol. 83, no. 3, Art. no. 032319, 2011.

[DAW 06] DAWSON C.M., NIELSEN M.A., "The Solovay-Kitaev theorem", *Quantum Information and Computation*, vol. 6, no. 1, pp. 81–95, 2006.

[DAY 08a] DAY J., *Patterns in Network Architecture: A Return to Fundamentals*, Prentice Hall, 2008.

[DAY 08b] DAY J., MATTA I., MATTAR K., "Networking is IPC: a guiding principle to a better Internet", *Proceedings of ACM SIGCOMM CoNext ReArch'08 Workshop*, 2008.

[DE 02] DE WOLF R., "Quantum communication and complexity", *Theoretical Computer Science*, vol. 287, no. 1, pp. 337–353, 2002.

[DE 05] DE BURGH M., BARTLETT S.D., "Quantum methods for clock synchronization: beating the standard quantum limit without entanglement", *Physical Review A*, vol. 72, Art. no. 042301, 2005.

[DE 12] DE GREVE K., YU L., MCMAHON P., *et al.*, "Quantum-dot spin-photon entanglement via frequency downconversion to telecom wavelength", *Nature*, vol. 491, no. 7424, pp. 421–425, 2012.

[DEE 01] DEERING S., Watching the waist of the protocol hourglass, Presentation at IETF 51, London, August 2001.

[DEH 03] DEHAENE J., VAN DEN NEST M., DE MOOR B., *et al.*, "Local permutations of products of Bell states and entanglement distillation", *Physical Review A*, vol. 67, no. 2, Art. no. 22310, 2003.

[DEN 02] DENNIS E., KITAEV A., LANDAHL A., *et al.*, "Topological quantum memory", *Journal of Mathematical Physics*, vol. 43, pp. 4452–4505, 2002.

[DEU 85] DEUTSCH D., "Quantum theory, the Church-Turing Principle, and the universal quantum computer", *Proceedings of the Royal Society A*, vol. 400, pp. 97–117, 1985.

[DEU 92] DEUTSCH D., JOZSA R., "Rapid solution of problems by quantum computation", *Proceedings of the Royal Society*, vol. 439, p. 553–558, 1992.

[DEU 96] DEUTSCH D., EKERT A., JOZSA R., *et al.*, "Quantum privacy amplification and the security of quantum cryptography over noisy channels", *Physical Review Letters*, vol. 77, no. 13, pp. 2818–2821, 1996.

[DEV 13] DEVITT S.J., MUNRO W.J., NEMOTO K., "Quantum error correction for beginners", *Reports on Progress in Physics*, vol. 76, no. 7, Art. no. 076001, 2013.

[D'HO 05a] D'HONDT E., Distributed quantum computation: a measurement-based approach, PhD Thesis, Vrije Universiteit Brussel, July 2005.

[D'HO 05b] D'HONDT E., PANANGADEN P., "The computational power of the W and GHZ states", *Quantum Information and Computation*, vol. 6, no. 2, pp. 13–183, 2005.

[DI 12] DI FRANCO C., BALLESTER D., "Optimal path for a quantum teleportation protocol in entangled networks", *Physical Review A*, vol. 85, Art. no. 010303, January 2012.

[DIA 07] DIANATI M., ALLÉAUME R., "Transport layer protocols for the SECOQC Quantum Key Distribution (QKD) network", *32nd IEEE Conference on Local Computer Networks, 2007 (LCN '07)*, IEEE, pp. 1025–1034, 2007.

[DIA 08] DIANATI M., ALLÉAUME R., GAGNAIRE M., *et al.*, "Architecture and protocols of the future European quantum key distribution network", *Security and Communication Networks*, vol. 1, no. 1, pp. 57–74, Wiley Online Library, 2008.

[DIF 76] DIFFIE W., HELLMAN M., "New directions in cryptography", *IEEE Transactions on Information Theory*, vol. 22, no. 6, pp. 644–654, 1976.

[DIJ 59] DIJKSTRA E., "A note on two problems in connexion with graphs", *Numerische Mathematik*, vol. 1, no. 1, pp. 269–271, 1959.

[DIV 98] DIVINCENZO D.P., "Quantum gates and circuits", *Proceedings of the Royal Society A*, 1998.

[DIX 08] DIXON A., YUAN Z., DYNES J., *et al.*, "Gigahertz decoy quantum key distribution with 1 Mbit/s secure key rate", *Optics Express*, vol. 16, no. 23, pp. 18790–18979, 2008.

[DOD 09] DODSON D., FUJIWARA M., GRANGIER P., *et al.*, Updating quantum cryptography report ver. 1, Arxiv preprint arXiv:0905.4325, 2009.

[DOY 05] DOYLE J.C., ALDERSON D.L., LI L., *et al.*, "The 'robust yet fragile' nature of the Internet", *Proceedings of the National Academy of Sciences*, vol. 102, no. 41, pp. 14497–14502, 2005.

[DUA 04] DUAN L.-M., BLINOV B.B., MOEHRING D.L., *et al.*, "Scalable trapped ion quantum computation with a probabilistic ion-photon mapping", *Quantum Information and Computation*, vol. 4, pp. 165–173, 2004.

[DÜR 99] DÜR W., BRIEGEL H.-J., CIRAC J.I., *et al.*, "Quantum repeaters based on entanglement purification", *Physical Review A*, vol. 59, no. 1, pp. 169–181, 1999.

[DÜR 00] DÜR W., VIDAL G., CIRAC J.I., "Three qubits can be entangled in two inequivalent ways", *Physical Review A*, vol. 62, no. 6, Art. no. 062314, 2000.

[DÜR 07] DÜR W., BRIEGEL H., "Entanglement purification and quantum error correction", *Reports on Progress in Physics*, vol. 70, pp. 1381–1424, 2007.

[EKE 91] EKERT A., "Quantum cryptography based on Bell's theorem", *Physical Review Letters*, vol. 67, no. 6, pp. 661–663, APS, 1991.

[ELK 13] ELKIN M., KLAUCK H., NANONGKAI D., *et al.*, "Quantum lower bounds for distributed network computing (full version)", *Proceedings of PODC*, 2013.

[ELL 02] ELLIOTT C., "Building the quantum network", *New Journal of Physics*, vol. 4, Art. no. 46, 2002.

[ELL 03] ELLIOTT C., PEARSON D., TROXEL G., "Quantum cryptography in practice", *Proceedings of SIGCOMM 2003*, August 2003.

[ELL 05a] ELLIOTT C., COLVIN A., PEARSON D., *et al.*, "Current status of the DARPA quantum network", *Proceedings of SPIE*, vol. 5815, Art. no. 138, 2005.

[ELL 05b] ELLIOTT C., "The DARPA quantum network", *Quantum Communications and Cryptography*, CRC Press, 2005.

[ENG 96] ENGLERT B.-G., "Fringe visibility and which-way information: an inequality", *Physical Review Letters*, vol. 77, pp. 2154–2157, 1996.

[ENK 97] VAN ENK S.J., CIRAC J.I., ZOLLER P., "Ideal quantum communication over noisy channels: a quantum optical implementation", *Physical Review Letters*, vol. 78, no. 22, pp. 4293–4296, June 1997.

[EVA 11] EVANS D., The Internet of things: how the next evolution of the Internet is changing everything, Cisco white paper, April 2011.

[FAR 01] FARHI E., GOLDSTONE J., GUTMANN S., *et al.*, "A quantum adiabatic evolution algorithm applied to random instances of an NP-complete Problem", *Science*, vol. 292, pp. 472–476, 2001.

[FED 09] FEDRIZZI A., URSIN R., HERBST T., *et al.*, "High-fidelity transmission of entanglement over a high-loss free-space channel", *Nature Physics*, vol. 5, no. 6, pp. 389–392, 2009.

[FEK 13] FEKETE J., DANIEL R., MATTEO C., *et al.*, "Ultranarrow-band photon-pair source compatible with solid state quantum memories and telecommunication networks", *Physical Review Letters*, vol. 110, no. 22, pp. 220–502, May 2013.

[FEY 63] FEYNMAN R.P., LEIGHTON R.B., SANDS M., *The Feynman Lectures on Physics*, Addison-Wesley, Reading, MA, 1963.

[FEY 02] FEYNMAN R.P., "Simulating physics with computers", in HEY A.J.G., (ed.), *Feynman and Computation*, Westview Press, 2002.

[FOR 07] FOREMAN S.M., HOLMAN K.W., HUDSON D.D., *et al.*, "Remote transfer of ultrastable frequency references via fiber networks", *Review of Scientific Instruments*, vol. 78, no. 2, pp. 021101–021101, 2007.

[FOW 09] FOWLER A., STEPHENS A., GROSZKOWSKI P., "High threshold universal quantum computation on the surface code", *Physical Review A*, vol. 80, Art. no. 052312, 2009.

[FOW 10] FOWLER A.G., WANG D.S., HILL C.D., *et al.*, "Surface code quantum communication", *Physical Review Letters*, vol. 104, no. 18, Art. no. 180503, May 2010.

[FRA 11] FRANKEL S., KRISHNAN S., "IP security (IPsec) and internet key exchange (IKE) document roadmap", RFC 6071, February 2011.

[FRE 79] FRENCH A.P., TAYLOR E.F., *An Introduction To Quantum Physics*, CRC Press, 1979.

[FUJ 09] FUJII K., YAMAMOTO K., "Entanglement purification with double selection", *Physical Review A*, vol. 80, no. 4, Art. no. 042308, October, 2009.

[FUR 98] FURUSAWA A., SØRENSEN J.L., BRAUNSTEIN S.L., *et al.*, "Unconditional quantum teleportation", *Science*, vol. 282, no. 5389, pp. 706–709, 1998.

[GAO 12] GAO W., FALLAHI P., TOGAN E., *et al.*, "Observation of entanglement between a quantum dot spin and a single photon", *Nature*, vol. 491, no. 7424, pp. 426–430, 2012.

[GAV 09] GAVOILLE C., KOSOWSKI A., MARKIEWICZ M., "What can be observed locally?", *Distributed Computing*, pp. 243–257, 2009.

[GAY 05] GAY S., "Quantum programming languages: survey and bibliography", *Bulletin of the European Association for Theoretical Computer Science*, June 2005.

[GEN 09] GENTRY C., "Fully homomorphic encryption using ideal lattices", *Proceedings of the 41st Annual ACM Symposium on Theory of Computing*, STOC '09, New York, NY, pp. 169–178, 2009.

[GEN 10] GENTRY C., "Computing arbitrary functions of encrypted data", *Communications of the ACM*, vol. 53, no. 3, pp. 97–105, March 2010.

[GEN 11] GENTRY C., HALEVI S., "Implementing Gentry's fully-homomorphic encryption scheme", *Advances in Cryptology–EUROCRYPT 2011*, Springer, pp. 129–148, 2011.

[GER 05] GERRY C.C., KNIGHT P.L., *Introductory Quantum Optics*, Cambridge University Press, 2005.

[GIL 64] GILBERT C., *The Design and Use of Electronic Analogue Computers*, Chapman and Hall, Ltd., 1964.

[GIL 08] GILDER L., *The Age of Entanglement: When Quantum Physics Was Reborn*, Vintage, 2008.

[GIO 01] GIOVANNETTI V., LLOYD S., MACCONE L., "Quantum-enhanced positioning and clock synchronization", *Nature*, vol. 412, no. 6845, pp. 417–419, 2001.

[GIS 02] GISIN N., RIBORDY G., TITTEL W., *et al.*, "Quantum cryptography", *Reviews of Modern Physics*, vol. 74, no. 1, pp. 145–195, 2002.

[GIS 07] GISIN N., THEW R., "Quantum communication", *Nature Photonics*, vol. 1, pp. 165–171, March 2007.

[GOE 08] GOEBEL A.M., WAGENKNECHT C., ZHANG Q., *et al.*, "Multistage entanglement swapping", *Physical Review Letters*, vol. 101, Art. no. 080403, August 2008.

[GOT 97] GOTTESMAN D., Stabilizer codes and quantum error correction, PhD Thesis, California Institute of Technology, May 1997.

[GOT 99] GOTTESMAN D., CHUANG I.L., "Demonstrating the viability of universal quantum computation using teleportation and single-qubit operations", *Nature*, vol. 402, pp. 390–393, 1999.

[GOT 00] GOTTESMAN D., "Theory of quantum secret sharing", *Physical Review A*, vol. 61, no. 4, Art. no. 42311, 2000.

[GOT 12] GOTTESMAN D., JENNEWEIN T., CROKE S., "Longer-baseline telescopes using quantum repeaters", *Physical Review Letters*, vol. 109, Art. no. 070503, 2012.

[GOV 02] GOVINDAN R., RADOSLAVOV P., An analysis of the internal structure of large autonomous systems, Report no. 02–777, CS Department, University of Southern California, 2002.

[GRA 86] GRANGIER P., ROGER G., ASPECT A., "Experimental evidence for a photon anticorrelation effect on a beam splitter: a new light on single-photon interferences", *EPL (Europhysics Letters)*, vol. 1, no. 4, pp. 173–179, 1986.

[GRA 09] GRASSL M., RÖTTELER M., "Quantum error correction and fault tolerant quantum computing", *Encyclopedia of Complexity and Systems Science*, Springer, pp. 7324–7342, 2009.

[GRE 88] GREENBERGER D.M., YASIN A., "Simultaneous wave and particle knowledge in a neutron interferometer", *Physics Letters A*, vol. 128, no. 8, pp. 391–394, 1988.

[GRE 89] GREENBERGER D.M., HORNE M.A., ZEILINGER A., "Going beyond Bell's theorem", KAFATOS M., (ed.), *Bell's Theorem, Quantum Theory, and Conceptions of the Universe*, Kluwer, pp. 69–72, available as arXiv:0712.0921v1 [quant-ph], 1989.

[GRO 96] GROVER L., "A fast quantum-mechanical algorithm for database search", *Proceedings of 28th Annual ACM Symposium on the Theory of Computation*, pp. 212–219, available at: http://arXiv.org/quant-ph/9605043, 1996.

[HAL 07] HALLGREN S., "Polynomial-time quantum algorithms for Pell's equation and the principal ideal problem", *Journal of the ACM*, vol. 54, no. 1, 2007.

[HÄN 06] HÄNSCH T.W., "Nobel lecture: passion for precision", *Reviews of Modern Physics*, vol. 78, no. 4, pp. 1297–1309, 2006.

[HAR 07] HARTMANN L., KRAUS B., BRIEGEL H.-J., *et al.*, "On the role of memory errors in quantum repeaters", *Physical Review A*, vol. 75, Art. no. 032310, 2007.

[HAR 09] HARROW A.W., HASSIDIM A., LLOYD S., "Quantum algorithm for linear systems of equations", *Physical Review Letters*, vol. 103, no. 15, Art. no. 150502, 2009.

[HAY 06] HAYASHI M., *Quantum Information*, Springer, 2006.

[HAY 07] HAYASHI M., IWAMA K., NISHIMURA H., *et al.*, "Quantum network coding", *24th International Symposium on Theoretical Aspects of Computer Science (STACS'7)*, Arxiv preprint quant-ph/0601088, 2007.

[HEC 02] HECHT E., *Optics*, 4th ed., Pearson Education/Addison-Wesley, 2002.

[HEI 06] HEIN M., DÜR W., EISERT J., *et al.*, Entanglement in graph states and its applications, Arxiv preprint quant-ph/0602096, 2006, Presented at 173rd International School of Physics "Enrico Fermi": Quantum Computers, Algorithms and Chaos, Varenna, Italy, 5–15 July 2005.

[HIN 13] HINKLEY N., SHERMAN J.A., PHILLIPS N.B., *et al.*, "An atomic clock with 10^{-18} instability", *Science*, vol. 341, no. 6151, pp. 1215–1218, 2013.

[HOL 02] HOLROYD A., "Inequalities in entanglement percolation", *Journal of Statistical Physics*, vol. 109, no. 1, pp. 317–323, 2002.

[HOR 12] HORSMAN C., FOWLER A., DEVITT S., *et al.*, "Surface code quantum computing by lattice surgery", *New Journal of Physics*, vol. 14, Art. no. 123011, 2012.

[IKU 11] IKUTA R., KUSAKA Y., KITANO T., *et al.*, "Wide-band quantum interface for visible-to-telecommunication wavelength conversion", *Nature Communications*, vol. 2, Art. no. 1544, 2011.

[IKU 13] IKUTA R., KATO H., KUSAKA Y., *et al.*, "High-fidelity conversion of photonic quantum information to telecommunication wavelength with superconducting single-photon detectors", *Physical Review A*, vol. 87, Art. no. 010301, January 2013.

[IMR 12] IMRE S., GYONGYOSI L., *Advanced Quantum Communications: An Engineering Approach*, Wiley-IEEE Press, 2012.

[INA 13] INAGAKI T., MATSUDA N., TADANAGA O., *et al.*, "Entanglement distribution over 300 km of fiber", *Optics Express*, vol. 21, no. 20, pp. 23241–23249, October 2013.

[ISE 97] ISENBERG D.S., "The rise of the stupid network", *Computer Telephony*, pp. 16–26, August 1997.

[IWA 06] IWAMA K., NISHIMURA H., RAYMOND R., *et al.*, Quantum network coding for general graphs, Arxiv preprint quant-ph/0611039, 2006.

[JAC 88] JACOBSON V., "Congestion avoidance and control", *SIGCOMM Computer Communication Review*, vol. 18, no. 4, pp. 314–329, August 1988.

[JAI 91] JAIN R., *The Art of Computer Systems Performance Analysis*, John Wiley & Sons, 1991.

[JEN 01] JENNEWEIN T., WEIHS G., PAN J.-W., *et al.*, "Experimental nonlocality proof of quantum teleportation and entanglement swapping", *Physical Review Letters*, vol. 88, no. 1, Art. no. 017903, December 2001.

[JIA 07a] JIANG L., TAYLOR J.M., LUKIN M.D., "Fast and robust approach to long-distance quantum communication with atomic ensembles", *Physical Review A*, vol. 76, Art. no. 012301, July 2007.

[JIA 07b] JIANG L., TAYLOR J.M., SØRENSEN A.S., *et al.*, "Distributed quantum computation based on small quantum registers", *Physical Review A*, vol. 76, Art. no. 062323, December 2007.

[JIA 09] JIANG L., TAYLOR J.M., NEMOTO K., *et al.*, "Quantum repeater with encoding", *Physical Review A*, vol. 79, no. 3, Art. no. 032325, 2009.

[JÖN 61] JÖNSSON C., "Elektroneninterferenzen an mehreren künstlich hergestellten Feinspalten", *Zeitschrift für Physik*, vol. 161, no. 4, pp. 454–474, 1961.

[JON 12a] JONES N.C., VAN METER R., FOWLER A.G., *et al.*, "A layered architecture for quantum computing using quantum dots", *Physical Review X*, vol. 2, no. 27, Art. no. 031007, 2012.

[JON 12b] JONES N., WHITFIELD J., MCMAHON P., *et al.*, "Faster quantum chemistry simulation on fault-tolerant quantum computers", *New Journal of Physics*, vol. 14, no. 11, p. 115023, 2012.

[JOZ 94] JOZSA R., "Fidelity for mixed quantum states", *Journal of Modern Optics*, vol. 41, no. 12, pp. 2315–2323, 1994.

[JOZ 00] JOZSA R., ABRAMS D., DOWLING J., *et al.*, "Quantum clock synchronization based on shared prior entanglement", *Physical Review Letters*, vol. 85, no. 9, pp. 2010–2013, 2000.

[JOZ 01] JOZSA R., ABRAMS D.S., DOWLING J.P., *et al.*, "Jozsa *et al.* reply", *Physical Review Letters*, vol. 87, Art. no. 129802, 2001.

[KAS 11] KASSAL I., WHITFIELD J.D., PERDOMO-ORTIZ A., *et al.*, "Simulating chemistry using quantum computers", *Annual Review of Physical Chemistry*, vol. 62, no. 1, pp. 185–207, 2011.

[KEN 05] KENT S., SEO S., Security architecture for the Internet protocol, RFC 4031, December 2005.

[KER 09] KERENIDIS I., "Quantum multiparty communication complexity and circuit lower bounds", *Mathematical Structures in Computer Science*, vol. 19, no. Special Issue 01, pp. 119–132, 2009.

[KIM 08] KIMBLE H.J., "The quantum Internet", *Nature*, vol. 453, pp. 1023–1030, June 2008.

[KIT 02] KITAEV A.Y., SHEN A.H., VYALYI M.N., *Classical and Quantum Computation*, American Mathematical Society, 2002.

[KIT 03] KITAEV A., "Fault-tolerant quantum computation by anyons", *Annals of Physics*, vol. 303, no. 1, pp. 2–30, 2003.

[KLE 10] KLEINJUNG T., AOKI K., FRANKE J., *et al.*, "Factorization of a 768-bit RSA modulus", *Advances in Cryptology–CRYPTO 2010*, Springer, pp. 333–350, 2010.

[KNI 96] KNILL E., LAFLAMME R., Concatenated Quantum Codes, available at: http://arXiv.org/quant-ph/9608012, August 1996.

[KNI 01] KNILL E., LAFLAMME R., MILBURN G.J., "A scheme for efficient quantum computation with linear optics", *Nature*, vol. 409, pp. 46–52, 2001.

[KÓM 13] KÓMÁR P., KESSLER E., BISHOF M., *et al.*, A quantum network of clocks, arXiv:1310.6045 [quant-ph], October 2013.

[KOR 56] KORN G.A., KORN T.M., *Electronic Analog Computers*, 2nd ed., McGraw-Hill, 1956.

[KOT 10] KOTLA R., ALVISI L., DAHLIN M., *et al.*, "Zyzzyva: speculative Byzantine fault tolerance", *ACM Transactions on Computer Systems*, vol. 27, pp. 7:1–7:39, January 2010.

[KRA 13] KRAUTER H., SALART D., MUSCHIK C., *et al.*, "Deterministic quantum teleportation between distant atomic objects", *Nature Physics*, vol. 9, pp. 400–404, 2013.

[KUR 12] KUROSE J.F., ROSS K.W., *Computer networking*, 6th ed., Pearson Education, 2012.

[KWI 95] KWIAT P.G., MATTLE K., WEINFURTER H., *et al.*, "New high-intensity source of polarization-entangled photon pairs", *Physical Review Letters*, vol. 75, pp. 4337–4341, 1995.

[LAD 06] LADD T.D., VAN LOOCK P., NEMOTO K., *et al.*, "Hybrid quantum repeater based on dispersive CQED interaction between matter qubits and bright coherent light", *New Journal of Physics*, vol. 8, Art. no. 184, 2006.

[LAD 10] LADD T., JELEZKO F., LAFLAMME R., *et al.*, "Quantum computers", *Nature*, vol. 464, pp. 45–53, March 2010.

[LAF 96] LAFLAMME R., MIQUEL C., PAZ J.P., *et al.*, "Perfect quantum error correcting code", *Physical Review Letters*, vol. 77, pp. 198–201, 1996.

[LAM 82] LAMPORT L., SHOSTAK R., PEASE M., "The Byzantine generals problem", *ACM Transactions on Programming Languages and Systems*, vol. 4, no. 3, pp. 382–401, 1982.

[LAN 07] LANDRY O., VAN HOUWELINGEN J., BEVERATOS A., *et al.*, "Quantum teleportation over the Swisscom telecommunication network", *JOSA B*, vol. 24, no. 2, pp. 398–403, 2007.

[LAN 11] LANYON B.P., HEMPEL C., NIGG D., *et al.*, "Universal digital quantum simulation with trapped ions", *Science*, vol. 334, no. 6052, pp. 57–61, 2011.

[LAP 09] LAPEYRE JR G., WEHR J., LEWENSTEIN M., "Enhancement of entanglement percolation in quantum networks via lattice transformations", *Physical Review A*, vol. 79, no. 4, Art. no. 042324, 2009.

[LEN 03] LENSTRA A., TROMER E., SHAMIR A., *et al.*, "Factoring estimates for a 1024-bit RSA modulus", *AsiaCrypt 2003*, Lecture Notes in Computer Science, New York, Springer-Verlag, 2003.

[LEU 10] LEUNG D., OPPENHEIM J., WINTER A., "Quantum network communication – the butterfly and beyond", *IEEE Transactions on Information Theory*, Piscataway, NJ, USA, vol. 56, no. 7, pp. 3478–3490, 2010.

[LIM 05] LIM Y.L., BARRETT S.D., BEIGE A., *et al.*, "Repeat-until-success quantum computing using stationary and flying qubits", *Physical Review Letters*, vol. 95, no. 3, Art. no. 30505, 2005.

[LLO 93] LLOYD S., "A potentially realizable quantum computer", *Science*, vol. 261, pp. 1569–1571, 1993.

[LLO 96] LLOYD S., "Universal quantum simulators", *Science*, vol. 273, pp. 1073–1078, 1996.

[LLO 04] LLOYD S., SHAPIRO J., WONG F., *et al.*, "Infrastructure for the quantum Internet", *ACM SIGCOMM Computer Communication Review*, vol. 34, no. 5, pp. 9–20, 2004.

[LLO 08] LLOYD S., "Enhanced sensitivity of photodetection via quantum illumination", *Science*, vol. 321, no. 5895, pp. 1463–1465, 2008.

[LLO 13] LLOYD S., MOHSENI M., REBENTROST P., Quantum algorithms for supervised and unsupervised machine learning, arXiv preprint arXiv:1307.0411, 2013.

[LO 08] LO H.-K., ZHAO Y., "Quantum cryptography", *Encyclopedia of Complexity and System Science*, Springer, 2008, arXiv:0803.2507v4 [quant-ph].

[LOO 06] VAN LOOCK P., LADD T.D., SANAKA K., *et al.*, "Hybrid Quantum Repeater Using Bright Coherent Light", *Physical Review Letters*, vol. 96, Art. no. 240501, 2006.

[LOW 99] LOW S.H., LAPSLEY D.E., "Optimization flow control-I: basic algorithm and convergence", *IEEE/ACM Transactions on Networking (TON)*, vol. 7, no. 6, pp. 861–874, 1999.

[LYN 96] LYNCH N.A., *Distributed Algorithms*, Morgan Kaufmann, 1996.

[MA 12] MA X., HERBST T., SCHEIDL T., *et al.*, "Quantum teleportation over 143 [thinsp] kilometres using active feed-forward", *Nature*, 2012.

[MAG 05] MAGNIEZ F., SANTHA M., SZEGEDY M., "Quantum algorithms for the triangle problem", *Proceedings of the 16th Annual ACM-SIAM Symposium on Discrete Algorithms*, Society for Industrial and Applied Mathematics, pp. 1109–1117, 2005.

[MAH 06] MAHADEVAN P., KRIOUKOV D., FALL K., *et al.*, "Systematic topology analysis and generation using degree correlations", *Proceedings of SIGCOMM 2006*, ACM, August 2006.

[MAR 04] MARKHAM D.J.H., Local distinguishability, entanglement and mixedness of quantum states, PhD Thesis, Imperial College, University of London, 2004.

[MAR 08] MARKHAM D., SANDERS B., "Graph states for quantum secret sharing", *Physical Review A*, vol. 78, no. 4, Art. no. 42309, 2008.

[MAT 96] MATTLE K., WEINFURTER H., KWIAT P.G., *et al.*, "Dense coding in experimental quantum communication", *Physical Review Letters*, vol. 76, pp. 4656–4659, 1996.

[MEA 89] MEAD C., *Analog VLSI and Neural Systems*, Addison Wesley, 1989.

[MED 02] MEDINA A., TAFT N., SALAMATIAN K., *et al.*, "Traffic matrix estimation: existing techniques and new directions", *ACM SIGCOMM Computer Communication Review*, vol. 32, no. 4, pp. 161–174, 2002.

[MEY 04] MEYER D.A., "Quantum communication in games", *AIP Conference Proceedings*, vol. 734, pp. 36–39, 2004.

[MIN 09] MINK A., FRANKEL S., PERLNER R., "Quantum key distribution (QKD) and commodity security protocols: introduction and integration", *International Journal of Network Security & Its Applications*, vol. 1, no. 2, 2009.

[MOC 88] MOCKAPETRIS P., DUNLAP K.J., "Development of the domain name system", *SIGCOMM Computer Communication Review*, vol. 18, no. 4, pp. 123–133, August 1988.

[MOC 07] MOCHON C., Quantum weak coin flipping with arbitrarily small bias, arXiv preprint arXiv:0711.4114, 2007.

[MOE 07] MOEHRING D., MAUNZ P., OLMSCHENK S., *et al.*, "Entanglement of single-atom quantum bits at a distance", *Nature*, vol. 449, no. 7158, pp. 68–71, 2007.

[MOL 07] MOLINA-TERRIZA G., TORRES J.P., TORNER L., "Twisted photons", *Nature Physics*, vol. 3, no. 5, pp. 305–310, 2007.

[MOR 13] MORIMAE T., FUJII K., "Blind quantum computation protocol in which Alice only makes measurements", *Physical Review A*, vol. 87, Art. no. 050301, 2013.

[MOS 09] MOSCA M., "Quantum algorithms", *Encyclopedia of Complexity Systems Science*, ROBERT M., (ed.), 2009.

[MOY 97] MOY J., "OSPF Version 2", RFC 2178, July 1997.

[MUN 01] MUNRO W., NEMOTO K., WHITE A., "The Bell inequality: a measure of entanglement?", *J. Modern Optics*, vol. 48, no. 7, pp. 1239–1246, June 2001.

[MUN 05] MUNRO W., NEMOTO K., SPILLER T., "Weak nonlinearities: a new route to optical quantum computation", *New Journal of Physics*, vol. 7, Art. no. 137, May 2005.

[MUN 08] MUNRO W.J., VAN METER R., LOUIS S. G.R., *et al.*, "High-bandwidth hybrid quantum repeater", *Physical Review Letters*, vol. 101, no. 4, Art. no. 040502, July 2008.

[MUN 10] MUNRO W., HARRISON K., STEPHENS A., *et al.*, "From quantum multiplexing to high-performance quantum networking", *Nature Photonics*, vol. 4, pp. 792–796, 2010.

[MUN 12] MUNRO W., STEPHENS A., DEVITT S., *et al.*, "Quantum communication without the necessity of quantum memories", *Nature Photonics*, 2012.

[NAG 09] NAGAYAMA S., VAN METER R., IKE for IPsec with QKD, Internet draft, draft-nagayama-ipsecme-ipsec-with-qkd-00; October 2009.

[NIE 00] NIELSEN M.A., CHUANG I.L., *Quantum Computation and Quantum Information*, Cambridge University Press, 2000.

[NIE 03] NIELSEN M.A., "Simple rules for a complex quantum world", *The Edge of Physics*, Scientific American, 2003.

[NIE 05] NIELSEN M.A., Cluster-state quantum computation, available at: http://arxiv.org/abs/quant-ph/0504097, April 2005.

[NIE 06] NIELSEN M.A., DOWLING M.R., GU M., *et al.*, "Quantum computation as geometry", *Science*, vol. 311, pp. 1133–1135, 2006.

[NÖL 13] NÖLLEKE C., NEUZNER A., REISERER A., *et al.*, "Efficient teleportation between remote single-atom quantum memories", *Physical Review Letters*, vol. 110, Art. no. 140403, 2013.

[NYG 10] NYGREN E., SITARAMAN R.K., SUN J., "The Akamai network: a platform for high-performance internet applications", *SIGOPS Operating Systems Review*, vol. 44, no. 3, pp. 2–19, 2010.

[ÖME 02] ÖMER B., "Classical Concepts in Quantum Programming", *Proceedings of Quantum Structures*, 2002.

[OI 06] OI D.K.L., DEVITT S.J., HOLLENBERG L.C.L., "Scalable error correction in distributed ion trap computers", *Physical Review A*, vol. 74, Art. no. 052313, 2006.

[OLM 09] OLMSCHENK S., MATSUKEVICH D.N., MAUNZ P., *et al.*, "Quantum teleportation between distant matter qubits", *Science*, vol. 323, no. 5913, pp. 486–489, 2009.

[PAN 03] PAN J.-W., GASPARONI S., URSIN R., *et al.*, "Experimental entanglement purification of arbitrary unknown states", *Nature*, vol. 423, pp. 417–422, May 2003.

[PAP 11] PAPPA A., CHAILLOUX A., DIAMANTI E., *et al.*, "Practical quantum coin flipping", *Physical Review A*, vol. 84, no. 5, Art. no. 052305, 2011.

[PEA 80] PEASE M., SHOSTAK R., LAMPORT L., "Reaching agreement in the presence of faults", *The Journal of the ACM*, vol. 27, no. 2, pp. 228–234, 1980.

[PEA 04] PEARSON D., "High-speed QKD reconciliation using forward error correction", *AIP Conference Proceedings*, vol. 734, pp. 299–302, 2004.

[PEE 09] PEEV M., PACHER C., ALLEAUME R., *et al.*, "The SECOQC quantum key distribution network in Vienna", *New Journal of Physics*, vol. 11, no. 7, Art. no. 075001, 2009.

[PEN 05] PENG C.-Z., YANG T., BAO X.-H., *et al.*, "Experimental free-space distribution of entangled photon pairs over 13 km: towards satellite-based global quantum communication", *Physical Review Letters*, vol. 94, p. 150501, 2005.

[PER 00] PERLMAN R., *Interconnections: Bridges, Routers, Switches, and Internetworking Protocols*, Pearson Education India, 2000.

[PER 10] PERSEGUERS S., CAVALCANTI D., LAPEYRE G.J., *et al.*, "Multipartite entanglement percolation", *Physical Review A*, vol. 81, no. 3, p. 032327, 2010.

[PER 13] PERSEGUERS S., JR G. J.L., CAVALCANTI D., *et al.*, "Distribution of entanglement in large-scale quantum networks", *Reports on Progress in Physics*, vol. 76, no. 9, Art. no. 096001, 2013.

[PET 11] PETERSON L.L., DAVIE B.S., *Computer Networks: A Systems Approach*, 5th ed., Elsevier, 2011.

[PFA 13] PFAFF W., TAMINIAU T.H., ROBLEDO L., *et al.*, "Demonstration of entanglement-by-measurement of solid state qubits", *Nature Physics*, vol. 9, no. 1, pp. 29–33, 2013.

[PRE 98a] PRESKILL J., *Lectures Notes on Quantum Computation, available at: http://www.theory.caltech.edu/˜preskill/ph219/index.html*, October 1998.

[PRE 98b] PRESKILL J., "Reliable quantum computers", *Proceedings of the Royal Society A*, vol. 454, pp. 385–410, 1998.

[RAI 01] RAIMOND J.M., BRUNE M., HAROCHE S., "Manipulating quantum entanglement with atoms and photons in a cavity", *Reviews of Modern Physics*, vol. 73, pp. 565–582, 2001.

[RAU 03] RAUSSENDORF R., BROWNE D.E., BRIEGEL H.J., "Measurement-based quantum computation on cluster states", *Physical Review A*, vol. 68, no. 022312, 2003.

[RAU 07a] RAUSSENDORF R., HARRINGTON J., "Fault-tolerant quantum computation with high threshold in two dimensions", *Physical Review Letters*, vol. 98, Art. no. 190504, 2007.

[RAU 07b] RAUSSENDORF R., HARRINGTON J., GOYAL K., "Topological fault-tolerance in cluster state quantum computation", *New Journal of Physics*, vol. 9, Art. no. 199, 2007.

[RAU 12] RAUSSENDORF R., "Key ideas in quantum error correction", *Philosophical Transactions of the Royal Society A: Mathematical, Physical and Engineering Sciences*, vol. 370, no. 1975, pp. 4541–4565, 2012.

[RAY 10] RAYNAL P., KALEV A., SUZUKI J., *et al.*, "Encoding many qubits in a rotor", *Physical Review A*, vol. 81, Art. no. 052327, May 2010.

[REG 02] REGEV O., "Quantum computation and lattice problems", *Proceedings of the 43rd Annual IEEE Symposium on Foundations of Computer Science*, IEEE, pp. 520–529, 2002.

[REI 06] REICHLE R., LEIBFRIED D., KNILL E., *et al.*, "Experimental purification of two-atom entanglement.", *Nature*, vol. 443, no. 7113, pp. 838–41, 2006.

[RFC 05] "Internet key exchange (IKEv2) protocol", RFC 4306, December 2005.

[RIE 04] RIEBE M., HÄFFNER H., ROOS C., *et al.*, "Deterministic quantum teleportation with atoms", *Nature*, vol. 429, no. 6993, pp. 734–737, 2004.

[RIT 12] RITTER S., NÖLLEKE C., HAHN C., *et al.*, "An elementary quantum network of single atoms in optical cavities", *Nature*, vol. 484, no. 7393, pp. 195–200, 2012.

[ROS 81] ROSEN E., Vulnerabilities of network control protocols: an example, Report no. RFC 789, July 1981.

[RUD 03] RUDOLPH T., GROVER L., "Quantum communication complexity of establishing a shared reference frame", *Physical Review Letters*, vol. 91, Art. no. 217905, 2003.

[SAL 84] SALTZER J.H., REED D.P., CLARK D.D., "End-to-end arguments in system design", *ACM Transactions on Computer Systems*, vol. 2, no. 4, pp. 277–288, 1984.

[SAL 10] SALVAIL L., PEEV M., DIAMANTI E., *et al.*, "Security of trusted repeater quantum key distribution networks", *Journal of Computer Security*, vol. 18, no. 1, pp. 61–87, 2010.

[SAN 08] SANGOUARD N., SIMON C., COUDREAU T., *et al.*, "Purification of single-photon entanglement with linear optics", *Physical Review A*, vol. 78, Art. no. 050301, 2008.

[SAN 11] SANGOUARD N., SIMON C., DE RIEDMATTEN H., *et al.*, "Quantum repeaters based on atomic ensembles and linear optics", *Reviews of Modern Physics*, vol. 83, no. 1, p. 33, 2011.

[SAT 12] SATOH T., LE GALL F., IMAI H., "Quantum network coding for quantum repeaters", *Physical Review A*, vol. 86, Art. no. 032331, 2012.

[SCH 90] SCHNEIDER F.B., "Implementing fault-tolerant services using the state machine approach: a tutorial", *ACM Comput. Surv.*, vol. 22, no. 4, pp. 299–319, 1990.

[SCH 96] SCHNEIER B., *Applied Cryptography*, 2nd ed., John Wiley, 1996.

[SCH 03] SCHUCH N., SIEWERT J., "Programmable networks for quantum algorithms", *Physical Review Letters*, vol. 91, Art. no. 027902, 2003.

[SCH 13] SCHAIBLEY J.R., BURGERS A.P., MCCRACKEN G.A., *et al.*, "Demonstration of quantum entanglement between a single electron spin confined to an InAs quantum dot and a photon", *Physical Review Letters*, vol. 110, Art. no. 167401, 2013.

[SCU 91] SCULLY M.O., ENGLERT B.-G., WALTHER H., "Quantum optical tests of complementarity", *Nature*, vol. 351, pp. 111–116, 1991.

[SHE 06] SHERSON J., KRAUTER H., OLSSON R., *et al.*, "Quantum teleportation between light and matter.", *Nature*, vol. 443, no. 7111, pp. 557–60, 2006.

[SHO 94] SHOR P.W., "Algorithms for quantum computation: discrete logarithms and factoring", *Proceedings of 35th Symposium on Foundations of Computer Science*, Los Alamitos, CA, IEEE Computer Society Press, pp. 124–134, 1994.

[SHO 95] SHOR P.W., "Scheme for reducing decoherence in quantum computer memory", *Physical Review A*, vol. 52, no. 4, pp. R2493–R2496, October 1995.

[SHO 97] SHOR P.W., "Polynomial time algorithms for prime factorization and discrete logarithms on a quantum computer", *SIAM Journal on Computing*, vol. 26, no. 5, pp. 1484–1509, 1997,

[SIP 12] SIPAHIGIL A., GOLDMAN M.L., TOGAN E., *et al.*, "Quantum interference of single photons from remote nitrogen-vacancy centers in diamond", *Physical Review Letters*, vol. 108, Art. no. 143601, 2012.

[SLO 13] SLODIČKA L., HÉTET G., RÖCK N., *et al.*, "Atom-atom entanglement by single-photon detection", *Physical Review Letters*, vol. 110, Art. no. 083603, 2013.

[SMI 01] SMITH A., Multi-party quantum computation, Master's Thesis, Massachusetts Institute of Technology, 2001.

[SMI 08] SMITH G., YARD J., "Quantum communication with zero-capacity channels", *Science*, vol. 321, no. 5897, pp. 1812–1815, 2008.

[SOU 05] SOULE A., LAKHINA A., TAFT N., *et al.*, "Traffic matrices: balancing measurements, inference and modeling", *ACM SIGMETRICS Performance Evaluation Review*, vol. 33, pp. 362–373, 2005.

[SPI 06] SPILLER T.P., NEMOTO K., BRAUNSTEIN S.L., *et al.*, "Quantum computation by communication", *New Journal of Physics*, vol. 8, Art. no. 30, February 2006.

[STE 96] STEANE A., "Error correcting codes in quantum theory", *Physical Review Letters*, vol. 77, pp. 793–797, 1996.

[STE 03] STEFFEN M., VAN DAM W., HOGG T., *et al.*, "Experimental implementation of an adiabatic quantum optimization algorithm", *Physical Review Letters*, vol. 90, Art. no. 067903, 2003.

[STE 13] STEFFEN L., SALATHE Y., OPPLIGER M., *et al.*, "Deterministic quantum teleportation with feed-forward in a solid state system", *Nature*, vol. 500, no. 319–322, August 2013.

[STU 11] STUCKI D., LEGRE M., BUNTSCHU F., *et al.*, "Long-term performance of the SwissQuantum quantum key distribution network in a field environment", *New Journal of Physics*, vol. 13, no. 12, Art. no. 123001, 2011.

[SVO 06] SVORE K.M., AHO A.V., CROSS A.W., *et al.*, "A layered software architecture for quantum computing design tools", *IEEE Computer*, pp. 74–83, January 2006.

[TAK 05] TAKAMOTO M., HONG F., HIGASHI R., *et al.*, "An optical lattice clock", *Nature*, vol. 435, no. 7040, pp. 321–324, 2005.

[TAK 13] TAKEDA S., MIZUTA T., FUWA M., *et al.*, "Deterministic quantum teleportation of photonic quantum bits by a hybrid technique", *Nature*, vol. 500, pp. 315–318, August 2013.

[TAN 05] TANI S., KOBAYASHI H., MATSUMOTO K., "Exact quantum algorithms for the leader election problem", *Proceedings of 22nd Annual Symposium on Theoretical Aspects of Computer Science (STACS'05)*, Lecture Notes in Computer Science, Springer-Verlag, vol. 3404, pp. 581–592, 2005.

[TAN 10] TANENBAUM A.S., *Computer Networks*, 5 ed., Prentice Hall, 2010.

[TAN 12] TANI S., KOBAYASHI H., MATSUMOTO K., "Exact quantum algorithms for the leader election problem", *The ACM Transactions on Computation Theory*, vol. 4, no. 1, pp. 1:1–1:24, 2012.

[TAS 10] TASHIMA T., KITANO T., ÖZDEMIR I.M.C.K., *et al.*, "Demonstration of local expansion toward large-scale entangled webs", *Physical Review Letters*, vol. 105, no. 21, Art. no. 210503, 2010.

[TAY 09] TAYLOR G.I., "Interference fringes with feeble light", *Proceedings of the Cambridge Philosophical Society*, vol. 15, pp. 114–115, 1909.

[TER 13] TERHAL B.M., Quantum error correction for quantum memories, arXiv:1302.3428 [quant-ph], 2013.

[THA 06] THAKER D.D., METODI T., CROSS A., *et al.*, "CQLA: matching density to exploitable parallelism in quantum computing", *Proceedings of 33rd Annual International Symposium on Computer Architecture*, June 2006.

[TIT 99] TITTEL W., BRENDEL J., GISIN N., *et al.*, "Long-distance Bell-type tests using energy-time entangled photons", *Physical Review A*, vol. 59, no. 6, pp. 4150–4163, 1999.

[TIT 01] TITTEL W., ZBINDEN H., GISIN N., "Experimental demonstration of quantum secret sharing", *Physical Review A*, vol. 63, no. 4, Art. no. 42301, 2001.

[TOG 11] TOGAN E., CHU Y., TRIFONOV A., *et al.*, "Quantum entanglement between an optical photon and a solid-state spin qubit", *Frontiers in Optics*, Optical Society of America, 2011.

[TON 89] TONOMURA A., ENDO J., MATSUDA T., *et al.*, "Demonstration of single-electron buildup of an interference pattern", *American Journal of Physics*, vol. 57, pp. 117–120, 1989. Accompanying movie available at http://www.hitachi.com/rd/portal/research/em/doubleslit.html.

[TOU 01] TOUCH J., "Dynamic internet overlay deployment and management using the x-bone", *Computer Networks*, pp. 117–135, 2001.

[TOU 06] TOUCH J., WANG Y., PINGALI V., A recursive network architecture, ISI Technical Report ISI-TR-2006-626, 2006, Presented at the IEEE Workshop on Computer Communications (CCW), Pittsburgh PA, February 2007.

[TOU 08] TOUCH J., PINGALI V., "The RNA metaprotocol", *Proceedings of IEEE International Conference on Computer Communications (ICCCN)*, 2008.

[TOU 10] TOUCH J., BALDINE I., DUTTA R., *et al.*, "A dynamic recursive unified internet design (DRUID)", *Computer Networks*, 2010.

[TUN 10] TUNICK A., MOORE T., DEACON K., *et al.*, "Review of representative free-space quantum communications experiments", 2010.

[TUR 98] TURCHETTE Q.A., WOOD C.S., KING B.E., *et al.*, "Deterministic entanglement of two trapped ions", *Physical Review Letters*, vol. 81, pp. 3631–3634, 1998.

[URS 07] URSIN F., TIEFENBACHER T., SCHMITT-MANDERBACH H., *et al.*, "Entanglement-based quantum communication over 144 km", *Nature Physics*, vol. 3, pp. 481–486, July 2007.

[VAN 06] VAN METER R., Architecture of a quantum multicomputer optimized for shor's factoring algorithm, PhD Thesis, Keio University, available as arXiv:quant-ph/0607065, 2006.

[VAN 08] VAN METER R., MUNRO W.J., NEMOTO K., *et al.*, "Arithmetic on a distributed-memory quantum multicomputer", *ACM Journal of Emerging Technologies in Computing Systems*, vol. 3, no. 4, Art. no. 17, January 2008.

[VAN 09] VAN METER R., LADD T.D., MUNRO W.J., *et al.*, "System design for a long-line quantum repeater", *IEEE/ACM Transactions on Networking*, vol. 17, no. 3, pp. 1002–1013, June 2009.

[VAN 11] VAN METER R., TOUCH J., HORSMAN C., "Recursive quantum repeater networks", *Progress in Informatics*, , no. 8, pp. 65–79, March 2011.

[VAN 13a] VAN METER R., HORSMAN C., "A blueprint for building a quantum computer", *Communications of the ACM*, vol. 53, no. 10, pp. 84–93, October 2013.

[VAN 13b] VAN METER R., SATOH T., LADD T.D., *et al.*, "Path selection for quantum repeater networks", *Networking Science*, pp. 1-14, 2013.

[VIL 08] VILLORESI P., JENNEWEIN T., TAMBURINI F., *et al.*, "Experimental verification of the feasibility of a quantum channel between Space and Earth", *New Journal of Physics*, vol. 10, Art. no. 033038, 2008.

[WAL 83] WALLS D.F., "Squeezed states of light", *Nature*, vol. 306, pp. 141–146, 1983.

[WAL 05] WALTHER P., RESCH K.J., RUDOLPH T., *et al.*, "Experimental one-way quantum computing", *Nature*, vol. 434, pp. 169–176, 2005.

[WAN 10] WANG D., FOWLER A., STEPHENS A., *et al.*, "Threshold error rates for the toric and surface codes", *Quantum Information and Computation*, vol. 10, pp. 456–459, 2010.

[WAN 13] WANG J.-Y., YANG B., LIAO S.-K., *et al.*, "Direct and full-scale experimental verifications towards ground-satellite quantum key distribution", *Nature Photonics*, Nature Publishing Group, vol. 7, no. 5, pp. 387–393, 2013.

[WEI 11] WEIGEL W., LENHART G., "Standardization of quantum key distribution in ETSI", *Wireless Personal Communications*, vol. 58, no. 1, pp. 145–157, 2011.

[WIE 83] WIESNER S., "Conjugate coding", *SIGACT News*, vol. 15, no. 1, pp. 78–88, 1983.

[WIL 99] WILLIAMS C.P., CLEARWATER S.H., *Ultimate Zero and One: Computing at the Quantum Frontier*, Copernicus Books, 1999.

[WIL 13] WILDE M.M., *Quantum Information Theory*, Cambridge University Press, 2013.

[WOO 82] WOOTTERS W.K., ZUREK W.H., "A single quantum cannot be cloned", *Nature*, vol. 299, pp. 802–803, October 1982.

[WRI 95a] WRIGHT G.R., STEVENS W.R., *TCP/IP Illustrated*, Addison-Wesley Professional, 1995.

[WRI 95b] WRIGHT G.R., STEVENS W.R., *TCP/IP Illustrated, Volume 2: The Implementation*, Addison-Wesley, Boston, MA, 1995.

[YAO 11] YAO A.M., PADGETT M.J., "Orbital angular momentum: origins, behavior and applications", *Advances in Optics and Photonics*, vol. 3, no. 2, pp. 161–204, 2011.

[YE 08] YE J., KIMBLE H., KATORI H., "Quantum state engineering and precision metrology using state-insensitive light traps", *Science*, vol. 320, no. 5884, pp. 1734–1738, 2008.

[YUA 88] YUANG M., "Survey of protocol verification techniques based on finite state machine models", *Proceedings of the Computer Networking Symposium*, pp. 164–172, 1988.

[ŻUK 93] ŻUKOWSKI M., ZEILINGER A., HORNE M.A., *et al.*, ""Event-ready-detectors" Bell experiment via entanglement swapping", *Physical Review Letters*, vol. 71, pp. 4287–4290, American Physical Society, December 1993.

[ZAL 99] ZALKA C., "Grover's quantum searching algorithm is optimal", *Physical Review A*, vol. 60, no. 4, pp. 2746–2751, 1999.

[ZHA 03] ZHAO Z., YANG T., CHEN Y., *et al.*, "Experimental realization of entanglement concentration and a quantum repeater", *Physical Review Letters*, vol. 90, no. 20, Art. no. 207901, 2003.

[ZHA 04] ZHAO Z., CHEN Y.-A., ZHANG A.-N., *et al.*, "Experimental demonstration of five-photon entanglement and open-destination teleportation", *Nature*, vol. 430, no. 6995, pp. 54–58, 2004.

[ZWE 12] ZWERGER M., DÜR W., BRIEGEL H.J., "Measurement-based quantum repeaters", *Physical Review A*, vol. 85, Art. no. 062326, 2012.

Index

A

address, 17
AS number, 61
Aspect, Alain, 52
attenuation length, 153
authentication, 62, 95
authorization, 62
autonomous system (AS), 61
avalanche photodiode (APD), 155

B

backbone, 66
Bell measurement, 49
  basis, 49
  pair, 5, 47
Bell, John, 52
Bennett, Charles H., 3, 79, 97
bias, 63
binary state, 180
blind quantum computation, 10, 129
Brassard, Giles, 3, 97, 114
Briegel, Hans, 13, 117, 185
broadcast, 55
buffer, 60, 69
  memory division multiplexing, 69

C

capability, 62
cavity QED, 157
Cerf, Vinton (Vint) Gray, 67
channel, 57
Cirac, J. Ignacio, 185
circuit switching, 68
Clark, David D. (Dave), 70
Clauser-Horner-Shimony-Holt (CHSH) inequality, 50
client, 61
Clifford group operations, 86
computational basis, 49
congestion, 59, 64
convergence of routing, 64
cryptoperiod, 103

D

dark count, 155
  integrity, 62
  packet, 56
  data privacy, 63
deadlock, 206
decibel (dB), 152
denial of service (DoS), 65, 77, 104
density matrix, 34
destination, 55
detection efficiency, 155
diameter of a network, 58
digital signature, 62
Dirac, Paul, 26
Dür, Wolfgang, 13, 185

E

eigenvalue, 27, 51
eigenvector, 27

Ekert, Artur, 100
european telecommunication standards institute (ETSI), 109
exterior gateway protocol (EGP), 73

F

fidelity, 7
Furusawa, Akira, 82

G, H

gateway, 66
Gentry, Craig, 129
Gerry, Christopher C., 149
GHZ state, 5, 87, 115, 145
Gisin, Nicolas, 121, 170, 193
Grangier, Philippe, 52
graph state, 5, 116
ground state, 156
Hadamard gate, 33
Harald Weinfurter, 151, 306
Haroche, Serge, 170
hierarchy, 65

I, K

Imoto, Nobuyuki, 154
interface, 60
interference, 25, 32, 39, 133, 141, 155, 161, 170, 171
interior gateway protocol (IGP), 73
internet assigned numbers authority (IANA), 71
internet engineering task force (IETF), 101
internet protocol (IP), 61, 63, 67
internetwork, 4
ion trap, 151, 156
IPsec, 75
Kahn, Robert (Bob), 67
key rollover, 103
Kimble, Jeff, 82
Knight, Peter L., 149
Kwiat, Paul, 151, 306

L

Lamport, Leslie, 65
local area network (LAN), 66
local operations and classical communication (LOCC), 49
local operations and classical communication (LOCC), 116
Lukin, Mikhail (Misha) D., 145, 185
Lynch, Nancy A., 71

M

man-in-the-middle attack, 95
Mattle, Klaus, 151, 306
measurement-based quantum computation (MBQC), 117, 129
Metcalfe's Law, 56
Monroe, Chris, 170
multicast, 55
multihop forwarding, 57
multiplexing, 19, 59
Munro, W.J. (Bill), 227, 238, 245

N

name resolution, 61
Nemoto, Kae, 238
network address translation (NAT), 75
network interface card (NIC), 196
network protocol, 60
next hop, 58
nitrogen vacancy (NV) center in diamond, 151, 156
node degree, 58
noise operator, 202
non-repudiation, 63
number field sieve, 95

O

one-time pad (OTP), 94
open shortest path first (OSPF), 73, 110
optical fiber, 152
orbital angular momentum (OAM), 151
overlay network, 75, 102

P

packet forwarding, 58
Pan, Jian-Wei, 115, 192, 203
parametric down conversion (PDC), 151
partial trace of a matrix, 39

path, 15, 57
Pauli frame correction, 89, 117, 118, 198, 223, 227
permission, 62
physical interface, 57
physical memory address, 62
Polzik, Eugene, 84
port number, 61
protocol layer, 66
purification, 7

Q, R

quantum coherences, 34
  dot, 151, 156
  error correction, 7
Raussendorf, Robert, 117
Reed, David P., 70
rekeying, 103
relay, 106
resource, 59
right, 62
Roger, Gérard, 52
routing, 57

S

Saltzer, Jerome J. (Jerry), 70
Schrödinger, Erwin, 28
SECOQC, 109
separable, 39
server, 61
singlet, 50, 132
socket, 16, 60
source, 55
spin, 156
squeezed states, 150
state variable, 156
state vector, 28
statistical multiplexing, 69
Steane, Andrew, 221
subnet, 61
superconducting Josephson junction flux qubit, 157
superconducting single photon detectors (SSPDs), 155
superposition, 29

T, U, V, W

tensor product, 27
time division multiplexing (TDM), 68
trace of a matrix, 28
traffic load, 57
transit, 74
tree, 66
tunnel, 75, 101
unicast, 55
Vernon cipher, 94
virtual memory address, 62
virtual private network (VPN), 75, 102
W state, 5, 116, 120
waveguide, 152
well-known port, 61
Werner state, 173, 180–182, 229
white noise, 157
wide area network (WAN), 66
Wineland, David, 170

Y, Z

Yamamoto, Yoshihisa, 154, 170
Zeilinger, Anton, 82, 84, 151, 153, 192, 203, 306
Zoller, Peter, 185

Printed and bound by CPI Group (UK) Ltd, Croydon, CR0 4YY
13/09/2026
14961076-0006